Advance in Friction Stir Processed Materials

Advance in Friction Stir Processed Materials

Editors

Józef Iwaszko
Jerzy Winczek

MDPI • Basel • Beijing • Wuhan • Barcelona • Belgrade • Manchester • Tokyo • Cluj • Tianjin

Editors

Józef Iwaszko
Department of Materials Engineering
Czestochowa University of Technology
Czestochowa
Poland

Jerzy Winczek
Department of Technology and Automation
Czestochowa University of Technology
Czestochowa
Poland

Editorial Office
MDPI
St. Alban-Anlage 66
4052 Basel, Switzerland

This is a reprint of articles from the Special Issue published online in the open access journal *Materials* (ISSN 1996-1944) (available at: www.mdpi.com/journal/materials/special_issues/friction_stir_processed_materials).

For citation purposes, cite each article independently as indicated on the article page online and as indicated below:

LastName, A.A.; LastName, B.B.; LastName, C.C. Article Title. *Journal Name* **Year**, *Volume Number*, Page Range.

ISBN 978-3-0365-4438-0 (Hbk)
ISBN 978-3-0365-4437-3 (PDF)

Contents

Editorial

Special Issue: Advance in Friction Stir Processed Materials

Józef Iwaszko [1,*] and Jerzy Winczek [2]

[1] Department of Materials Engineering, Faculty of Production Engineering and Materials Technology, Czestochowa University of Technology, 19 Armii Krajowej Ave., 42-201 Czestochowa, Poland

[2] Department of Technology and Automation, Faculty of Mechanical Engineering and Computer Science, Czestochowa University of Technology, 21 Armii Krajowej Ave., 42-201 Czestochowa, Poland; jerzy.winczek@pcz.pl

* Correspondence: iwaszko@wip.pcz.pl

1. Introduction

In recent years, on the basis of FSP/FSW technologies, a number of new solutions, methods and variants have been developed, constituting not only proof of the continuous evolution of FSP/FSW technologies, but also of the huge scientific and application potential hidden in these methods. The key idea of these works was to induce more beneficial changes in the microstructure of materials and their properties, as well as to ensure better control of the processes taking place at individual stages of machining [1,2]. As is known, the main condition to obtain the assumed material characteristics is appropriate selection of the process parameters and knowledge of the changes in the microstructure of the material and its properties as a function of the machining parameters [3], shape and dimensions of the tool [4,5] or, for example, the sample cooling method [6–9]. That is why systematic research in this area and experimental verification of developed concepts and ideas are so important. The key objective of this Special Issue was to present the current state of knowledge and the recent advances in developing the microstructure and properties of materials using modern FSP and related technologies. The Special Issue made it possible to collect valuable articles presenting the results of research on changes in the microstructure and properties of the material caused by the application of the above technologies, as well as presenting the cognitive and application potential in addition to underlining the implementation problems and challenges faced by the users of FSP and related technologies.

Citation: Iwaszko, J.; Winczek, J. Special Issue: Advance in Friction Stir Processed Materials. *Materials* **2022**, *15*, 3742. https://doi.org/10.3390/ma15113742

Received: 16 May 2022
Accepted: 23 May 2022
Published: 24 May 2022

2. Discussion of Research Results

An important factor influencing the degree of grain refinement, i.e., the key parameter characterizing the microstructure formed during FSP, is the method and intensity of sample cooling. The paper [10] presents a solution in which a stream of compressed air, additionally cooled to the temperature of −11 °C with a cooling jet nozzle, was used to cool the 7075 aluminium alloy during FSP. The authors applied two variants of air blowing, i.e., at angles of 45° and 90° to the sample surface and carried out an analysis of the impact of FSP performed under accelerated cooling conditions on the microstructure and properties of the alloy, taking a sample treated with FSP in still air as the reference material. In relation to the material cooled in still air, greater homogeneity of the microstructure of the material and a higher degree of grain refinement were obtained. In the case of the naturally cooled sample, the average grain size in the near-surface zone was 7.6 μm, while in the case of the air-cooled sample it was 1.4 μm in the variant with air blowing at the angle of 45° and 3.2 μm in the variant with air blowing at the angle of 90° with reference to the surface. The consequences of the more favourable changes in the microstructure were higher hardness and resistance to wear resistance of the alloy. The paper also assessed the impact of the blowing angle of the air stream on the microstructure and material properties and indicate that the solution in which the nozzle is inclined at the angle of 45° is particularly favourable because the air stream cools both the material subjected to friction modification and the tool itself. It was shown that such a cooling method leads to stronger grain refinement and better material properties because in such a case there is both direct cooling of the material with the air stream and indirect cooling by means of the tool cooled down by this stream. According to the authors, the proposed cooling method may be a competitive solution compared to other methods because it combines the advantages of compressed air cooling and cryogenic factors.

In turn, Ahmed et al. [11] focused on the use of bobbin tool friction stir welding (BT-FSW) to produce defect-free lap joints in welding AA1050-H14 alloy sheets. The BT-FSW method is an expansion of classic FSW and is increasingly used in industrial practice owing to the number of advantages and benefits of its application. In paper [11], a newly designed and fabricated bobbin tool with different pin geometries (cylindrical, square, and triangular) and concave shoulder profile was used, and then an assessment of the impact of BT-FSW pin geometries on the joint properties was carried out in the context of assessing the possibility of obtaining defect-free lap joints. The authors showed a clear relationship between the temperature at the weld centre, and the pin geometry along with the traverse speed. The authors found, that a square pin leads to a higher BT-FSW stir zone temperature, and the temperature measured at the advancing side is higher than that at the retreating side in all welding conditions. In the case of a tool with cylindrical and square pin geometries, defect-free welds at all the researched traverse welding speeds were obtained.

In paper [12], a friction stir deposition technique (FSD) in the additive manufacturing of AA2011 parts was used. FSD is a solid-state additive manufacturing technique that can be used to deposit different materials. The authors assessed the impact of the feeding speed and the temper condition of the AA2011 alloy on the properties and microstructures of the deposited materials. The conducted research showed that the use of FSD allows the production of sound, continuous multi-layered AA2011 parts without any physical discontinuities or interfacial defects between the layers in all the applied processing parameters. The microstructure of the deposited material was characterized by equiaxed grains, whose size was much smaller in comparison to the grain size measured in the initial material. Despite the intense grain refinement in the case of the AA2011-T6 alloy, lower hardness of the friction stir deposited parts in comparison to the base material was found. An improvement in hardness was obtained only in the case of the AA2011-O friction stir deposited parts and it reached even 163% of the starting material hardness at the applied feeding speed of 1 mm/min.

Analysis of the factors affecting the properties of joints obtained with friction stir lap welding (FSLW) was the subject of research conducted by Choy et al. [13]. An element of a torsion beam shaft of a car chassis produced by joining a pipe made of A357 aluminium alloy with a steel pipe made of FB590 steel of increased strength was tested. The impact of the process parameters on the mechanical parameters of the obtained joint was analysed, as well as on the thickness of the intermetallic compound layer (IMCL) formed in the intermediate layer between the aluminium and steel. The authors placed great emphasis on analysis of the IMCL due to its significant impact on the properties of the joint. The authors found that the thickness of the IMCL, but also the tensile shear load, is primarily impacted by the tool penetration depth. It was found that the thickness of the IMCL decreases with the increase in the tool penetration depth, while the impact of the tool penetration depth on the tensile shear load is more complex. Choy et al. [14] also determined the impact of various process factors on the tensile shear load using definitive screening design. They used four types of tools with different pin lengths. The scope of the experiment adopted by the authors, expressed in its multi-variant nature, allowed a number of relationships to be determined, useful in the control of friction stir lap welding, and facilitating achievement of the assumed goals. The authors discovered, for instance, that the plunge depth has a significant impact on the tensile shear load in addition to the tool penetration depth and also indicated that the plunge depth negatively affects the magnitude of tensile shear load.

They also found that the tool penetration depth has the greatest impact on the size of the hooking part in the lap welding of the pipe.

In many industrial applications, aluminium alloys are reinforced with hard ceramic particles to enhance the mechanical properties of aluminium metal matrix composites (Al-MMC). Ali et al. [15] conducted studies on AA6061/SiC/B_4C composites subjected to friction stir welding. The matrix of the composite was the AA6061 aluminium alloy, while SiC and B_4C powders were used as reinforcement, whose total share in the composite was 13%. Microstructural investigations were carried out using SEM, as well as radiographic tests, tensile strength tests, and hardness and wear resistance measurements. The microstructural studies revealed strong grain refinement and even particle distribution in the joint area. The use of the reinforcement phase improved the tensile properties of the friction stir welded Al-MMCs. The highest ultimate tensile stress was exhibited by the sample with 10% SiC and 3% B_4C. It was found that with the increase in B_4C and decrease in the SiC content in the friction stir welded Al-MMCs samples, the tensile strength declined. The elongation decreased as the percentage of SiC was reduced and B_4C was raised. The hardness measurements showed growth in its value directly proportional to the B_4C share and inversely proportional to the SiC content. The coefficient of friction was higher for the samples whose SiC content ranged from 2% to 4%. The conducted research indicated that FSW is an appropriate process for welding Al-MMC, and the obtained results prove the ability to meet the requirements of the aviation and automotive industry. In the opinion of the authors of the work, this technology and the products manufactured thanks to it can also be used in navy and civil navigation owing to the high efficiency of welding.

In [16], an assessment was carried out for the possibility of using the FSW as a welding technique to fill the grooves for welding duplex stainless steel (DSS). For this purpose, three different groove geometries without a root gap were designed and machined in DSS plates with the thickness of 6.5 mm. For comparative purposes, the DSS plate was also welded using the gas–tungsten arc welding (GTAW) method. The obtained joints were subjected to a comparative assessment employing radiographic inspection, optical microscopy, electron back scattering diffraction, as well as hardness and tensile testing. Studies indicated that FSW can be successfully used for welding 2205 duplex stainless steel. It was found that defect-free and sound joints can be produced using a 60° V-shaped groove with a 2 mm root face without a root gap. Compared to the joint obtained by the GTAW method, the FSW joint was characterized by higher yield strength, ultimate tensile strength, and elongation by 21%, 41%, and 66%, respectively. In addition, a clearly higher hardness of the weld joint produced by FSW was found than that produced by GTAW. The maximum hardness of material in the stirring zone in the case of the FSW joint was 280 HV, while in the case of the GTAW joint it was 265 HV.

In papers [17,18], the authors focused on selection of the parameters in friction stir spot-welding (FSSW) processes. Ataya et al. [17] joined a low-carbon steel sheet (A283M-Grade C) with a brass sheet (CuZn40) by means of FSSW using different rotational speeds and dwell times to explore the effective range of parameters, enabling joints to be obtained with a high load-carrying capacity. The quality of the joints was assessed by visual examination, macro- and micro-structural investigations, EDS analysis, the tensile lap shear test, and hardness measurements. Determination of the effect of the total number of revolutions on the heat generated per spot weld was also an important aspect of the work. It was found that the heat input ranged from 11 kJ to 1.5 kJ and it was linearly proportional to the number of revolutions per spot joint. The authors also revealed that when the number of revolutions per spot ranged from 250 to 500, it was possible to produce joints with a high load-carrying capacity from 4 kN to 7.5 kN. In turn, in work [18], FSSW was used to join AA6082-T6 aluminium alloy sheets. Sheets with thicknesses of 1 mm and 2 mm were welded using different rotation speeds, while the dwell time was constant in all the variants. Verification of the effects of FSSW was based on macro- and microstructural investigations, hardness testing and a tensile-shear test. An important element of the characterisation of the process and the manufactured joint was the determination of the heat input generated during FSSW and the peak temperature in the stirring zone. It was found that the highest heat input energy of 3 kJ occurs at the highest rotational speed, and the heat input grows with increasing rotational speed. A similar trend was found in the case of the peak temperature measured in the stirring zone, namely with the increase in the rotation speed, the maximum temperature rose from 236 ± 4 °C to 367 ± 3 °C. The conducted research also showed that the spot joints welded at 600 rpm are characterised by highest the hardness and shear load.

3. Summary

Knowledge of the processes, phenomena, and relationships occurring during material processing using FSP/FSW and related technologies is the key to achieving the assumed research and application objectives. Therefore, in this aspect, this Special Issue has become a kind of a platform for exchanging practical experience and knowledge, as well as a source of valuable information that can ultimately become an inspiration to undertake one's own research in this area. The articles included in this Special Issue indicates that FSP/FSW and related technologies are constantly evolving, and, thus, the possibilities in shaping the microstructure and properties of materials are also growing, which consequently generates new research challenges.

Author Contributions: Conceptualization, J.I. and J.W.; formal analysis, J.I. and J.W.; resources, J.I. and J.W.; writing—original draft preparation, J.I. and J.W.; writing—review and editing, J.I. and J.W.; supervision, J.I. and J.W.; project administration, J.I. and J.W. All authors have read and agreed to the published version of the manuscript.

Funding: This research received no external funding.

Acknowledgments: The Guest Editors would like to thank all the authors from all over the world (Korea, Egypt, India, Saudi Arabia, Tunisia, and Poland) who contributed with their valuable works to the accomplishment of the Special Issue. Special thanks are due to the Reviewers for their constructive comments.

Conflicts of Interest: The authors declare no conflict of interest.

References

1. Almazrouee, A.I.; Al-Fadhalah, K.J.; Alhajeri, S.N. A New Approach to Direct Friction Stir Processing for Fabricating Surface Composites. *Crystals* **2021**, *11*, 638. [CrossRef]
2. Iwaszko, J.; Kudła, K.; Fila, K. Technological Aspects of Friction Stir Processing of AlZn5.5MgCu Aluminum Alloy. *Bull. Pol. Acad. Sci. Tech. Sci.* **2018**, *66*, 713–719.
3. Bagheri, B.; Abdollahzadeh, A.; Sharifi, F.; Abbasi, M. The role of vibration and pass number on microstructure and mechanical properties of AZ91/SiC composite layer during friction stir processing. *Proc. Inst. Mech. Eng. C J. Mech. Eng. Sci.* **2021**, *236*, 2312–2326. [CrossRef]
4. Janeczek, A.; Tomków, J.; Fydrych, D. The Influence of Tool Shape and Process Parameters on the Mechanical Properties of AW-3004 Aluminium Alloy Friction Stir Welded Joints. *Materials* **2021**, *14*, 3244. [CrossRef] [PubMed]
5. Paidar, M.; Bokov, D.; Mehrez, S.; Ojo, O.O.; Ramalingam, V.V.; Memon, S. Improvement of mechanical and wear behavior by the development of a new tool for the friction stir processing of Mg/B_4C composite. *Surf. Coat. Technol.* **2021**, *426*, 127797. [CrossRef]
6. Khalaf, H.I.; Al-Sabur, R.; Abdullah, M.E.; Kubit, A.; Derazkola, H.A. Effects of Underwater Friction Stir Welding Heat Generation on Residual Stress of AA6068-T6 Aluminum Alloy. *Materials* **2022**, *15*, 2223. [CrossRef] [PubMed]
7. Memon, S.; Tomków, J.; Derazkola, H.A. Thermo-Mechanical Simulation of Underwater Friction Stir Welding of Low Carbon Steel. *Materials* **2021**, *14*, 4953. [CrossRef] [PubMed]
8. Kumar, A.; Kumar Godasu, A.; Pal, K.; Mula, S. Effects of in-process cryocooling on metallurgical and mechanical properties of friction stir processed Al7075 alloy. *Mater. Charact.* **2018**, *144*, 440–447. [CrossRef]
9. Ramaiyan, S.; Chandran, R.; Santhanam, S.K.V. Effect of Cooling Conditions on Mechanical and Microstructural Behaviours of Friction Stir Processed AZ31B Mg Alloy. *Mod. Mech. Eng.* **2017**, *7*, 144–160. [CrossRef]
10. Iwaszko, J.; Kudła, K. Evolution of Microstructure and Properties of Air-Cooled Friction-Stir-Processed 7075 Aluminum Alloy. *Materials* **2022**, *15*, 2633. [CrossRef] [PubMed]
11. Ahmed, M.M.Z.; Habba, M.I.A.; El-Sayed Seleman, M.M.; Hajlaoui, K.; Ataya, S.; Latief, F.H.; EL-Nikhaily, A.E. Bobbin Tool Friction Stir Welding of Aluminum Thick Lap Joints: Effect of Process Parameters on Temperature Distribution and Joints' Properties. *Materials* **2021**, *14*, 4585. [CrossRef] [PubMed]

12. Ahmed, M.M.Z.; El-Sayed Seleman, M.M.; Elfishawy, E.; Alzahrani, B.; Touileb, K.; Habba, M.I.A. The Effect of Temper Condition and Feeding Speed on the Additive Manufacturing of AA2011 Parts Using Friction Stir Deposition. *Materials* **2021**, *14*, 6396. [CrossRef] [PubMed]
13. Choy, L.; Kang, M.; Jung, D. Effect of Microstructure and Tensile Shear Load Characteristics Evaluated by Process Parameters in Friction Stir Lap Welding of Aluminum-Steel with Pipe Shapes. *Materials* **2022**, *15*, 2602. [CrossRef] [PubMed]
14. Choy, L.; Kim, S.; Park, J.; Kang, M.; Jung, D. Effect of Process Factors on Tensile Shear Load Using the Definitive Screening Design in Friction Stir Lap Welding of Aluminum–Steel with a Pipe Shape. *Materials* **2021**, *14*, 5787. [CrossRef] [PubMed]
15. Ali, K.S.A.; Mohanavel, V.; Vendan, S.A.; Ravichandran, M.; Yadav, A.; Gucwa, M.; Winczek, J. Mechanical and Microstructural Characterization of Friction Stir Welded SiC and B4C Reinforced Aluminium Alloy AA6061 Metal Matrix Composites. *Materials* **2021**, *14*, 3110. [CrossRef] [PubMed]
16. Ahmed, M.M.Z.; Abdelazem, K.A.; El-Sayed Seleman, M.M.; Alzahrani, B.; Touileb, K.; Jouini, N.; El-Batanony, I.G.; Abd El-Aziz, H.M. Friction Stir Welding of 2205 Duplex Stainless Steel: Feasibility of Butt Joint Groove Filling in Comparison to Gas Tungsten Arc Welding. *Materials* **2021**, *14*, 4597. [CrossRef] [PubMed]
17. Ataya, S.; Ahmed, M.M.Z.; El-Sayed Seleman, M.M.; Hajlaoui, K.; Latief, F.H.; Soliman, A.M.; Elshaghoul, Y.G.Y.; Habba, M.I.A. Effective Range of FSSW Parameters for High Load-Carrying Capacity of Dissimilar Steel A283M-C/Brass CuZn40 Joints. *Materials* **2022**, *15*, 1394. [CrossRef] [PubMed]
18. Ahmed, M.M.Z.; El-Sayed Seleman, M.M.; Ahmed, E.; Reyad, H.A.; Touileb, K.; Albaijan, I. Friction Stir Spot Welding of Different Thickness Sheets of Aluminum Alloy AA6082-T6. *Materials* **2022**, *15*, 2971. [CrossRef] [PubMed]

 MDPI

Article

Evolution of Microstructure and Properties of Air-Cooled Friction-Stir-Processed 7075 Aluminum Alloy

Józef Iwaszko [1,*] and Krzysztof Kudła [2]

[1] Department of Materials Engineering, Faculty of Production Engineering and Materials Technology, Czestochowa University of Technology, 19 Armii Krajowej Ave., 42-200 Czestochowa, Poland

[2] Department of Technology and Automation, Faculty of Mechanical Engineering and Computer Science, Czestochowa University of Technology, 21 Armii Krajowej Ave., 42-200 Czestochowa, Poland; krzykudla@gmail.com

* Correspondence: iwaszko@wip.pcz.pl

Abstract: A rolled plate of 7075 aluminum alloy was friction-stir-processed (FSP) with simultaneous cooling by an air stream cooled to −11 °C with a jet cooling nozzle. Two variants of air blowing were used: at an angle of 45° to the sample surface and at an angle of 90°. The reference material was a sample subjected to analogous treatment but naturally cooled in still air. The microstructural tests revealed strong grain refinement in all the samples, with higher grain refinement obtained in the air-cooled friction-stir-processed samples. For the naturally cooled samples, the average grain size in the near-surface area was 7.6 μm, while for the air-cooled sample, it was 1.4 μm for the 45° airflow variant and 3.2 μm for the 90° airflow variant. A consequence of the greater grain refinement was that the hardness of the air-cooled friction-stir-processed samples was higher than that of the naturally cooled samples. The improvement in abrasive wear resistance was achieved only in the case of the friction-stir-processed specimens with air cooling. It was found that the change in the air blowing angle affects not only the degree of grain refinement in the stirring zone, but also the geometrical structure of the surface. In all the samples, FSP caused redistribution of the intermetallic precipitates combined with their partial dissolution in the matrix.

Keywords: friction stir processing; 7075 aluminum alloy; air cooling; jet cooling nozzle; microstructure evolution; hardness; tribological properties

Citation: Iwaszko, J.; Kudła, K. Evolution of Microstructure and Properties of Air-Cooled Friction-Stir-Processed 7075 Aluminum Alloy. *Materials* **2022**, *15*, 2633. https://doi.org/10.3390/ma15072633

Academic Editor: Thomas Niendorf

Received: 9 March 2022
Accepted: 31 March 2022
Published: 2 April 2022

1. Introduction

Friction stir processing (FSP) is a novel grain refinement technique and one of the most promising methods of modifying the microstructure and properties of engineering materials. This solid-state processing technique was developed by Mishra et al. [1,2], but the basic principles and idea of friction stir processing are derived from friction stir welding (FSW) technology [3]. The differences between FSP and FSW mainly concern their purpose because FSW technology is used to join materials [4,5], while FSP is employed to modify the microstructure of the material [6–9]. As with FSW technology, FSP uses a special cylindrical tool with a pin that rotates and plunges into the material to be processed, and then moves along designed paths. The nature and scope of changes occurring during FSP or FSW depend, among others, on the shape of the tool pin [10], the rotational speed [11] or the number of tool passes [12]. The friction of the tool against the surface of the modified material generates a large amount of heat, which makes the material plastic. The plasticized material flows to the back of the pin, where it is extruded and forged behind the tool, consolidated and cooled under hydrostatic pressure conditions [13]. FSP technology is also successfully used to eliminate defects and material loss in the sample [14] and to produce surface composites [15–17] or to modify the microstructure of composites [18,19]. The production of surface composites consists in introducing the reinforcing phase in the form of particles [20–22] or fibers [23] into the plasticized matrix. Currently, there are a number of solutions and methods allowing effective introduction of the reinforcing phase during FSP and the production of a surface composite [24]. For this purpose, the most common method is the groove method [25,26] or the hole method [27,28], less often used are direct friction stir processing [29–31], the sandwich method [23,32,33] or other solutions.

FSP technology is constantly evolving as new equipment and methodological solutions are constantly being developed, enabling more favorable changes to be obtained in the microstructure of the material, especially a higher degree of grain refinement. FSP is most often performed under conditions of natural cooling of the sample, i.e., in still air, and then the degree of grain refinement, and thus the properties of the material, are the result of the applied processing parameters. One of the newest trends in FSP technology is treatment with additional cooling in order to increase the temperature gradient in the material, and thus raise the cooling rate. The cooling rate of the material determines the degree of refinement of the microstructure, and therefore its mechanical properties; hence, the use of solutions increasing the cooling intensity is absolutely justified. The grain size affects not only the mechanical properties of the material [34], but also, for example, the corrosion resistance as smaller grains result in a lower value of the corrosion potential and the corrosion current [35]. A no less important aspect is the possibility of extending the tool life by reducing the amount of heat accumulated in the tool during FSP. The tool is exposed not only to abrasive wear, but also to high temperature resulting from friction against the surface of the material, as well as to cyclical changes in this temperature.

Currently, accelerated cooling is carried out by spraying coolant on the tool–workpiece interface [36,37], cooling the modified sample with compressed air [38,39] and conducting FSP on a sample immersed in a cooling agent [40,41] or intensive cooling of the sample before processing [42]. Special heat sinks with an internal cooling system are also utilized, in which the modified material is placed [43].

Assessment of the effect of additional cooling on the microstructure and material properties has been the subject of research by many research teams. For example, Luo et al. [12] analyzed the microstructure and properties of a material subjected to one-pass and two-pass submerged friction stir processing, using water as a coolant. In turn, Ai et al. [36] cooled samples of cast aluminum alloy A356 with a stream of water. The effect of cooling with water was greater grain refinement than in the analogous sample cooled with air, but at the same time, lower hardness of the alloy was obtained due to limitation of the growth of secondary-phase particles and their dissolution in the alloy matrix. Cooling with a stream of water was also used by Chen et al. [37]. In turn, Heidarpour et al. [43] used a water heat sink in surface modification of the AZ31 magnesium alloy. Derazkola et al. [44] tested three different coolants in the friction stir welding of aluminum AA3003 and A441 AISI steel sheet, namely, CO_2, water and air, and found that the smallest grain size in the sample was obtained after using CO_2, but simultaneously, the poorest joint with notable segregation at interface was formed. The water-cooled sample, in turn, was characterized by the greatest tensile strength, but at the same time, it was characterized by a lower hardness than in the case of cooling with CO_2. In turn, Yazdipour et al. [45] employed a mixture of dry ice and ethanol to modify the Al5083 aluminum alloy. Analogous cooling was used by Satyanarayana et al. [46] in processing of the 6061 aluminum alloy. In turn, Moaref et al. [40] applied the water submerged processing technique in the friction stir processing of pure copper and a copper–zinc alloy, and Feng et al. [41] utilized the same method for treatment of the 2219-T6 aluminum alloy. Cooling of the AZ31B magnesium alloy by spraying liquid nitrogen was used by Ammouri et al. [42]. In turn, Alavi Nia [47] used a die made of copper with internal grooves for the flow of water and additional cooling by means of compressed air in the FSP of the AZ31 magnesium alloy.

By analyzing the literature data, it can be concluded that the application of accelerated cooling promotes refinement of the microstructure [48,49] and its homogenization [36,47] and usually improves the mechanical properties of the material [43,50]. However, it should be noted that some of the proposed solutions are difficult to implement in industry or their application requires additional operations and adaptation activities or even significant reorganization of the workplace. In the case of the most commonly used water cooling, it is necessary to dry the material after processing and remove the coolant from the working area, in addition to its management during and after FSP. Moreover, water as a coolant should not be used for materials that tend to corrode in an aqueous environment. Cryogenic cooling with liquid nitrogen makes it possible to obtain a higher cooling rate of the sample, but it is a problematic coolant owing to its intense evaporation in contact with a warmer material. The gas cloud formed in this case has insulating properties, which in turn reduces the effective cooling rate of the sample. The application of liquid nitrogen cooling is also troublesome because of the need to insulate all the components of the cooling system, which still does not completely exclude the risk of nitrogen gas formation in the cooling system. Another cryogenic coolant employed in FSP is dry ice, i.e., solid carbon dioxide. This material is usually utilized in the form of a mixture of dry ice and ethanol/methanol, which is applied to the surface of the sample behind the tool [45,48,51]. The least troublesome and most convenient cooling medium is compressed air because the sample in this case does not require cleaning or drying, and the coolant does not need to be removed from the working area. In addition, air is a high-purity coolant and can be used to process virtually any material. An additional advantage of this method of cooling is the continuous exchange of the cooling agent during the procedure, thanks to which the "cooling parameters" of the air (temperature, flow velocity) do not change during the cooling process, as can be the case with liquid coolants. It should be noted, however, that air has a lower heat capacity than water or cryogenic coolants; hence, the application of compressed air generates lower cooling rates. Nonetheless, it is possible to eliminate this disadvantage and increase the intensity of air cooling, for example, by cooling it with a jet cooling nozzle.

The novelty of the work is an innovative method of cooling the sample during FSP, namely by means of an air stream cooled to -11 °C with a jet cooling nozzle. The proposed solution is located between cooling with compressed air and cooling with cryogenic agents and combines the main advantages of air (cleanliness, no need to remove the coolant from the working area or its disposal/management after treatment) and the advantages of cooling with cryogenic agents (high cooling intensity) [52]. The advantage of the solution used in this study over air cooling results from the much lower air temperature generated by the jet cooling nozzle. To the best knowledge of the authors of this study, there are no similar studies in the world literature on the use of a jet nozzle in FSP, as well as studies on the influence of the angle of the air stream on the size and nature of microstructural changes and material properties. It is worth adding that jet cooling nozzles are offered by suppliers or manufacturers of milling machines as additional equipment for devices; therefore, their availability and potential implementation in industrial conditions is not a problem, which is an additional argument for this solution. The efficiency of the process, the economic effects and the ease of conducting this type of treatment outweigh the previously used technologies of microstructure refinement employed in the process of light metal surface modification using FSP technology.

The main purpose of the work was to analyze the microstructure and properties of the 7075 aluminum alloy subjected to FSP with simultaneous cooling by means of an air stream cooled with a jet nozzle. The reference material was an analogously processed but naturally cooled alloy in still air. The scope of the research also included assessment of the effect of the air blowing method on the size and nature of microstructural changes in the processed zone and the properties of the aluminum alloy.

2. Materials and Experiment Procedures

The base material used for the experiment was the 7075 (Al-5.5Zn-2.4Mg-1.6Cu-0.20Cr) aluminum alloy in the T6 state (supersaturated solution treated and artificially aged alloy). This alloy, due to its low density, high specific strength and high fracture toughness, is a material widely applied in many industries, among others, in the production of particularly responsible structures in the aviation and automotive industries [53]. The chemical composition of the aluminum alloy is presented in Table 1.

Table 1. Chemical composition of 7075 aluminum alloy.

Alloy	Element Content, wt%								
7075	Zn	Fe	Cu	Mn	Mg	Cr	Si	Ti	Al
	5.5	0.3	1.6	0.15	2.4	0.2	0.2	0.1	89.55

The material was supplied in the form of 15 mm thick rolled plates. Samples with dimensions of 90 × 70 × 15 mm were cut from the plate, the surface of which was degreased with acetone and then subjected to FSP. Friction stir processing was performed by means of a vertical CNC milling machine (AVIA FNE 50, AVIA S.A., Warsaw, Poland). The tool was made of hardened and tempered X37CrMoV5-1 (H11) hot work tool steel with a hardness of 51 ± 1 HRC. The shoulder diameter of the tool was 18 mm. The tool was equipped with a cone-shaped pin, 4.3 mm long and with diameters of 6 mm (cone base) and 4 mm (cone top), respectively. The side surface of the cone was threaded. The tool utilized in the work is shown in Figure 1. The shoulder tilt angle was fixed at 2°.

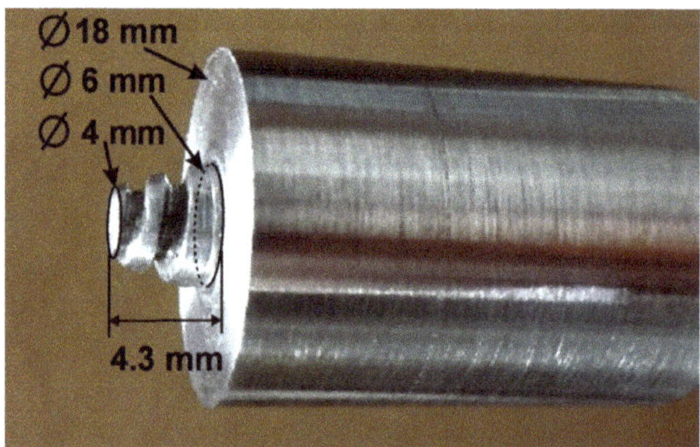

Figure 1. Tool used in friction stir processing of 7075 aluminum alloy.

In order to intensify the cooling process of the friction-modified zone, the stand was equipped with an additional cooling system to be employed during processing of the samples. The cooling medium was air-cooled to a temperature of about −11 °C with a jet cooling nozzle. The structure of the nozzle and the principles of its operation are shown in Figure 2.

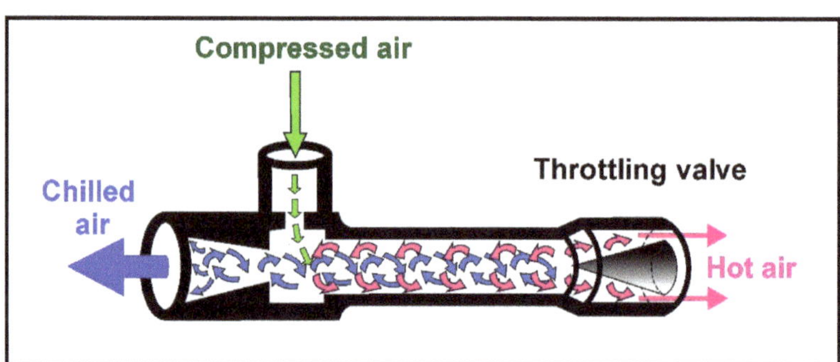

Figure 2. Jet cooling nozzle.

The jet cooling nozzle is supplied with compressed air, which is set into rotary motion in the nozzle chamber, and then two streams, external and internal, are separated from it. Hot air from the external stream is led outside the nozzle through one of its ends, while the air in the internal vortex loses heat and escapes through the other end of the nozzle in the form of a strongly cooled stream [54]. The temperature of the individual streams depends primarily on the pressure and temperature of the air supplied to the nozzle. The lower the temperature of the supplied air and the higher its pressure, the lower the temperature the air can be cooled to. As a result, the temperature of the cooled air at the outlet can be even several dozen degrees lower than the temperature of the air supplied to the nozzle, which enables high cooling efficiency and energy savings, as the nozzle does not require additional power for its operation. It is worth adding that cooling nozzles are simple, inexpensive and reliable devices that do not require maintenance and can be easily implemented in industrial conditions, and their use does not interfere with the production process. Therefore, nozzles appear to be the ideal solution for spot cooling in FSP and FSW processes.

As part of this work, a commercial cooling nozzle was utilized, offered as an accessory for machine tools, which, according to the manufacturer's data, can generate a cold air stream with a temperature of up to −46 °C. The auxiliary nozzle, which supplied air previously cooled by the jet cooling nozzle (Exair, Cincinnati, OH, USA), was integrated with the working tool holder, thanks to which its position and distance from the tool and the sample surface were constant during the entire processing. The air pressure supplied to the cooling nozzle was 8 bar.

Two variants of cooling the sample were employed; in the first case, the auxiliary nozzle supplying the cooling stream was inclined at an angle of 45° to the sample surface. This sample was denoted as AC45. The stream of cooled air cooled both the material during friction stir processing and the already modified material located just behind the tool. The cooling stream also cooled the tool shoulder. The distance of the end of the cooling nozzle from the edge of the pin was 20 mm, which allowed for precise directing of the air stream to the place subjected to the friction stir processing, as well as to the tool working in this place. In this way, a double cooling effect was obtained, namely, direct cooling of the modified material by means of an air stream and indirect cooling of the alloy by means of a tool cooled during FSP.

In the second case, the nozzle was positioned perpendicular to the surface of the sample, and the cooling stream mainly cooled the material located behind the tool, and to a lesser extent, the tool shoulder and the material undergoing frictional treatment at the moment. This sample was denoted as AC90. The FSP variants are shown in Figure 3 in the form of diagrams and photos of the stand. The distance of the end of the cooling nozzle from the edge of the pin in this case was 45 mm, so it was greater than in the case of the AC45 sample. The use of a different distance resulted from the adopted strategy, in the case of the AC90 sample, the focus was on direct cooling of the material behind the pin, i.e., already subjected to friction stir processing, while the indirect cooling of the material with the tool was limited. It should be noted, however, that the key parameter, i.e., the distance of the nozzle tip from the sample surface, was identical in both samples AC45 and AC90, thanks to which the FSP-subjected material was cooled under uniform conditions in terms of air flow parameters.

The use of the different variants in the location of the coolant supply nozzle aimed to demonstrate whether and to what extent a change in the angle of the coolant stream affects the degree of grain refinement in the surface layer of the material. The reference material was an FSPed sample naturally cooled in still air. This sample was denoted as NC.

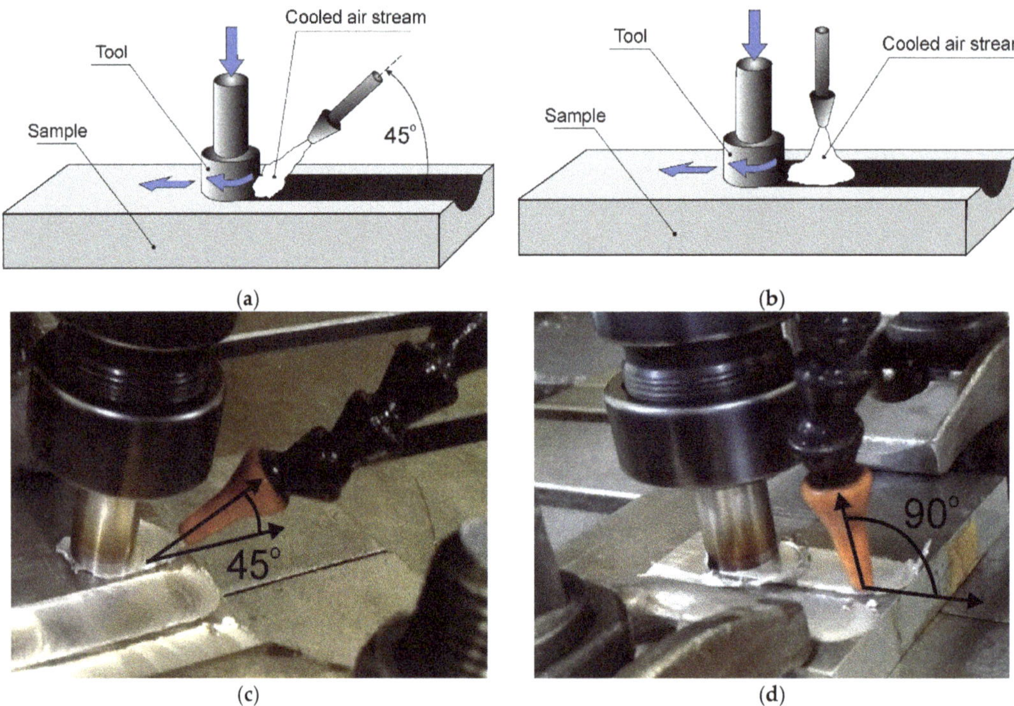

Figure 3. Schemes of FSP variants (**a**,**b**), photos of stands (**c**,**d**).

The detailed processing parameters are presented in Table 2. The selection of FSP parameters was based on the practical knowledge gained during previous tests conducted with the use of 7075 aluminum alloy. The processing was carried out using parameters, at which no previously disqualifying changes in the geometric structure of the surface or unsatisfactory changes in the microstructure of the material were observed.

Table 2. Processing parameters of 7075 aluminum alloy.

Parameter	NC	AC90	AC45
Tool rotational speed [rpm]	400	400	400
Tool traverse speed [mm/min]	30	30	30
Plunge time [s]	2	2	2
Plunge speed [mm/min]	6	6	6
Pin length [mm]	4.3	4.3	4.3
Plunge depth [mm]	5.3	5.3	5.3
Auxiliary nozzle inclination angle [°]	–	90°	45°
Distance of end of cooling nozzle from edge of pin [mm]	–	45	20

The obtained samples were subjected to macroscopic and profilometric tests in order to assess the sample surface and analyze its geometric structure. Macroscopic examinations were performed using an Olympus SZ61 stereoscopic microscope (Olympus, Tokio, Japan), and profilometric examinations with a Taylor Hobson TALYSURF 120 profilometer (Taylor Hobson Ltd., Leicester, UK). Microstructural examinations were performed by means of an Olympus GX41 light microscope (Olympus, Tokio, Japan) and a JEOL JSM-6610LV scanning electron microscope (JEOL Ltd., Tokio, Japan). The microstructural investigations were carried out on cross-sections perpendicular to the tool traverse direction. The metallographic specimens were etched with Keller's reagent (2 mL HF + 3 mL HCl + 5 mL HNO_3 + 190 mL distilled water). To determine the grain size, a measurement technique was used based on determination of the number of grains per unit area (Jeffries method). For each of the samples, the grain size variability was also measured. A representative area located in the near-surface layer was analyzed. Chemical composition analysis was performed using an Oxford Instruments EDS microanalyzer (Oxford Instruments Plc., Abingdon, UK). The hardness was measured on sample cross-sections utilizing a Shimadzu hardness tester (Shimadzu Corp., Kioto, Japan) with a load of 980.7 mN. To evaluate the tribological properties of the material, a laboratory pin-on-disc tribometer, T01-M (ITEE, Radom, Poland), was used under unlubricated sliding contact against a rotating steel ring (HRC 58-63).

3. Results and Discussion

3.1. Macroscopic and Profilometric Investigations

The macroscopic effect of FSP is shown in Figure 4a. The main element characterizing the geometric structure of each surface is its roughness, which is the set of irregularities resulting from processing. In order to assess the geometric structure of the samples subjected to surface treatment employing the various FSP variants, three main roughness parameters were measured, namely, the Ra parameter (arithmetic mean deviation of the roughness profile from the mean line), Rz parameter (maximum roughness profile height) and the Rc parameter (mean height of the roughness profile elements). All the samples were subjected to roughness analysis. Seven segment measurements were taken on each band to visualize the variability of the surface geometry at different locations of the band, and in particular to visualize the differences in the values of the roughness parameters in the central part of the band and in the peripheral regions of the band. The measuring section length was 6 mm and the distance between adjacent measuring sections was 2 mm. The adopted measurement methodology is presented in Figure 4b.

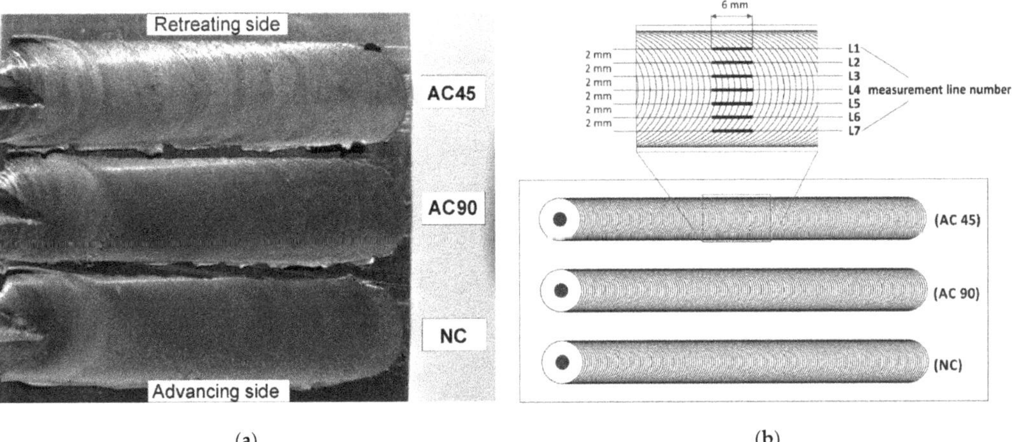

(a) (b)

Figure 4. Macroscopic effect of FSP (**a**), scheme for measuring surface roughness of bands (**b**).

The obtained values of the individual roughness parameters as a function of the measuring line location are shown in Figure 5, and the mean values of the Ra, Rz, Rc parameters together with the calculated confidence intervals are given in Table 3.

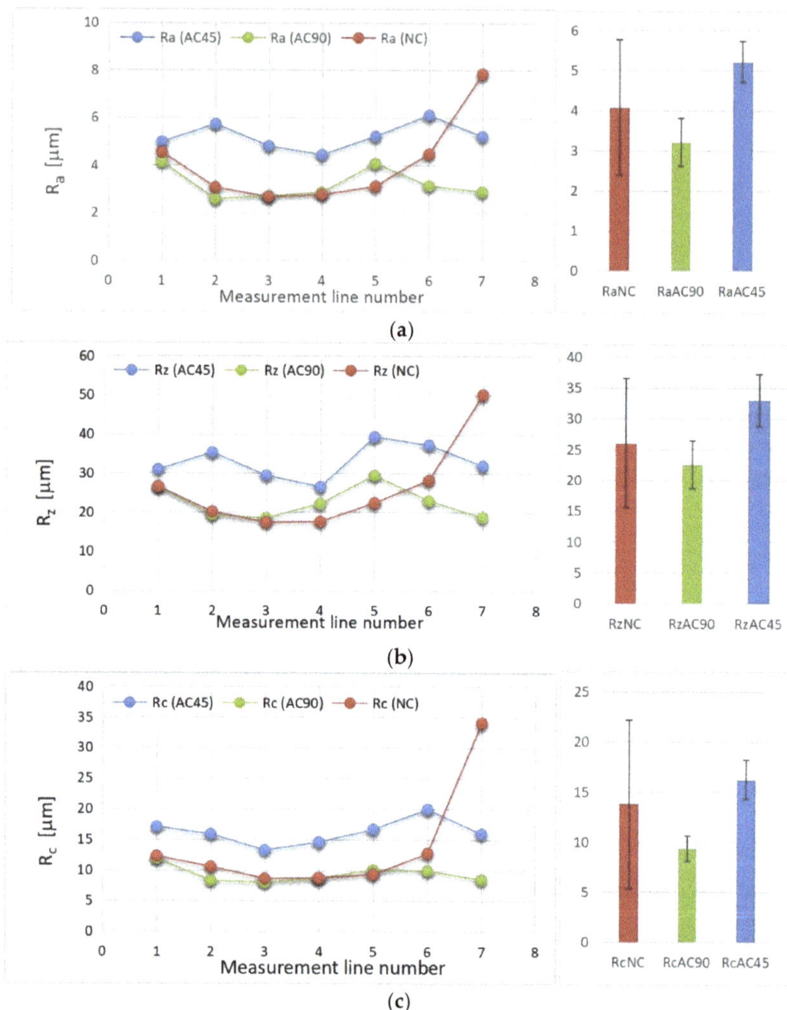

Figure 5. Roughness parameter values: Ra (**a**), Rz (**b**) and Rc (**c**).

Table 3. Mean values of Rz, Ra, Rc parameters with the determined confidence intervals.

Roughness Parameter	Mean Values			Confidence Interval +/−		
	AC45	AC90	NC	AC45	AC90	NC
Ra	5.23	3.21	4.08	0.52	0.60	1.70
Rz	33.01	22.53	26.06	4.22	3.89	10.57
Rc	16.23	9.37	13.79	1.95	1.27	8.42

By analyzing the obtained results, it can be seen that the geometric structure of the AC45 sample was characterized by higher Ra, Rz and Rc parameters than the naturally-cooled sample or the AC90 sample. This fact should be explained by the different kinetics of cooling the material during processing. In the case of the AC45 sample, the cooling stream reduced the thermal effect directly affecting the material at the time of processing, thus influencing the process of shaping the geometric structure of the surface. The lower process temperature resulted in less plasticization of the alloy, which brought about greater surface roughness after FSP. In the case of the AC90 sample, the air stream mainly cooled the material already subjected to FSP, located just behind the pin, and to a lesser extent contributed to a reduction in the thermal effect generated in the tool–material interaction zone and decisive for the degree of plasticization of the material. In this case, the influence of cooling on the process of shaping the geometric structure of the material was smaller because the air stream cooled the material, the geometric structure of which was largely already formed. By analyzing the roughness parameter values as a function of the measuring section location, it can also be noticed that the lowest values of the roughness parameters occurred in the central part of the band and grew as they approached the band boundary, while the rise in the case of the naturally cooled sample was very large on the advancing side. In the case of the remaining samples, no such significant differences were observed on the advancing or retreating side.

The results presented in Table 3 show that the largest spread (the widest confidence interval) relates to FSP without additional cooling (NC) and is due to an unfavorable, strong increase in the tested parameters (Rz, Ra, Rc) on the advancing side. The deterioration of the surface shaping conditions under free cooling conditions (without additional cooling) is an important argument for introducing additional cooling treatments to the friction stir processing of heat-sensitive materials, including the solutions proposed in this article.

3.2. Microstructure Evolution

3.2.1. Microstructural Investigations of 7075 Aluminum Alloy in Initial State

In order to illustrate the size and scope of changes in the microstructure of the aluminum alloy caused by FSP, detailed studies of the material in its initial state were carried out, paying particular attention to those features and microstructure elements, the evolution of which was to be expected after friction stir processing. The microstructure of the 7075 alloy in its initial state is shown in Figure 6. The 7075 aluminum alloy was characterized by a microstructure typical of materials subjected to directional plastic processing; this microstructure was formed by strongly deformed grains of the α primary solution, with a clearly elongated shape, whose orientation corresponded to the direction of rolling. The average grain length was about 240 μm.

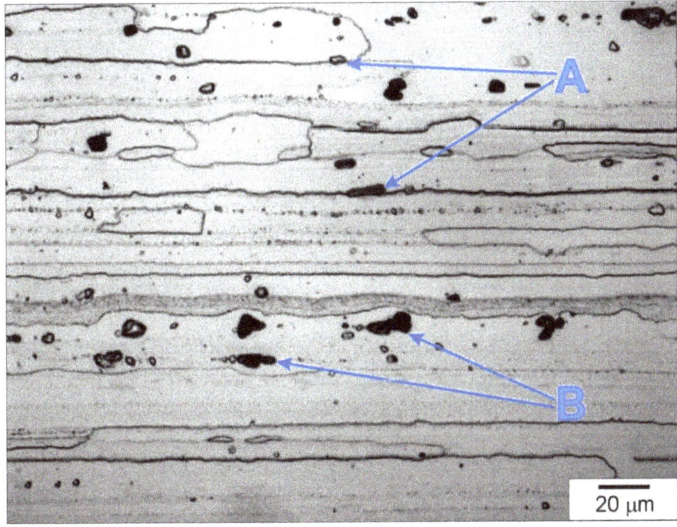

Figure 6. Microstructure of 7075 aluminum alloy in initial state, light microscope, etched specimen.

A characteristic feature of the 7075 alloy microstructure was the presence of numerous intermetallic phases located both at the boundaries (examples of precipitates are marked with the symbol A in Figure 6) and inside the deformed grains (examples of precipitates are marked with the symbol B in Figure 6). The precipitates exhibited distinct banding resulting from the applied plastic processing and were present both in the form of clusters and individual particles. The size of the precipitates varied and ranged from approximately 1 μm to 18 μm, with precipitates of dimensions from 4 to 8 μm dominating. The EDS analysis revealed the presence of several different forms of precipitates. The presence of numerous precipitates rich in Al, Zn and Cu was found, as well as numerous precipitates rich in Al, Fe and Cu and less numerous phases rich in Al, Si and Zn, as well as Al, Zn and Mg. The results of the EDS analysis of the exemplary intermetallic phases are presented in Figure 7 and Table 4. The site subjected to EDS analysis is marked with a frame in the figures. Moreover, EDS analysis of a representative part of BM (the base material) was performed in order to compare this composition with the results of the EDS analysis of the material subjected to FSP, which is discussed later in the study. The result of the EDS analysis of BM is shown in Figure 8. The height of the spectral lines of the alloying elements reflects their content in the alloy.

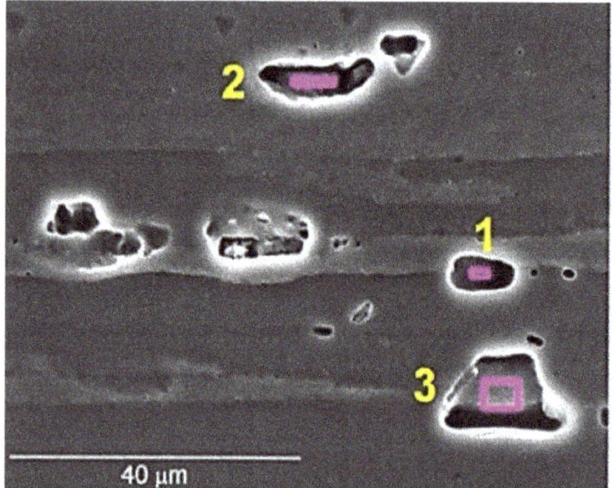

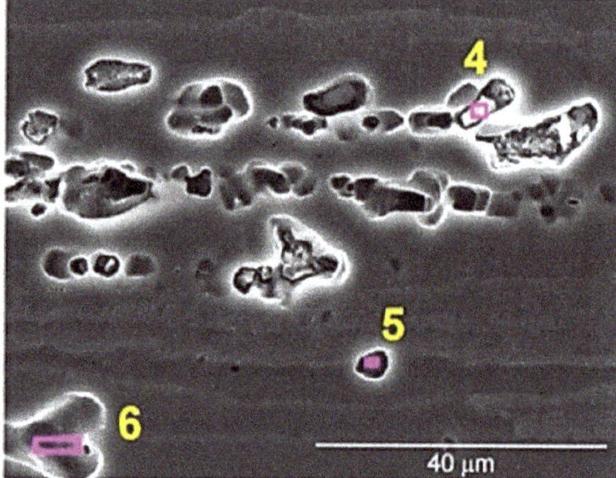

Figure 7. Precipitates of intermetallic phases in 7075 aluminum alloy in initial state, SEM.

Table 4. Results of EDS analysis of intermetallic phases.

Element	Mass Fraction of Elements in Intermetallic Phases, wt%					
	No. 1	**No. 2**	**No. 3**	**No. 4**	**No. 5**	**No. 6**
Al	80.31	27.30	88.43	44.90	83.55	78.00
Mg	1.93	0.72	2.52	2.03	2.25	1.34
Zn	10.85	3.16	7.94	4.22	11.14	15.33
Fe	–	–	–	12.80	–	–
Cu	3.95	1.20	–	29.92	3.06	5.33
Cr	0.95	–	–	0.34	–	–
O	2.01	32.78	1.11	5.79	–	–
Si	–	34.85	–	–	–	–

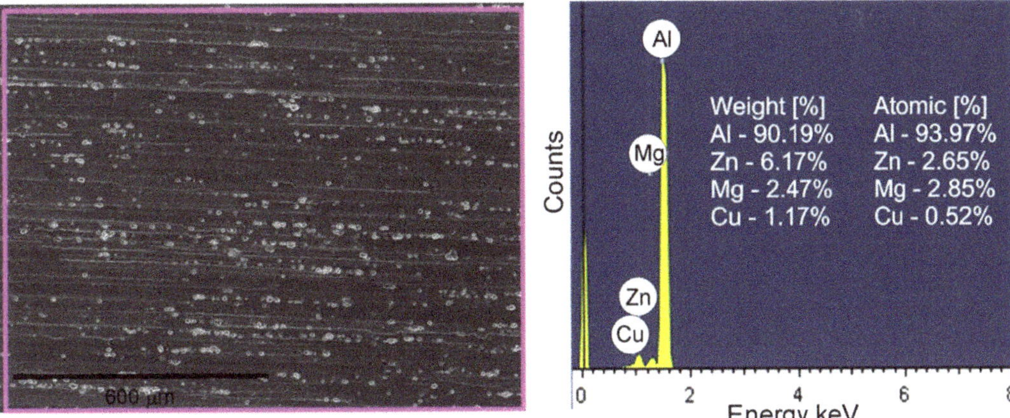

Figure 8. EDS of BM results.

3.2.2. Macro- and Microstructural Investigations of Friction-Stir-Processed 7075 Aluminium Alloy

The friction-stir-processed cross-sections of the samples are shown in Figure 9. The width and depth of the processed zone were measured on each sample. It was found that these dimensions correlate with the diameter of the shoulder of the tool (in terms of bandwidth) and the length of the pin (in terms of the depths of the processed zone). These zones for the samples cooled with the jet nozzle were slightly smaller than for the naturally-cooled samples. This regularity mainly concerned the bandwidth, and to a lesser extent, the depth of the processed zone. This fact should be explained by the differences in the cooling intensity of the samples and the resulting differences in the size of the temperature gradient, which is formed in individual samples. It should be noted, however, that the differences in the dimensions of the zone of microstructural changes were not significant. Differences in the widths of the stirring zone (SZ) between water-cooled and naturally cooled samples were found, among others, by Chen et al. [37]. In the case of using water cooling, the authors registered a reduction in the width of SZ by over 15%.

In Figure 9, the advancing and retreating sides are highlighted. As can be seen, there are clear differences in the shape and course of the processed zone border on the advancing and retreating sides. In the case of the advancing side, the border was more clearly marked than in the case of the retreating side.

The following zones were found in the surface layer of the samples subjected to FSP: the stirring zone (SZ) located in the central part of the microstructural changes zone, the thermomechanically affected zone (TMAZ) and the heat-affected zone (HAZ). The stirring zone was clearly the dominant zone. In this zone, greatly refined grains were present, mostly equiaxed or close to equiaxial in shape. Both the strong grain refinement and the grain shape in SZ prove that dynamic recrystallization (DRX) occurred during FSP, which shaped the microstructure of the material in the stirring zone. Dynamic recrystallization is a predominant mechanism for grain refinement in FSPed alloys. The occurrence of dynamic recrystallization is a consequence of the strong deformation of the material and the influence of high temperature that accompany the treatment. As indicated in the literature, dynamic recrystallization in aluminum alloys can proceed according to various mechanisms, namely, the evolution of grains can be caused by discontinuous dynamic recrystallization (DDRX), continuous dynamic recrystallization (CDRX) or geometric dynamic recrystallization (GDRX) [55]. DDRX is based on the formation of recrystallization grains and the subsequent migration of grain boundaries [56], while CDRX involves the formation of new grains by a gradual misorientation increase of the subgrains [55], while the main evidence for GDRX was related to the few retained high-aspect ratio fibrous grains [55,57]. The mechanism according to which the dynamic recrystallization of the material takes place is a function of the applied FSP parameters, but also of the intensity of the material cooling. This is confirmed by the research of Zeng et al. [55], these authors showed that different DRX mechanisms may occur in aluminum alloys subjected to FSW, depending on the welding parameters and the cooling method used.

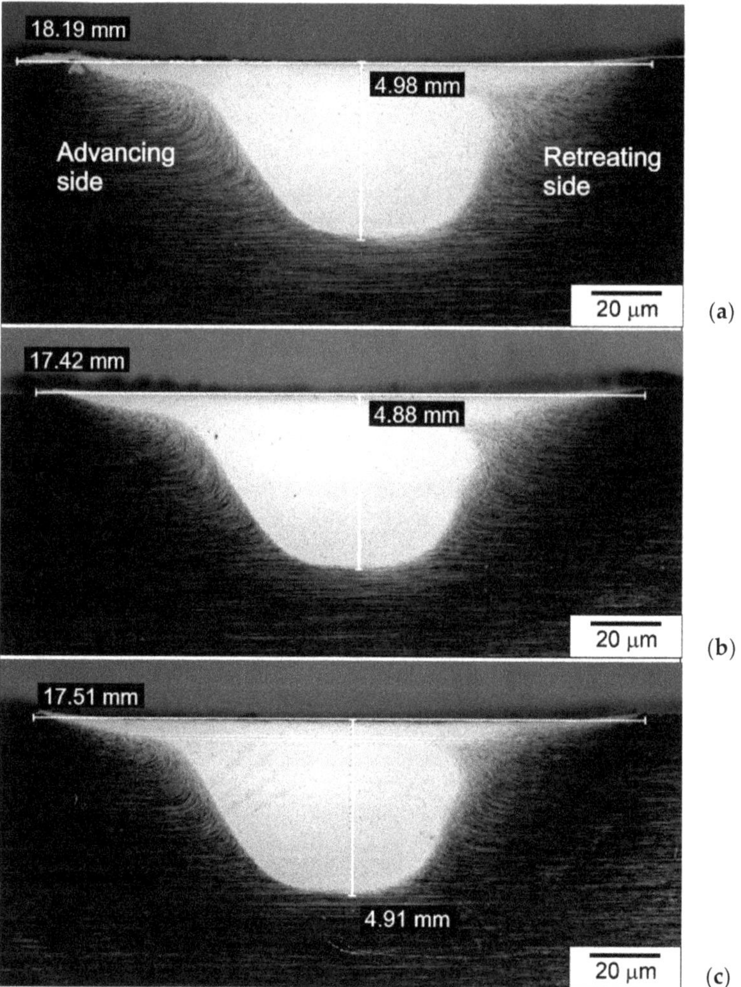

Figure 9. Processed zones of NC (**a**), AC45 (**b**), AC90 (**c**), light microscope, etched specimens.

The microstructural changes in SZ also prove that the recrystallization temperature was exceeded during FSP, which in the case of aluminum alloys is about 0.6–0.8 Tm, i.e., 300–450°C. In contrast to the SZ, in the thermomechanically affected zone, non-equiaxed grains dominated. This fact proves that the material in TMAZ was plastically deformed, but no recrystallization occurred in this zone because the temperature required for its occurrence was not reached. The non-equiaxed shape of grains in TMAZ is a consequence of lower strains and strain rates, as well as lower peak temperatures experienced by the material in this area [58]. The zone adjacent to the base material was the heat-affected zone (HAZ). In this zone, the material did not undergo plastic deformation, but is only subjected to a thermal cycle. The changes in the microstructure and properties of the material in this zone were only a consequence of the thermal influence.

The main measure of the effectiveness of the applied cooling method is, of course, the degree of grain refinement obtained in individual samples. The grain size was measured in the near-surface zone in three areas located at a distance of 0 to 100 μm from the surface, 200 to 300 μm and 400 to 500 μm, according to the diagram shown in Figure 10.

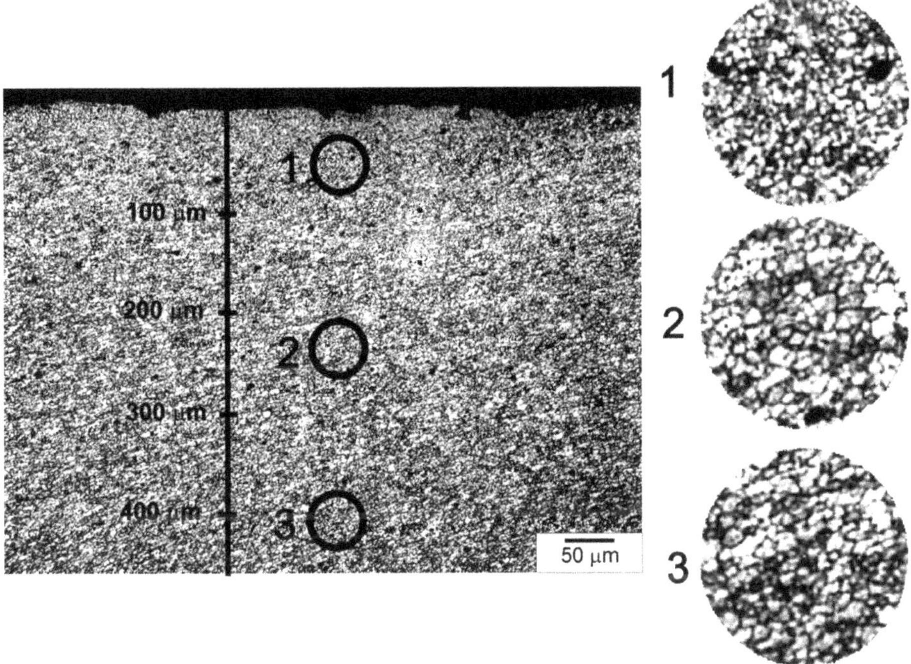

Figure 10. Differences in degree of grain refinement as function of distance from surface. AC90 sample. Sample etched. Light microscopy.

The microstructural observations revealed that the highest degree of grain refinement occurred in Zone 1, i.e., the zone closest to the surface. The grain sizes in Zones 2 and 3 were similar to each other. In the further discussion, to illustrate the differences in grain size in the different samples, the average grain size determined in the zone closest to the surface, i.e., at a distance of 0 to 100 μm, was used. Examples of microstructures observed in Zone 1 in individual samples are presented in Figure 11, and the grain size distributions are shown in Figure 12. As can be seen, the differences in the grain size are very clear, while in the case of the NC sample the average grain size was about 7.6 μm ± 1 μm, then in the case of the sample cooled with the jet cooling nozzle, the average grain size was 3.2 μm ± 1 μm for the AC90 sample and 1.4 μm ± 0.5 μm for the AC45 sample. Obviously, the degree of grain refinement is influenced by the material cooling rate; when the cooling intensity grows, the number of nucleation sites increases, and consequently, the number of grain boundaries rises, while the grain size decreases.

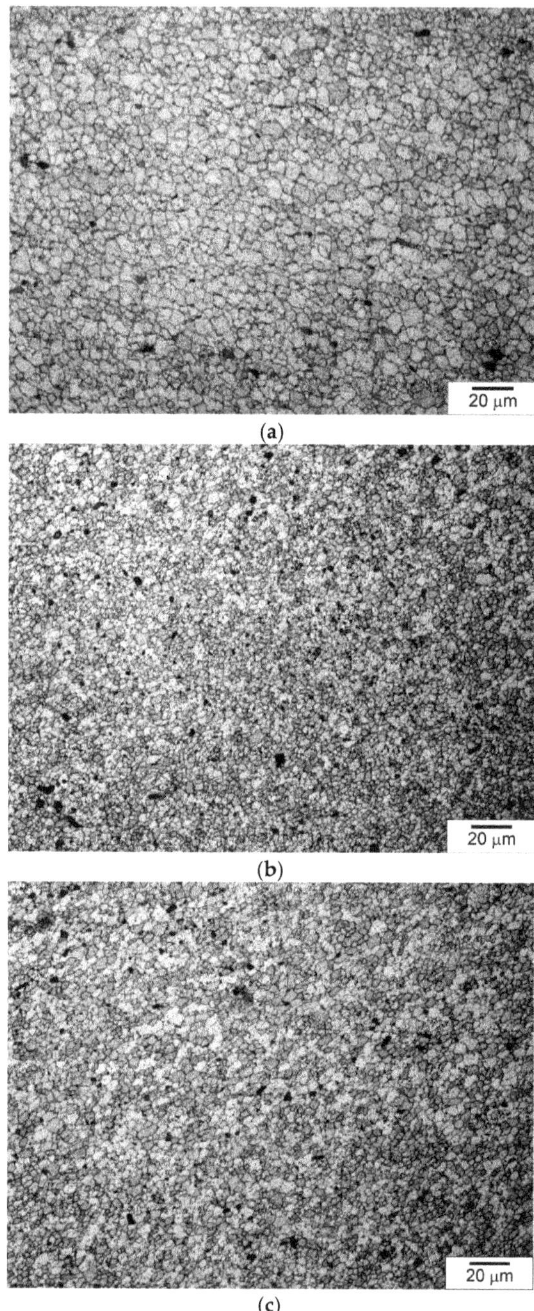

Figure 11. Microstructure of material in near-surface zone: (**a**) NC sample, (**b**) AC45 sample, (**c**) AC90 sample. Light microscope, etched specimens.

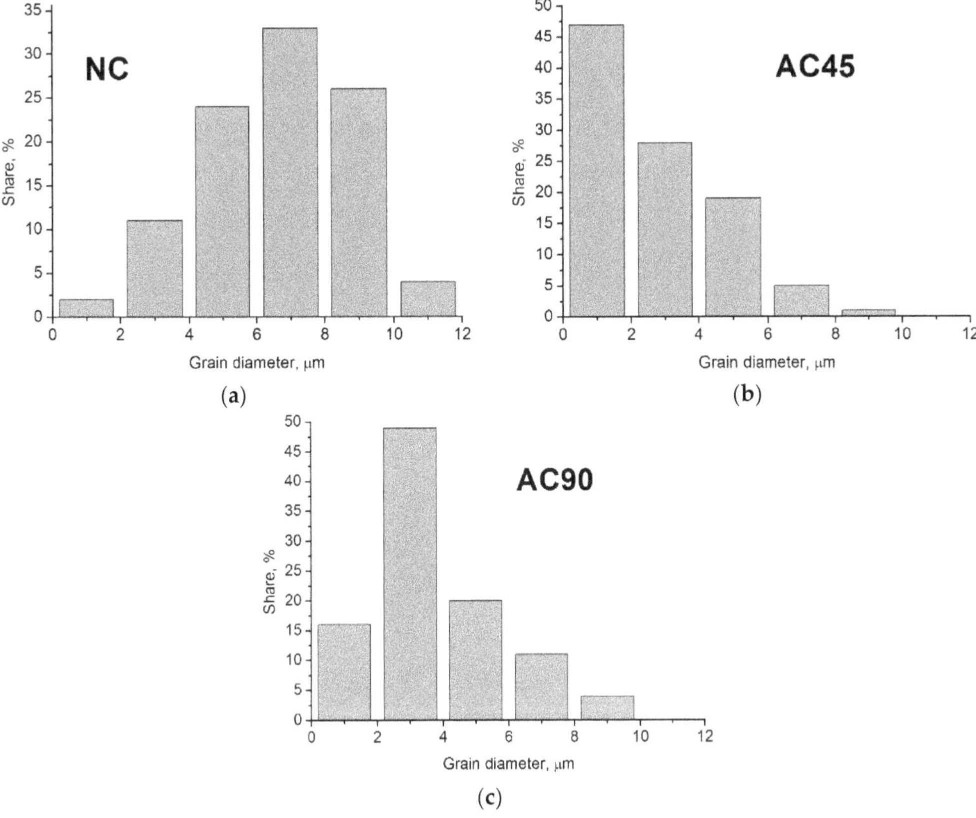

Figure 12. Grain size distributions in NC sample (**a**), AC45 sample (**b**), AC90 sample (**c**).

On the basis of the degree of obtained grain refinement, it can therefore be concluded that the most effective solution is the variant in which the nozzle is positioned at an angle of 45° in relation to the sample surface. It is then possible to obtain the highest degree of grain refinement. It is also worth noting that in this variant, the tool is also cooled, which contributes to an increase in the tool life because then the maximum temperature to which the tool is heated during FSP decreases, and thus the temperature range of the thermal cycle in which the tool works is reduced. This aspect is an additional argument in favor of just such a cooling variant. It should be noted, however, that both variants demonstrated the possibility of obtaining greater grain refinement than with natural cooling. Cooling with a jet nozzle is therefore an effective method of reducing the peak temperature during FSP, which is known to have a significant effect on grain evolution and the final grain refinement of a material. As indicated in the literature, the mechanism of the formation of recrystallized grain depends on the method of cooling and the intensity of cooling the material [55]. The factors which, apart from the cooling rate and the time of exposure to high temperature generated in the material during FSP, also affect the size of the recrystallized grain in the material, are the size of deformation and the rate of deformation of the material, which are a direct consequence of the applied process parameters. It is worth noting that the use of additional cooling also gives the operator the ability to carry out the FSP process with the use of lower traverse speed or higher rotational speeds of the tool. In the absence of additional cooling, both reducing the traverse speed or increasing the rotational speed of the tool would inevitably lead to an unfavorable increase in the grain size in the friction-stir-processed material due to higher heat input and higher peak temperature.

It is also worth adding that regardless of the cooling variant, the use of jet cooling nozzle affects not only the course and dynamics of the processes shaping the microstructure of the material, but also effectively cools the sample after FSP treatment, thanks to which the treatment is carried out in a multi-lane variant, i.e., when a series of bands covering the entire surface of the sample are made, can be performed faster as it is not necessary to cool down the sample after each cycle.

With increasing cooling intensity, the temperature gradient in the material increases, and thus the thermal stresses increase, which in the extreme case may lead to loss of cohesion by the material. In none of the analyzed samples, however, was the presence of microcracks or other material defects found, the presence of which could result from the use of air cooled with a jet cooling nozzle as a cooling agent.

The conducted microscopic examinations revealed high homogeneity of the microstructure of the material and a lack of banding of the precipitates so characteristic of the rolled material. Precipitates were present in the material subjected to FSP, but they were evenly redistributed in the volume of the processed zone and partially dissolved in the alloy matrix. Moreover, no material discontinuities were found in any of the analyzed samples, which is worth emphasizing.

EDS analyses were carried out in order to determine the differences in the chemical composition of the alloy as a function of the distance from the sample surface. The content of the elements was measured in three places, namely in the near-surface zone, in the central part of the processed zone and in the lower part of this zone. The contents of the main elements, i.e., Al, Mg, Zn, Cu, were analyzed, and the results of these analyses are presented in Table 5.

Table 5. EDS analysis results.

Sample	Element	Mass Fractions of Individual Elements, wt%			Average Value
		Near-Surface Zone	Central Part of SZ	Lower Part of SZ	
NC	Al	89.93	89.91	89.64	89.83
	Mg	2.06	2.32	2.32	2.23
	Zn	5.91	6.15	6.44	6.17
	Cu	2.10	1.62	1.61	1.78
AC45	Al	88.38	89.36	90.32	89.35
	Mg	2.45	2.05	2.18	2.23
	Zn	7.36	6.39	5.85	6.53
	Cu	1.82	2.20	1.66	1.89
AC90	Al	89.97	89.44	90.42	89.94
	Mg	2.47	2.21	2.32	2.33
	Zn	6.26	6.58	5.47	6.10
	Cu	1.30	1.77	1.79	1.62
BM	Al	90.19			
	Mg	2.47			
	Zn	6.17			
	Cu	1.17			

When analyzing the EDS results, one can notice slight differences in the shares of the main alloying elements in relation to their content in BM. The samples treated with FSP using a cooling nozzle had a higher concentration of Zn and Cu in the near-surface layer compared to the content of these elements in the core, with the differences being greater in the case of the AC45 sample. In the case of the NC sample, only a clearly increased concentration of Cu and a lower concentration of Zn in the near-surface zone were found in relation to the content of these elements in BM. According to Pang et al. [59], who analyzed the changes in the microstructure of the AA7075 aluminum alloy subjected to FSP in air and water, the higher concentration of Cu and Zn that was noted in the water-cooled samples may be due to the higher cooling rate, which prevented the redissolution of the alloying elements in the Al matrix. The results of the research carried out as part of this study confirm this hypothesis. It is worth noting, however, that the observed regularities do not apply to Mg. The content of this element in the subsurface zone was either lower than in BM (in the case of the NC samples), or the differences were marginal (in the case of the AC45 and AC90 samples). The analysis of the variability of the content of the main alloying elements as a function of the distance from the surface does not allow unequivocal conclusions to be drawn. While in the case of the AC 45 sample the Zn content clearly decreased with the distance from the surface, in the NC sample this tendency was exactly the opposite, and in the AC90 sample the highest content was found in the central part of the processed zone. There was no clear regularity in the case of Cu content either.

3.3. Hardness Measurements

As is known, the grain size has a significant effect on the properties of the material; along with a decrease in the average grain size, the yield strength of the material and the hardness increase, and the plastic properties of the material decline. The influence of the grain size on the hardness H_V of the material is described by the Hall–Petch equation [60]:

$$H_V = H_o + k \cdot d^{-1/2}$$

where H_o and k are appropriate constants and d is average grain diameter.

Since H_V is proportional to $d^{-1/2}$, the finer the grain size is, the greater the hardness value is [61]. Along with the reduction of the grain size, the number of grain boundaries increases, which constitute a barrier to dislocation, and act as pinning points inhibiting further dislocation propagation. This mechanism is the basis of the so-called "grain-boundary strengthening" or "Hall–Petch strengthening".

Taking into account the Hall–Petch equation and changes in the microstructure of the material caused by the FSP treatment, and especially the strong grain refinement in relation to the starting material, significant growth in the hardness of the material should be expected. On the other hand, the Hall–Petch equation does not take into account the numerous microstructural factors that affect or can affect the hardness of a material. In the case of the 7075 aluminum alloy, such a factor is intermetallic phase precipitates, which strengthen the material and improve its mechanical properties [62]. The potential decrease in hardness of the 7075 aluminum alloy may be a result of the dissolution of the precipitates during FSP. As can be seen from the microstructures presented in Section 3.2.2, a significant number of these precipitates dissolved in the aluminum solid solution during FSP. Nonetheless, the hardness measurement indicated that despite the dissolution of a certain part of the intermetallic phase precipitates, an increase in hardness in the subsurface layer was, nevertheless, noted. Therefore, it can be concluded that the increase in hardness resulting from the strong grain refinement, the disappearance of banding of the precipitates combined with the redistribution of undissolved precipitates effectively compensated for the decrease in hardness resulting from the reduction in the number of intermetallic phase precipitates.

Another key question was whether the use of the jet cooling nozzle contributed to a rise in the hardness of the material in relation to the naturally cooled sample. For this purpose, the average hardness of the material located in the 1 mm thick near-surface zone was calculated. The following hardness values were obtained: 153.7 HV0.1, 151.2 HV0.1, 133.1 HV0.1 for the AC45, AC90 and NC samples, respectively, with an average hardness of BM of about 129 HV0.1. Thus, in the subsurface zone, the hardness of the samples modified with the cooling nozzle in the AC90 variant was higher by about 13.6% compared to the hardness of the naturally cooled material and by about 15.4% in the AC45 variant. In relation to the hardness of BM, the hardness of the AC45 sample was higher by almost 19.2%, and the hardness of the AC90 sample was higher by almost 17.1%. Examples of hardness distributions as a function of the distance from the sample surface are shown in Figure 13, and in Table 6, the average value of hardness as a function of the distance from the processed surface along with the variability range of the examined parameter.

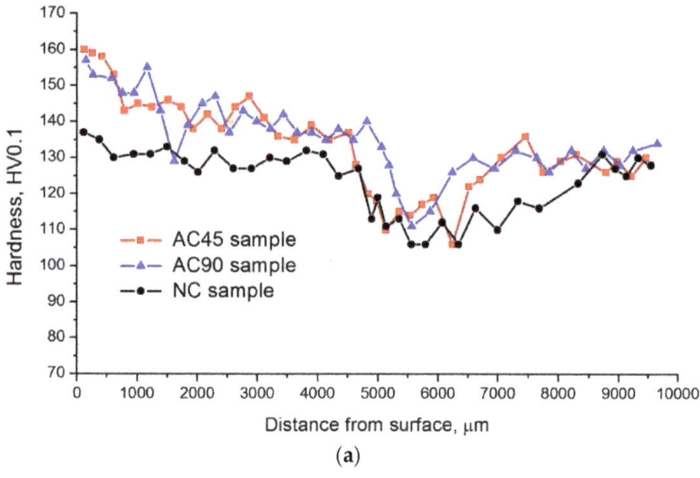

(**a**)

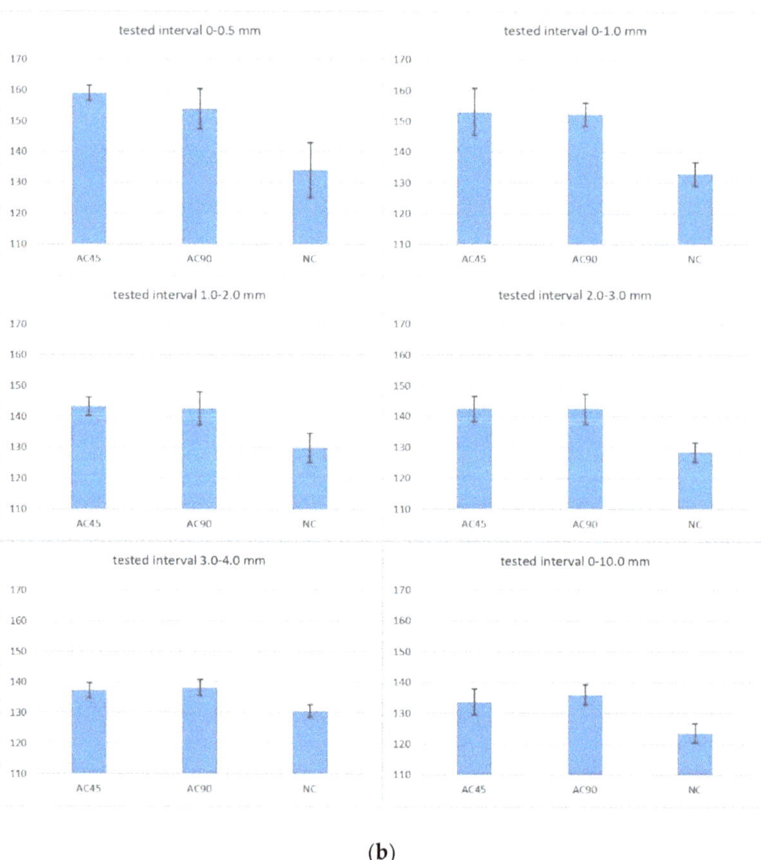

(**b**)

Figure 13. Hardness distributions as function of distance from surface (**a**), graphs of average hardness values and confidence intervals (**b**).

Table 6. Mean values of hardness as a function of distance from the surface and ranges of variability of the tested parameter.

Test Interval [mm]	Mean Value			Confidence Interval +/−		
	AC45	AC90	NC	AC45	AC90	NC
0–1 mm	153.7	151.2	133.1	7.8	3.8	3.8
1–2 mm	143.2	142.6	129.8	3.0	5.4	4.8
2–3 mm	142.4	142.4	128.4	4.2	4.9	3.1
3–4 mm	137.2	138.2	130.5	2.4	2.6	2.1
0–max (10 mm)	133.5	135.9	123.5	4.3	3.3	3.1

The analysis of the hardness distributions as a function of the distance from the surface shows that the hardness decreases in all samples as they move away from the surface, and this effect is observed to a depth of approximately 5–5.5 mm, then the hardness increases, and its stabilization is observed in the final part of the hardness distribution diagram. Therefore, in the hardness distributions, several ranges can be distinguished, differing in both the amplitude of hardness changes and the level of obtained hardness. The first zone is located at the surface, i.e., to a depth of about 1 mm from it. The material hardness in this zone is the highest and this regularity is observed in all analyzed samples. The high hardness in this zone is of course a consequence of the particularly fine grain refinement, especially in the case of samples cooled with a jet cooling nozzle. It can also be seen that the amplitude of changes in hardness in this zone, expressed by the degree of hardness decrease as a function of the distance from the surface, is clearly greater in the samples cooled with the jet cooling nozzle than in the NC sample. This fact should be explained by the fact that the very intensive cooling obtained thanks to the use of the jet cooling nozzle leads to a very strong refinement of the grain directly at the surface, which results in a clearly higher hardness of the material at this place. However, the scope of the coolant influence is limited, which results in a greater variability in hardness in this zone than in the case of the NC sample. The next zone is an area 1 to about 5–5.5 mm from the surface. In this area, the effect of stabilizing the hardness of the material is visible, the amplitude of changes in hardness in this area is definitely lower than in the near-surface zone, especially in the case of the NC sample, and in the case of AC45 and AC90 samples, a milder decrease in hardness is observed than in the near-surface zone. These results correspond with the results of microscopic tests which showed that in this zone, the grain size was characterized by relatively low variability. At a distance greater than 5–5.5 mm from the surface, i.e., at the boundary of the microstructural changes caused by the surface treatment, a clear drop in hardness to a level below the hardness of the base material can be observed, while in the case of the NC sample, the zone of reduced hardness is clearly wider than in the case of AC45 and AC90 samples, proving that the temperature gradient developed in the material during FSP was smaller, and thus the heat flow rate to BM was smaller. At a distance of more than 6 mm from the surface, an increase in hardness to the level characteristic for base material was observed in all samples. As can be seen, the amplitude of changes in hardness in this zone is relatively small.

By analyzing the literature data on the influence of FSP or FSW on the hardness of the 7075 aluminum alloy, it can be noticed that the increase in hardness is not a regularity. Despite the strong grain refinement favoring the increase in hardness, there is parallel partial dissolution of the precipitate particles, which may lead to a decrease in hardness. The hardness of the 7075 FSPed aluminum alloy is therefore the result of these processes. The lack of a significant increase in the hardness of the 7075 alloy subjected to FSP in relation to the hardness of BM was found, among others, by Sudhakar et al. [63]. Similar effects were also obtained by Kumar et al. [64].

Based on the data presented in Table 6, it can be concluded that with the assumed 95% probability, the unknown hardness falls within the confidence intervals given in this table. It should be stated that the use of additional cooling affects the change of hardness in the entire tested range (0–10 mm) and it is particularly important in the area of the surface layer and zones reaching even 3 mm from the processed surface.

3.4. Wear Resistance Tests

Wear resistance tests were carried out using the pin-on-disc method under dry friction conditions. The specimens for the tribological tests were made from the central part of individual bands. The specimens for the tribological tests had the shape of a cylinder with a diameter of 4 mm and a length equal to the thickness of the sample, i.e., 15 mm. In order to eliminate the influence of differences in the geometrical structure of the surface on the amount of wear of the material, the surface layer, with a thickness of about 25 μm, was removed. This layer was removed with the tribology tester, stopping the abrasion of the specimen when the tester showed a 25 μm linear material loss. Then, the surfaces of the specimens were cleaned of crushed particles and degreased, after which the actual test was started. The test lasted 90 min and was carried out in ambient temperature, which was 23 ± 1 °C. The disc rotation speed was 150 rpm and was identical for all the specimens. The counter-specimen was ring-shaped and made of EN 100Cr6 hard bearing steel (E52100 according to AISI). The wear resistance was determined by measuring the linear wear of the tested specimen. Both friction-stir-processed specimens and the starting material were tested for wear resistance to illustrate the effect of processing on the wear resistance of alloys and possible differences in the wear resistance of the samples modified using different variants of sample cooling. The results of the tribological tests are summarized in Table 7 and presented in Figure 14.

Table 7. Linear losses recorded during tribological test.

Sample	Linear Loss (μm)
BM	116
NC sample	121
AC45 sample	99.6
AC90 sample	100.5

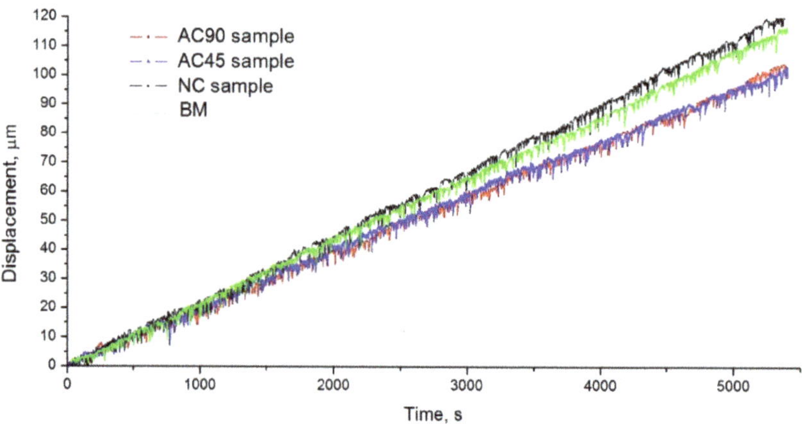

Figure 14. Results of tribological test.

The obtained results prove that the AC45 and AC90 specimens were characterized by a comparable wear intensity, but the wear was slightly lower than that of the analogous naturally cooled material and BM. These results are therefore consistent with the Archard wear equation, according to which the wear rate of the material is inversely proportional to the hardness of the material [65]. Notwithstanding, in the case of the NC specimen, the linear material loss was greater than in the case of BM, even though the hardness of the NC specimen was greater than that of BM. Hence, there was no regularity resulting from the Archard wear equation for the NC specimen. Similar results were obtained, among others, by Gholami et al. [66]; the alloy subjected to FSP analyzed by them was characterized by a lower wear rate than the base material, and the registered wear rates were 1.32 and 1.67 $mm^3 \cdot N^{-1} \cdot m^{-1}$, respectively. The lower wear resistance of the NC specimen than BM is mainly a consequence of the partial dissolution of the intermetallic phase precipitates. Although in the case of the AC45 and AC90 samples, there was also a dissolving effect of the intermetallic phase precipitates, the higher degree of grain refinement achieved in these cases largely compensated for the decrease in wear resistance resulting from the reduced number of intermetallic particles.

The samples after the tribological tests were then subjected to SEM examinations in order to visualize their geometric structure and determine the wear mechanism. The conducted observations revealed that the material abraded during the test by micro-cutting and grooving, which proves the abrasive wear of the material in this case. The characteristic scratches and grooves formed on the surface during the tribological test of the material are shown in Figure 15. No significant differences were found in the geometric structure of individual samples subjected to tribological tests. An important parameter useful in the comparative evaluation of the wear resistance of samples with an identical wear mechanism is the width of the grooves. A smaller width of the grooves proves a lesser deformation of the material, and thus also indicates a greater wear resistance of the tested material [67]. SEM tests of the surfaces of NC, AC45 and AC90 samples showed that the width of the grooves in the analyzed samples was similar, therefore, these results correspond to the results of the tribological test, which showed only slight differences in the wear resistance of individual samples. In addition, all the specimens exhibited the local presence of spalled regions, which proves that oxidative wear also occurred during the tribological tests. Under conditions of high friction and the accompanying thermal effect, the risk of oxidation of the sample surface increases, and the intensity of this process also depends on the physicochemical properties of the material, especially its affinity for oxygen. Aluminum is a metal with a strong affinity for oxygen; hence, the likelihood of an oxidation layer on the surface of the sample is very high. It is also worth adding that the oxidized places are more brittle than the metallic substrate; they do not form a permanent bond with the substrate, and therefore can easily detach from the substrate, leaving a characteristic spalled region in the place of detachment.

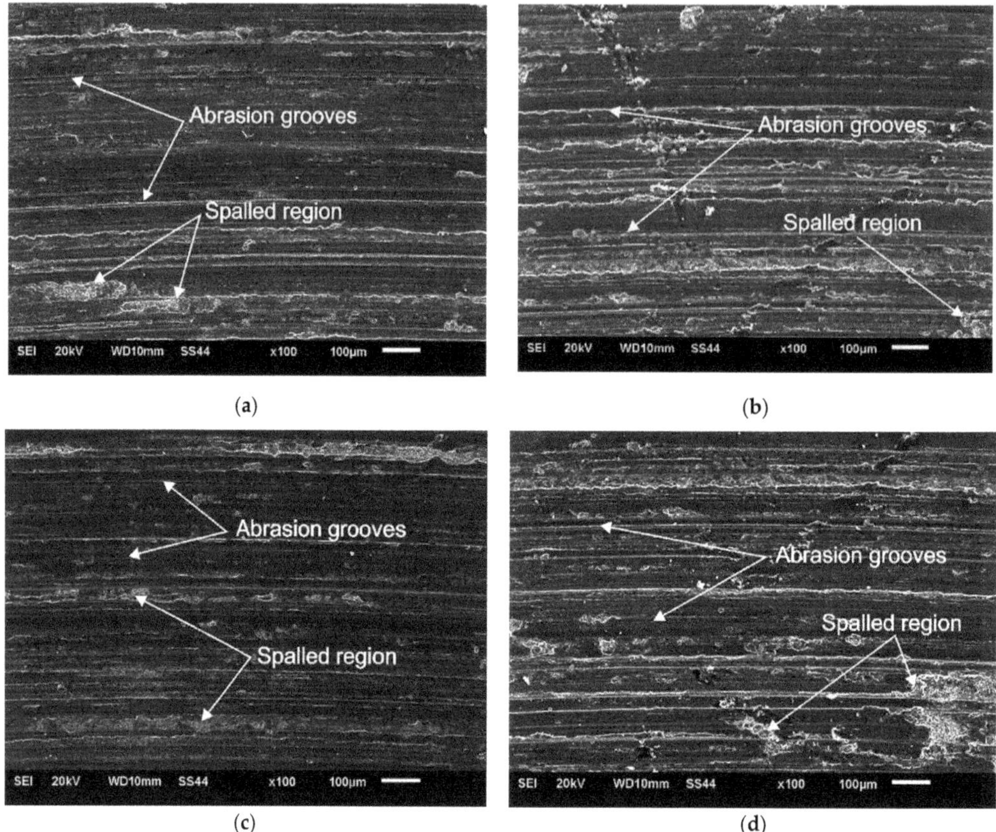

Figure 15. Worn surface morphologies of as-cast specimen (**a**), NC specimen (**b**), AC45 specimen (**c**), AC90 specimen (**d**). SEM.

4. Conclusions

The paper presents an analysis of the microstructure and selected properties of the 7075 alloy subjected to FSP with the use of various cooling variants. The novelty of the work is an innovative method of cooling the sample during FSP, namely, by means of an air stream cooled to −11 °C with a jet cooling nozzle. Therefore, an important aspect of the work was also the assessment of the suitability of the applied sample cooling method in shaping the microstructure and material properties.

Microstructural tests and selected material properties showed high efficiency of the new cooling method in relation to cooling in still air, which is expressed by more favorable changes in the microstructure, greater hardness and resistance to wear. The conducted research shows that the use of a jet cooling nozzle, intensifying the cooling process, leads to greater grain refinement, a more homogeneous microstructure and redistribution of intermetallic phase precipitates. The grain size decreases from 240 μm in BM to 7.6 μm in the NC sample and to 3.2 μm and 1.4 μm in AC90 and AC45, respectively. The consequence of microstructural changes caused by processing with air cooling was, in turn, an improvement in hardness, and in the case of the AC45 and AC90 samples, also the abrasion resistance of the material in relation to BM, or the sample subjected to FSP, but modified without employing a cooling system. The application of additional cooling allows the obtained microstructural effects to be influenced, e.g., by changing the position of the cooling nozzle, changing the distance of the nozzle from the frictional site, or changing the air pressure supplied to the cooling nozzle, determining the target air temperature at the outlet of the jet nozzle. The most effective solution is to cool both the surface of the processed material and the tool itself with a stream of air, which was done in the case of the AC45 sample. In the opinion of the authors of this paper, the presented solution has application potential and can be successfully employed in industrial practice, being a competitive solution to the currently used material cooling techniques.

Author Contributions: Conceptualization, J.I. and K.K.; methodology, J.I. and K.K.; material research, J.I. and K.K.; writing, J.I.; review and editing, J.I. and K.K.; supervision, J.I. All authors have read and agreed to the published version of the manuscript.

Funding: This research received no external funding.

Institutional Review Board Statement: Not applicable.

Informed Consent Statement: Not applicable.

Data Availability Statement: The data that support the findings of this study are available from the corresponding author, J. Iwaszko, upon reasonable request.

Conflicts of Interest: The authors declare no conflict of interest.

References

1. Mishra, R.S.; Mahoney, M.W.; McFadden, S.X.; Mara, N.A.; Mukherjee, A.K. High strain rate superplasticity in a friction stir processed 7075 Al alloy. *Scr. Mater.* **1999**, *42*, 163–168. [CrossRef]
2. Mishra, R.S.; Mahoney, M.W. Friction stir processing: A new grain refinement technique to achieve high strain rate superplasticity in commercial alloys. *Mater. Sci. Forum* **2001**, *357–359*, 507–514. [CrossRef]
3. Thomas, W.M. Friction Stir Butt Welding. International Patent Application No. PCT/GB92/02203, 6 December 1991.
4. Kosturek, R.; Śnieżek, L.; Torzewski, J.; Wachowski, M. Research on the friction stir welding of Sc-modified AA2519 extrusion. *Metals* **2019**, *9*, 1024. [CrossRef]
5. Kubit, A.; Bucior, M.; Kluz, R.; Święch, Ł.; Ochał, K. Application of the 3D digital image correlation to the analysis of deformation of joints welded with the FSW method after shot peening. *Adv. Mater. Sci.* **2019**, *19*, 57–66. [CrossRef]
6. Patel, V.; Li, W.; Vairis, A.; Badheka, V. Recent development in friction stir processing as a solid-state grain refinement technique: Microstructural evolution and property enhancement. *Crit. Rev. Solid State Mater. Sci.* **2018**, *44*, 377–425. [CrossRef]
7. Tripathi, H.; Bharti, A.; Saxena, K.K.; Kumar, N. Improvement in mechanical properties of structural AZ91 magnesium alloy processed by friction stir processing. *Adv. Mater. Process. Technol.* **2021**, 1–14. [CrossRef]
8. Dani, M.S.; Dave, I.B. Grain refinement and improvement in microhardness of AZ91 Mg alloy via friction stir processing. In *Recent Advances in Mechanical Infrastructure*; Springer: Singapore, 2020; pp. 197–209. [CrossRef]
9. Leszczyńska-Madej, B.; Madej, M.; Hrabia-Wiśnios, J.; Węglowska, A. Effects of the processing parameters of friction stir processing on the microstructure, hardness and tribological properties of SnSbCu bearing alloy. *Materials* **2020**, *13*, 5826. [CrossRef]
10. Janeczek, A.; Tomków, J.; Fydrych, D. The influence of tool shape and process parameters on the mechanical properties of AW-3004 aluminium alloy friction stir welded joints. *Materials* **2021**, *14*, 3244. [CrossRef]
11. Iwaszko, J.; Kudła, K.; Fila, K.; Strzelecka, M. The effect of friction stir processing (FSP) on the microstructure and properties of AM60 magnesium alloy. *Arch. Metall. Mater.* **2016**, *61*, 1209–1214. [CrossRef]
12. Luo, X.; Cao, G.; Zhang, W.; Qiu, C.; Zhang, D. Ductility improvement of an AZ61 magnesium alloy through two-pass submerged friction stir processing. *Materials* **2017**, *10*, 253. [CrossRef]
13. Gan, Y.X.; Solomon, D.; Reinbolt, M. Friction stir processing of particle reinforced composite materials. *Materials* **2010**, *3*, 329–350.
14. Sajed, M.; Seyedkashi, S.M.H. Multilayer friction stir plug welding: A novel solid-state method to repair cracks and voids in thick aluminum plates. *CIRP J. Manuf. Sci. Technol.* **2020**, *31*, 467–477. [CrossRef]
15. Velmurugan, T.; Subramanian, R.; Suganya Priyadharshini, G.; Raghu, R. Experimental investigation of microstructure, mechanical and wear characteristics of Cu-Ni/ZrC composites synthesized through friction stir processing. *Arch. Metall. Mater.* **2020**, *65*, 565–574.
16. Vaira Vignesh, R.; Padmanaban, R.; Govindaraju, M. Synthesis and characterization of magnesium alloy surface composite (AZ91D—SiO_2) by friction stir processing for bioimplants. *Silicon* **2020**, *12*, 1085–1102.
17. Deepan, M.; Pandey, C.; Saini, N.; Mahapatra, M.M.; Mulik, R.S. Estimation of strength and wear properties of Mg/SiC nanocomposite fabricated through FSP route. *J. Braz. Soc. Mech. Sci. Eng.* **2017**, *39*, 4613–4622. [CrossRef]
18. Iwaszko, J.; Kudła, K. Effect of friction stir processing (FSP) on microstructure and hardness of AlMg10/SiC composite. *Bull. Pol. Acad. Sci. Tech. Sci.* **2019**, *67*, 185–192.
19. Stawiarz, M.; Kurtyka, P.; Rylko, N.; Gluzman, S. Influence of FSP process modification on selected properties of Al-Si-Cu/SiC composite surface layer. *Compos. Theory Pract.* **2019**, *19*, 161–168.

20. Mohammed, M.H.; Subhi, A.D. Exploring the influence of process parameters on the properties of SiC/A380 Al alloy surface composite fabricated by friction stir processing. *Eng. Sci. Technol. Int. J.* **2021**, *69*, 1272–1280. [CrossRef]
21. Bagheri, B.; Abbasi, M. Development of AZ91/SiC surface composite by FSP: Effect of vibration and process parameters on microstructure and mechanical characteristics. *Adv. Manuf.* **2020**, *8*, 82–96. [CrossRef]
22. Iwaszko, J.; Kudła, K. Characterization of Cu/SiC surface composite produced by friction stir processing. *Bull. Pol. Acad. Sci. Tech. Sci.* **2020**, *68*, 555–564.
23. Mertens, A.; Simar, A.; Montrieux, H.-M.; Halleux, J.; Lecomte-Beckers, J. Friction stir processing of magnesium matrix composites reinforced with carbon fibres: Influence of the matrix characteristics and of the processing parameters on microstructural developments. In Proceedings of the 9th International Conference on Magnesium Alloys and their Applications, Vancouver, BC, Canada, 8–12 July 2012; pp. 845–850.
24. Iwaszo, J.; Sajed, M. Technological aspects of producing surface composites by friction stir processing—A review. *J. Compos. Sci.* **2021**, *5*, 323. [CrossRef]
25. Papantoniou, I.G.; Markopoulos, A.P.; Manolakos, D.E. A new approach in surface modification and surface hardening of aluminum alloys using friction stir process: Cu-reinforced AA5083. *Materials* **2020**, *13*, 1278. [CrossRef] [PubMed]
26. Morisada, Y.; Fujii, H.; Nagaoka, T.; Fukusumi, M. Effect of friction stir processing with SiC particles on microstructure and hardness of AZ31. *Mater. Sci. Eng. A* **2006**, *433*, 50–54. [CrossRef]
27. Lv, Y.; Ding, Z.; Sun, X.; Li, L.; Sha, G.; Liu, R.; Wang, L. Gradient microstructures and mechanical properties of Ti-6Al-4V/Zn composite prepared by friction stir processing. *Materials* **2019**, *12*, 2795. [CrossRef]
28. Iwaszko, J.; Kudła, K.; Fila, K. Technological aspects of friction stir processing of AlZn5.5MgCu aluminum alloy. *Bull. Pol. Acad. Sci. Tech. Sci.* **2018**, *66*, 713–719.
29. Huang, Y.; Wang, T.; Guo, W.; Wan, L.; Lv, S. Microstructure and surface mechanical property of AZ31 Mg/SiCp surface composite fabricated by direct friction stir processing. *Mater. Des.* **2014**, *59*, 274–278. [CrossRef]
30. Almazrouee, A.I.; Al-Fadhalah, K.J.; Alhajeri, S.N. A new approach to direct friction stir processing for fabricating surface composites. *Crystals* **2021**, *11*, 638. [CrossRef]
31. Takhakh, A.M.; Abdulla, H.H. Improving the mechanical properties of Al7075-T651 welded joint using direct friction stir processing. In Proceedings of theThe Second Conference of Post Graduate Researches (CPGR'2017) College of Engineering, Baghdad, Iraq, 4 October 2017.
32. Ratna Sunil, B. Different strategies of secondary phase incorporation into metallic sheets by friction stir processing in developing surface composites. *Int. J. Mech. Mater. Eng.* **2016**, *11*, 12.
33. Houshyar, M.; Nourouzi, S.; Aval, H.J. Sandwich method: Strategy to fabricate Al/SiC composites by FSP. *Trans. Indian Inst. Met.* **2019**, *72*, 3249–3259. [CrossRef]
34. Thapliyal, S.; Dwivedi, D.K. Study of the effect of friction stir processing of the sliding wear behavior of cast NiAl bronze: A statistical analysis. *Tribol. Int.* **2016**, *97*, 124–135. [CrossRef]
35. Ura-Bińczyk, E.; Bałkowiec, A.Z.; Mikołajczyk, Ł.; Lewandowska, M.; Kurzydłowski, K.J. Wpływ wielkości ziarna na odporność korozyjną stopu aluminium 7475. *Ochr. Przed Korozją* **2011**, *2*, 44–47.
36. Ai, X.; Yue, Y. Microstructure and mechanical properties of friction stir processed A356 cast Al under air cooling and water cooling. *High Temp. Mater. Process.* **2018**, *37*, 693–699. [CrossRef]
37. Chen, Y.; Jiang, Y.; Zhang, F.; Ding, H.; Zhao, J.; Ren, Z. Water cooling effects on the microstructural evolution and mechanical properties of friction-stir-processed Al-6061 alloy. *Trans. Indian Inst. Met.* **2018**, *71*, 3077–3087. [CrossRef]
38. Patel, V.; Badheka, V.; Li, W.; Akkireddy, S. Hybrid friction stir processing with active cooling approach to enhance superplastic behavior of AA7075 aluminum alloy. *Arch. Civ. Mech. Eng.* **2019**, *19*, 1368–1380. [CrossRef]
39. Akbari, M.; Khalkhali, A.; Keshavarz, S.M.E.; Sarikhani, E. The effect of in-process cooling conditions on temperature, force, wear resistance, microstructural, and mechanical properties of friction stir processed A356. *J. Mat. Des. Appl.* **2018**, *232*, 429–437. [CrossRef]
40. Moaref, A.; Rabiezadeh, A. Microstructural evaluation and tribological properties of underwater friction stir processed CP-copper and its alloy. *Trans. Nonferr. Met. Soc. China* **2020**, *30*, 972–981. [CrossRef]
41. Feng, X.; Liu, H.; Lippold, J.C. Microstructure characterization of the stir zone of submerged friction stir processed aluminum alloy 2219. *Mater. Charact.* **2013**, *82*, 97–102. [CrossRef]
42. Ammouri, A.H.; Kridli, G.T.; Ayoub, G.; Hamade, R.F. Investigating the effect of cryogenic pre-cooling on the friction stir processing of AZ31B. In Proceedings of the World Congress on Engineering, London, UK, 2–4 July 2014; Volume 2.
43. Heidarpour, A.; Ahmadifard, S.; Rohani, N. FSP pass number and cooling effects on the microstructure and properties of AZ31. *J. Adv. Mater. Process.* **2018**, *6*, 47–58.
44. Aghajani Derazkola, H.; García, E.; Eyvazian, A.; Aberoumand, M. Effects of rapid cooling on properties of aluminum-steel friction stir welded joint. *Materials* **2021**, *14*, 908. [CrossRef]
45. Yazdipour, A.; Shafiei, A.; Dehghani, M.K. Modeling the microstructural evolution and effect of cooling rate on the nanograins formed during the friction stir processing of Al5083. *Mater. Sci. Eng. A* **2009**, *527*, 192–197. [CrossRef]
46. Satyanarayana, M.V.N.V.; Reddy, P.; Kumar, A. Preparation of bulk-area stir zone in aluminium 6061 alloy via cryogenic friction stir processing. *Mater. Today Proc.* **2021**, *41*, 987–990. [CrossRef]

47. Alavi Nia, A.; Omidvar, H.; Nourbakhsh, S.H. Investigation of the effects of thread pitch and water cooling action on the mechanical strength and microstructure of friction stir processed AZ31. *Mater. Des.* **2013**, *52*, 615–620. [CrossRef]
48. Yang, X.; Dong, P.; Yan, Z.; Cheng, B.; Zhai, X.; Chen, H.; Zhang, H.; Wang, W. AlCoCrFeNi high-entropy alloy particle reinforced 5083Al matrix composites with fine grain structure fabricated by submerged friction stir processing. *J. Alloys Compd.* **2020**, *836*, 155411. [CrossRef]
49. Orozco-Caballero, A.; Cepeda-Jiménez, C.M.; Hidalgo-Manrique, P.; Rey, P.; Gesto, D.; Verdera, D.; Ruano, O.A.; Carreño, F. Lowering the temperature for high strain rate superplasticity in an Al–Mg–Zn–Cu alloy via cooled friction stir processing. *Mater. Chem. Phys.* **2013**, *142*, 182–185. [CrossRef]
50. Ramaiyan, S.; Chandran, R.; Santhanam, S.K.V. Effect of cooling conditions on mechanical and microstructural behaviours of friction stir processed AZ31B Mg alloy. *Mod. Mech. Eng.* **2017**, *7*, 144–160. [CrossRef]
51. Su, J.-Q.; Nelson, T.W.; Sterling, C.J. Friction stir processing of large-area bulk UFG aluminum alloys. *Scr. Mater.* **2005**, *52*, 135–140. [CrossRef]
52. Iwaszko, J.; Kudła, K. Microstructure, hardness, and wear resistance of AZ91 magnesium alloy produced by friction stir processing with air-cooling. *Int. J. Adv. Manuf. Technol.* **2021**, *116*, 1309–1323. [CrossRef]
53. Sunar, T.; Tuncay, T.; Özyürek, D.; Gürü, M. Investigation of mechanical properties of AA7075 alloys aged by various heat treatments. *Phys. Met. Metallogr.* **2020**, *121*, 1440–1446. [CrossRef]
54. Vortex Tubes. Available online: https://www.pneuma.pl/produkty/rurki-wirowe-i-produkty-do-chlodzenia-miejscowego (accessed on 25 January 2022).
55. Zeng, X.H.; Xue, P.; Wu, L.H.; Ni, D.R.; Xiao, B.L.; Wang, K.S.; Ma, Z.Y. Microstructural evolution of aluminum alloy during friction stir welding under different tool rotation rates and cooling conditions. *J. Mater. Sci. Technol.* **2019**, *35*, 972–981. [CrossRef]
56. Jata, K.V.; Semiatin, S.L. Continuous dynamic recrystallization during friction stir welding of high strength aluminum alloys. *Scripta Mater.* **2000**, *43*, 743–749. [CrossRef]
57. Prangnell, P.B.; Heason, C.P. Grain structure formation during friction stir welding observed by the 'stop action technique'. *Acta Mater.* **2005**, *53*, 3179–3192. [CrossRef]
58. McNelley, T.R.; Swaminathan, S.; Su, J.Q. Recrystallization mechanisms during friction stir welding/processing of aluminum alloys. *Scr. Mater.* **2008**, *58*, 349–354. [CrossRef]
59. Pang, J.J.; Liu, F.C.; Liu, J.; Tan, M.J.; Blackwood, D.J. Friction stir processing of aluminium alloy AA7075: Microstructure, surface chemistry and corrosion resistance. *Corros. Sci.* **2016**, *106*, 217–228. [CrossRef]
60. Sato, Y.S.; Park, S.H.C.; Kokawa, H. Microstructural factors governing hardness in friction-stir welds of solid-solution-hardened Al alloys. *Metall. Mater. Trans.* **2001**, *32*, 3033–3042. [CrossRef]
61. Nakata, K.; Kim, Y.G.; Fujii, H.; Tsumura, T.; Komazaki, T. Improvement of mechanical properties of aluminum die casting alloy by multi-pass friction stir processing. *Mater. Sci. Eng.* **2006**, *437*, 274–280.
62. Iwaszko, J.; Kudła, K. Surface remelting treatment of 7075 aluminum alloy—Microstructural and technological aspect. *Mater. Res. Express* **2020**, *7*, 016523. [CrossRef]
63. Sudhakar, I.; Madhu, V.; Madhusudhan Reddy, G.; Srinivasa Rao, K. Enhancement of wear and ballistic resistance of armour grade AA7075 aluminium alloy using friction stir processing. *Def. Technol.* **2015**, *11*, 10–17. [CrossRef]
64. Kumar, A.; Kumar, S.S.; Pal, K.; Mula, S. Effect of process parameters on microstructural evolution, mechanical properties and corrosion behavior of friction stir processed Al 7075 alloy. *J. Mater. Eng. Perform.* **2017**, *26*, 1122–1134. [CrossRef]
65. Archard, J.F. Contact and rubbing of flat surface. *J. Appl. Phys.* **1953**, *24*, 981–988. [CrossRef]
66. Gholami, S.; Emadoddin, E.; Tajally Borhani, E. Friction stir processing of 7075 Al alloy and subsequent aging treatment. *Trans. Nonferr. Met. Soc. China* **2015**, *25*, 2847–2855. [CrossRef]
67. Saini, N.; Pandey, C.; Thapliyal, S.; Dwivedi, D.K. Mechanical properties and wear behavior of Zn and MoS_2 reinforced, surface composite Al- Si alloys using friction stir processing. *Silicon* **2018**, *10*, 1979–1990. [CrossRef]

 *materials* 

Article

Mechanical and Microstructural Characterization of Friction Stir Welded SiC and B_4C Reinforced Aluminium Alloy AA6061 Metal Matrix Composites

Kaveripakkam Suban Ashraff Ali [1], Vinayagam Mohanavel [2], Subbiah Arungalai Vendan [3], Manickam Ravichandran [4], Anshul Yadav [5], Marek Gucwa [6] and Jerzy Winczek [6,*]

1 Department of Mechanical Engineering, C. Abdul Hakeem College of Engineering & Technology, Vellore 632509, India; ashraffali1986@gmail.com
2 Centre for Materials Engineering and Regenerative Medicine, Bharath Institute of Higher Education and Research, Chennai 600073, India; mohanavel2k16@gmail.com
3 Department of Electronics & Communication, Dayananda Sagar University, Bangalore 506114, India; arungalaisv@yahoo.co.in
4 Department of Mechanical Engineering, K. Ramakrishnan College of Engineering, Trichy 621112, India; smravichandran@hotmail.com
5 Membrane Science and Separation Technology Division, CSIR-Central Salt and Marine Chemicals Research Institute, Bhavnagar 364002, India; anshuly@csmcri.res.in
6 Department of Technology and Automation, Częstochowa University of Technology, 42-201 Częstochowa, Poland; mgucwa@spaw.pcz.pl
* Correspondence: winczek@imipkm.pcz.pl

Citation: Ali, K.S.A.; Mohanavel, V.; Vendan, S.A.; Ravichandran, M.; Yadav, A.; Gucwa, M.; Winczek, J. Mechanical and Microstructural Characterization of Friction Stir Welded SiC and B_4C Reinforced Aluminium Alloy AA6061 Metal Matrix Composites. *Materials* **2021**, *14*, 3110. https://doi.org/10.3390/ma14113110

Academic Editor: Hideki Hosoda

Received: 25 April 2021
Accepted: 3 June 2021
Published: 5 June 2021

Abstract: This study focuses on the properties and process parameters dictating behavioural aspects of friction stir welded Aluminium Alloy AA6061 metal matrix composites reinforced with varying percentages of SiC and B_4C. The joint properties in terms of mechanical strength, microstructural integrity and quality were examined. The weld reveals grain refinement and uniform distribution of reinforced particles in the joint region leading to improved strength compared to other joints of varying base material compositions. The tensile properties of the friction stir welded Al-MMCs improved after reinforcement with SiC and B_4C. The maximum ultimate tensile stress was around 172.8 ± 1.9 MPa for composite with 10% SiC and 3% B_4C reinforcement. The percentage elongation decreased as the percentage of SiC decreases and B_4C increases. The hardness of the Al-MMCs improved considerably by adding reinforcement and subsequent thermal action during the FSW process, indicating an optimal increase as it eliminates brittleness. It was seen that higher SiC content contributes to higher strength, improved wear properties and hardness. The wear rate was as high as 12 ± 0.9 g/s for 10% SiC reinforcement and 30 N load. The wear rate reduced for lower values of load and increased with B_4C reinforcement. The microstructural examination at the joints reveals the flow of plasticized metal from advancing to the retreating side. The formation of onion rings in the weld zone was due to the cylindrical FSW rotating tool material impression during the stirring action. Alterations in chemical properties are negligible, thereby retaining the original characteristics of the materials post welding. No major cracks or pores were observed during the non-destructive testing process that established good quality of the weld. The results are indicated improvement in mechanical and microstructural properties of the weld.

Keywords: stir casting; boron carbide; silicon carbide; AA6061 aluminium alloy; tensile strength; friction stir welding

1. Introduction

In many industrial applications, aluminium alloys are reinforced with hard ceramic particles to enhance the mechanical properties of aluminium metal matrix composites (Al-MMCs) [1–3]. Aluminium MMCs are lightweight material accompanied with good thermal and electrical conductivity, high stiffness, hardness, strength, melting point and wear resistance [4–6]. Due to their higher processing prices, Al-MMCs are deployed only for military weapons and aerospace applications. Aluminium MMCs have further found applications in automobile products, such as pistons, engine, disk brakes, cylinder liners and drum brakes [7,8]. Fusion welding of Al-MMCs creates brittle intermetallic components within the weld region's matrix and the reinforcement particles. The induced stress in the weld reduces the joint efficiency and reveals porosity and voids at the joint [9,10].

In the friction stir welding (FSW) process, a non-consumable rotating tool having higher toughness than the base material is pitched into the faying/butt ends of the plates to be welded. They are subject to appropriate axial force developed along the line of the joint. Pin and shoulder are the two key portions of the tool. The material experiencing the rotational movement of the tool pin is unstiffened by the frictional heat generated by the tool during the spinning action. The rotating tool drives the plastically distorted material from the front towards the reverse side of the tool. Subsequent forging facilitates the achievement of the weld. Though a solid-state joining process with the absence of material melting, FSW reveals a lack of stimulated melt solidified structure in the weld zone. This addresses the technical lacuna's encountered during the fusion welding of composites [11–16]. A few impactful literature related to the FSW of Al-MMCs has been reported [17–22]. Vijay and Murugan [22] reported that fine grain size and high tensile strength was achieved for the square pin profile of friction stir welded AA6061/TiB_2 composite. Xu et al. [21] stated that the high tensile strength values are obtained for high tool rotating speed. Topcu et al. [20] emphasized that the hardness increases with an increase in the B_4C reinforcement. Ali et al. [18] investigated the hardness and tensile properties of weld specimen and linked them with microstructural variation in AA6061/SiC/B_4C composites. Palanivelet et al. [17] revealed that the nugget zone and the grain growth are affected when the time of exposure of the FSW tool is high. Wook et al. [19] achieved uniform distribution of reinforced particles in weldment due to friction of FSW tool, and SiC increases the hardness of the Al-MMCs.

Limited literature reports are available on appropriate mixing percentages of SiC and B_4C reinforcements on Al-MMCs for which industrially acceptable mechanical property ranges and microstructural integrities for reliability are established. In particular, reports on weldability studies on these Al-MMC's are sporadic, lacking an interdisciplinary treatment. This provides a broader scope for exploration regarding the thermal implications of FSW on microstructures, process parametric influences on weld efficiency and mechanical behavioural analysis. Hence, this work is undertaken to fabricate Al-MMC reinforced with SiC and B_4C particles using the stir casting technique in various percentage compositions. FSW is performed on these composites by varying the process parameters to the weld's changes in properties and behaviours. The energy generated during the FSW process in terms of heat is also estimated to examine the thermal behaviour of the weld in different regions of the joint, which governs the microstructural grain sizes and the consequential, mechanical properties.

2. Experimental Procedure

The experimentation is carried out in two phases: the first phase involves stir casting to fabricate Al-MMC's with varying reinforcement composition. The second phase involves the FSW of the stir cast samples, followed by testing and characterization.

2.1. Fabrication of Composites (Phase-I)

A graphite crucible in electrical furnace (SwamEquip, Chennai, India) was used to prepare the Al-MMCs reinforced B_4C and SiC. Hexachloroethane tablets (M/s Madrad Fluorine Private Ltd., Chennai, India) were used for degassing (eradicate dross) from the composites. At 650 °C, B_4C and SiC were pre-oxidized for 2 h and transferred into the liquid matrix (AA6061) with constant stirring. To improve the metal bonding, heat treatment was given to B_4C to form a layer on SiC. After adding B_4C and SiC at an optimal speed

(500 rpm) and time (10 min), the melt was then transferred into an iron die mould. There were no traces of casting defects. To improve the wettability of the reinforced particles, magnesium was added during the stirring and melting of the alloy.

2.2. Experimental Setup

The Al-MMCs were prepared using the stir casting unit (Figure 1). The scientico electrical furnace of dimension 600 mm × 600 mm × 600 mm of the outer shell and furnace height 0.75 m was used. The crucible (dimension 200 mm × 300 mm × 50 mm) made of Zirconia coated AISI310 stainless steel material was employed with speed variations ranging between 300 and 500 rpm. The furnace operated at 230 V, 6 kW, and the molten metal working temperature was 990 ± 0.1 °C with a maximum sustainable temperature of 1200 °C. The stir casting furnace specifications are given in Table 1.

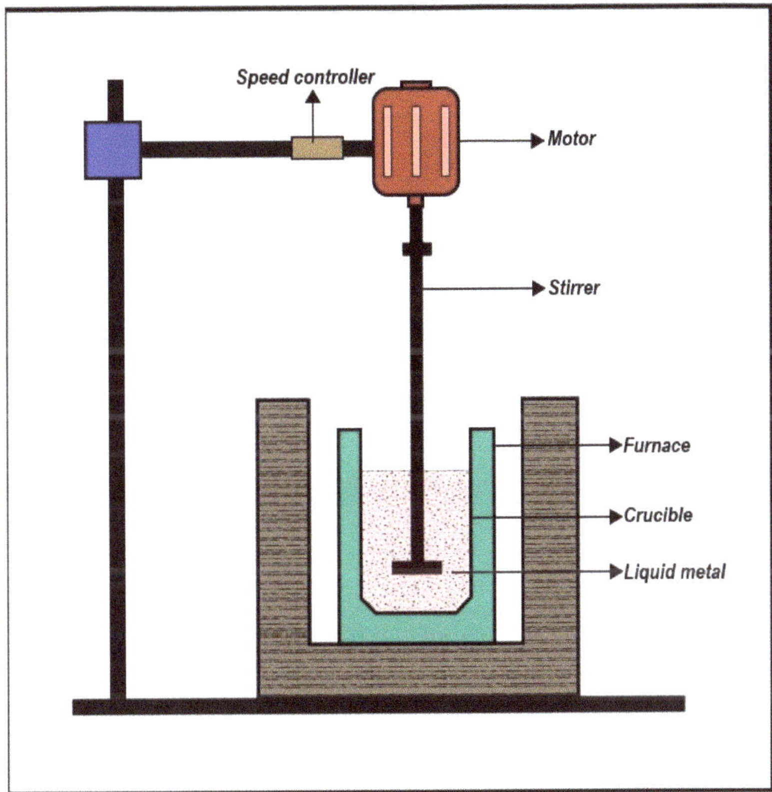

Figure 1. Schematic of stir casting setup.

Table 1. Process parameters for stir casting.

S. No.	Parameters	Value
1	Spindle speed	300–500 rpm
2	Stirring time	10 min
3	Temperature of melt	990 °C
4	Preheated temperature of SiC and B4C particles	650 °C
5	Powder feed rate	0.75–1.0 g/s

2.3. Experimental Procedure for FSW of Al-MMC's (Phase-II)

To strengthen Al-MMCs, SiC (325 nm mesh size) and B_4C (30 nm) particles were reinforced in the metal matrix (Figure 2). Hexachloroethane tablets was used to improve SiC and B4C powder's wettability and enhance their behaviour in Al melts [23]. The composition percentages of different samples fabricated in this study are listed in Table 2.

Figure 2. Scanning electron microscope micrograph of reinforcement particles: (**a**) SiC, (**b**) B_4C.

The workpieces of dimensions 150 mm × 150 mm × 6 mm were prepared in butt welding configuration (Figure 3). The FSW tool was made of tool steel (H-13) with a shoulder diameter of 25 mm. The cylindrical tool dimensions of 6 mm diameter and 5.7 mm height were considered (Figure 4). After a few trial-and-error experiments using FSW 3T 300NC, the optimal process parameters ranges were identified. The subsequent trials were based on the design of experiments (DOE) with different levels of process parameters accounting for 81 trials. However, the detailed information of the DOE based trials is not discussed here as the scope of the paper focuses on properties and behaviours of the weld. The FSW tool rotating speed, travelling speed and plunge force used for the trials were 1000 rpm, 75 mm/min and 10 KN, respectively. The welded specimens are cut along the cross-sections for tensile test specimens per ASTM (E8M) standards.

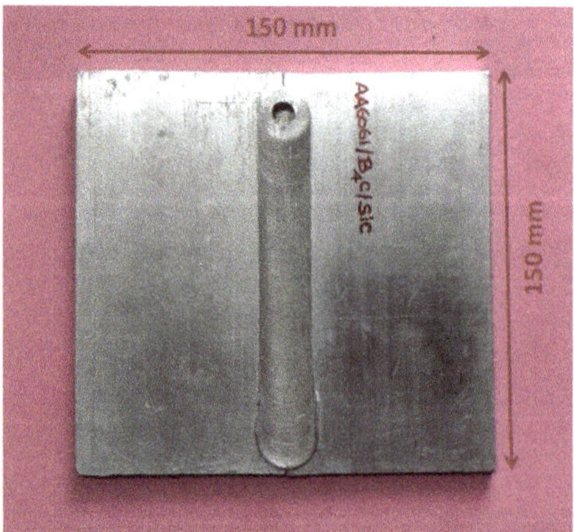

Figure 3. Friction stir welded specimen.

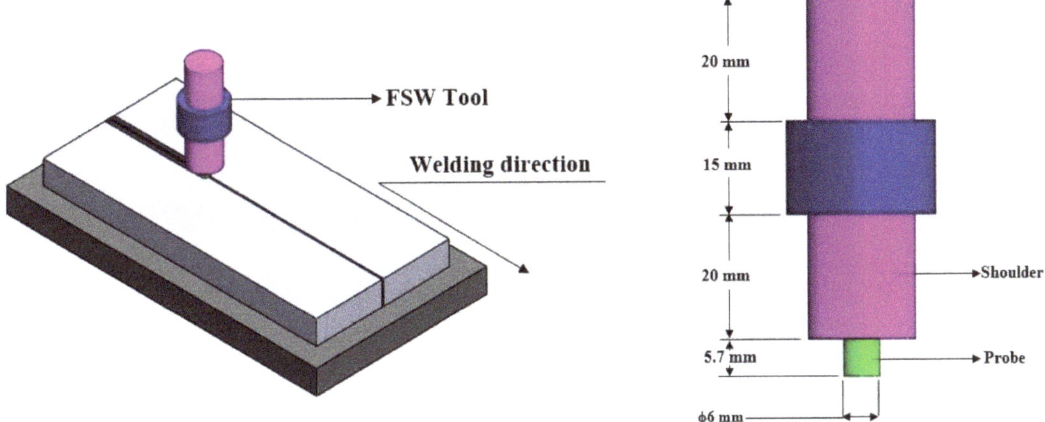

Figure 4. Schematic representation of FSW process and tool.

Table 2. Composition of various samples in wt.%.

Sample No.	Al %	SiC %	B4C %	Mg %
1	85	10	3	2
2	85	8	5	2
3	85	5	8	2
4	85	3	10	2

3. Estimation of Energy Generated in FSW for Heat Input Examinations

The amount of heat generated in FSW is the power transformed to the total amount of heat energy and expressed by Equation (1).

$$Q = \eta P \tag{1}$$

where η—power transformation coefficient, and P—power transferred.

The total amount of heat energy transferred is the sum of the translational (Q_t) and rotational (Q_r) heat energies as shown in Equation (2).

$$Q = Q_t + Q_r \tag{2}$$

The total amount of mechanical power derived is dependent on the amount of torque (M) and angular frequency (ω) and is expressed by Equations (3) and (4).

$dP = \omega dM = \omega r dF = \omega r \tau dA$

$$P = \omega M \tag{3}$$

$$dP = \omega dM = \omega r dF = \omega r \tau dA \tag{4}$$

where r—radial distance, F—force exerted by the FSW tool, τ—sheer contact stress in the material, and A—area of the weld.

By estimating the coefficients, the total heat input generated by the FSW tool is given by Equations (5) and (6):

$$Q = \int_0^{2\pi} \int_0^{R} \omega r^2 \tau d\theta dr \tag{5}$$

$$Q = \frac{2\pi\eta}{3S} \times \mu \times F_N \times \omega \times R \tag{6}$$

Heat input in Equation (6) is the amount of energy per unit length of the weld.

4. Results and Discussion

4.1. Mechanical Testing for Strength Assessment of Welds

4.1.1. Tensile Tests

The tensile tests were carried out at room temperature using a 40 tonnes capacity universal testing machine (UTM) INSTRON 8801 (High Wycombe, UK). The specimens were prepared according to the ASTM (E8M) standard. It was observed that sample #1 had the lowest B_4C content had the highest tensile strength (172 MPa) compared to the other samples (Figure 5). This is attributed to the good plasticity and density of Al [13]. With the increase in B_4C and a decrease in SiC content in the friction stir welded Al-MMCs samples, the tensile strength decreases. This can be attributed to the even distribution of fine Al-SiC particles. Higher percentages of SiC contributed to particle agglomeration of Al-SiC during the solidification that caused the decrease in strength of composites. The maximum value of yield stress was 142.58 MPa for sample #1, within the acceptable range of most applications. The strengthening of the composites joints after the FSW process may be attributed to dislocation density of SiC and B_4C particles, plastic deformation, and interactions between dislocations. The high yield strength of the Al-MMCs was due to the high amount of dislocations in its matrix. During the synthesis of Al-composite, additional dislocations were introduced by SiC and B_4C particles due to the induced internal stresses. It is attributed to the different cooling rates of the reinforcement particles from the weld process temperature due to the thermal expansion coefficients mismatch. The percentage elongation drastically reduced with a decrease in SiC and an increase in B_4C content. SiC exhibits high strength to weight ratio and high strength retention at elevated temperatures. With this, composite may also improve chemical stability and eliminate catastrophic failures when added to a metal matrix. However, it is important to account that processing of SiC at high temperature is preferable only when the reinforcement is to be processed at high temperature. There is also maximization of monolithic ceramics causing fragility in the composites. During the FSW process, the thermal expansion coefficient between reinforcement and the matrix contributes to the thermal cooling stress from processing temperature.

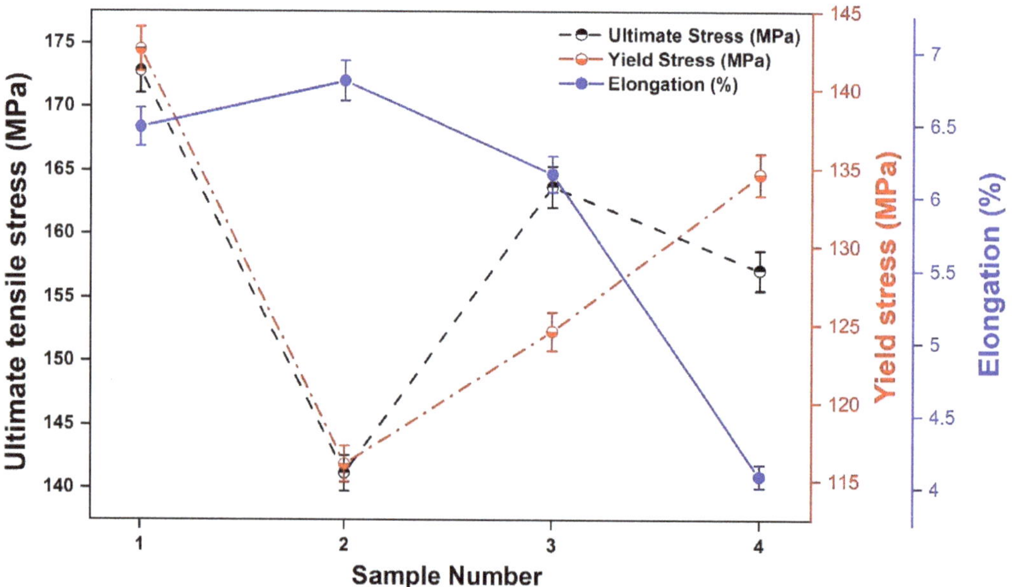

Figure 5. Tensile test properties measured for different sample compositions.

4.1.2. Hardness Test

The interface bonding strength between the Al matrix and reinforcement particles was recorded using the microhardness tests (Figure 6). It was observed that sample #4 had higher hardness followed by samples #3, #2 and #1. This observation emphasizes that the hardness is directly proportional to the B_4C and inversely proportional to the SiC content. Unreinforced Al alloy had an average hardness value of 23 HV, which improved considerably by adding reinforcement and subsequent thermal action during the FSW process, indicating an optimal increase as it eliminates brittleness. SiC and B_4C particles are relatively harder and tightly bonded, constricting dislocation movements in the Al matrix, thus enhancing the hardness. SiC has lower thermal expansion and can withstand high temperature due to its high creep and oxidation resistance reflected in the mechanical properties when added as reinforcement in Al-MMCs. Hence, a higher SiC content results in a decline in hardness. The hardness observed at the weld zone was higher than other zones for all the samples. It indicates the strengthening of Al-MMCs after it was subjected to the FSW process. This is due to the friction action in FSW, which aids in the uniform distribution of the reinforcement particles in the Al matrix. B_4C exhibits low density and high hardness, which improved the structural integrity and hardness of the Al-MMCs. It also possesses excellent stability with high thermal loading, wettability and abrasive capacity leading to prominent improvement in the mechanical properties.

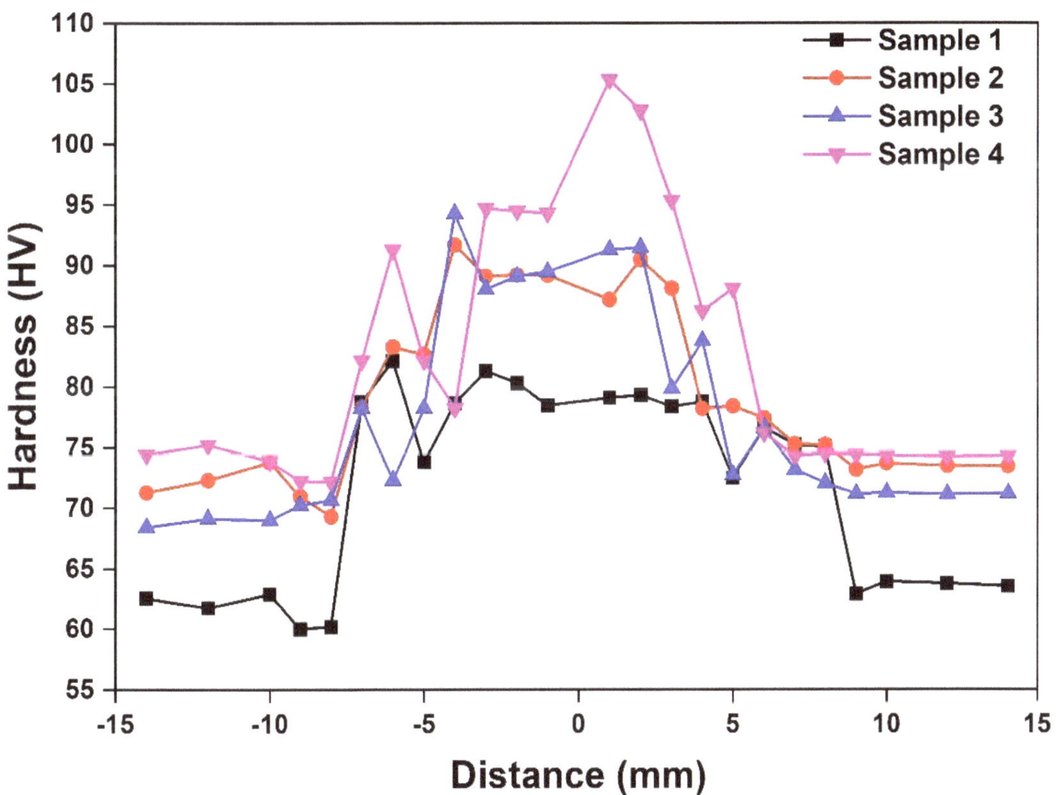

Figure 6. Comparison of hardness for friction stir welded Al-MMCs samples.

4.1.3. Wear Analysis

Wear tests were performed on SiC and B_4C reinforced Al-MMCs to study the wear behaviour of Al-MMCs. B_4C particles showed an insignificant impact on the wear properties, and consequently, the study was restricted to only the SiC content. SiC exhibits excellent wear and corrosion resistance for a wide range of temperature reflected in the Al MMC's tribological properties. For varying SiC content, the friction coefficient was recorded. The coefficient of friction was higher for samples whose SiC content ranged between 2 to 4% (Figure 7). It was also observed that the friction coefficient increased with an increase in load from 10 N to 20 N (while maintaining a constant velocity of 1 m/s). However, the friction coefficient decreased when the load was further increased to 30 N. This may be attributed to the increased brittleness of the samples [5]. The wear rate prominently increased with higher loading, and this observation was for all loading (Figure 8). The wear resistance gets transformed from abrasion to adhesion, reducing the average friction coefficient [5]. The wear rate of composites increased with the increase in load. This is due to the direct metal to metal contact, which leads to wear. The reinforcement particles reduced the plastic deformation by constricting the movement of dislocations. With the increased normal load, the reinforcement particles eroded from the surface leads to base Al resulted in a higher wear rate.

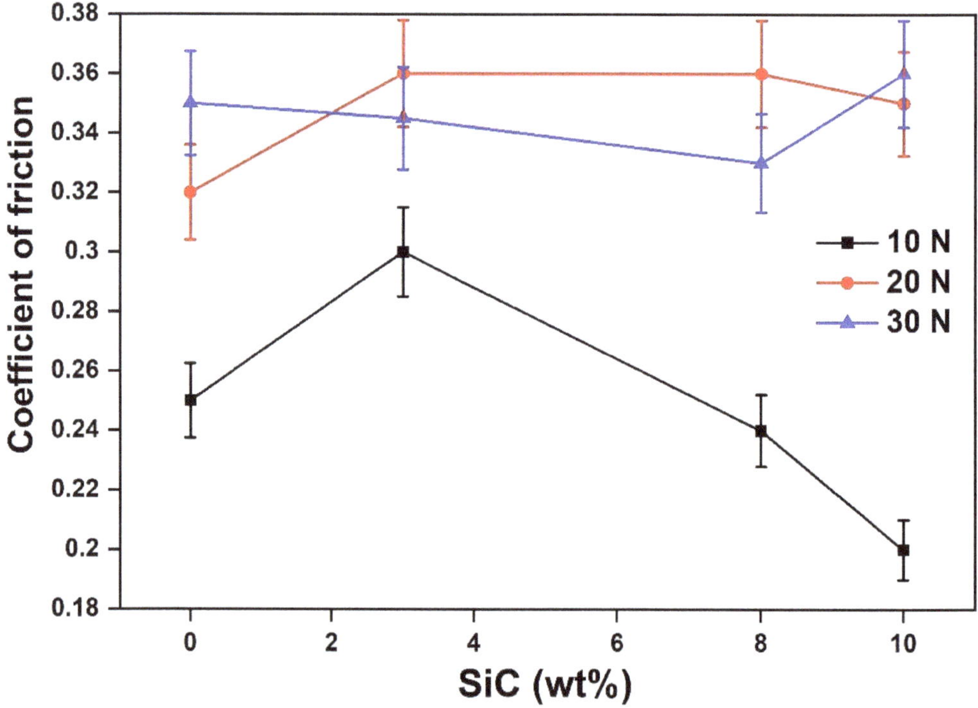

Figure 7. Variation of coefficient of friction of Al-MMCs with SiC content at different load and constant velocity of 1 m/s.

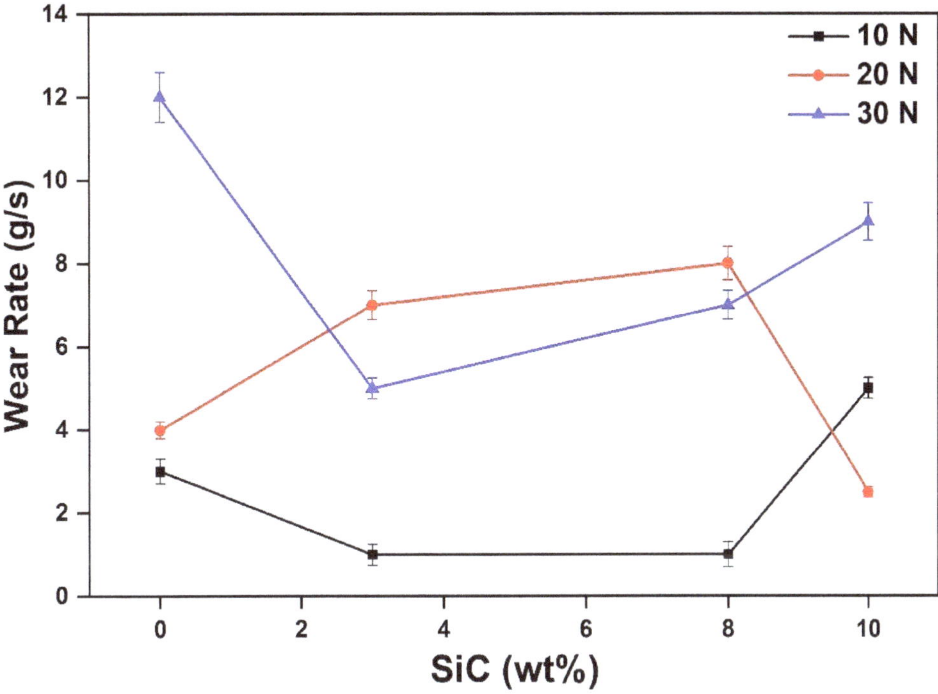

Figure 8. Variation of wear rate of Al-MMCs with SiC content at different load and constant velocity of 1 m/s.

SEM analysis was used to investigate the material loss from the surface. The SEM micrographs of worn surface are shown in Figure 9a,b in 10 N and 20 N loads, respectively. The presence of wear debris and ploughing marks are visible, which are insignificant to cause any behavioural changes of the material. The reinforced particles agglomerated in the weld region due to the frictional heat produced during welding. The SEM micrographs of 30 N load are shown in Figure 9c,d. The corresponding EDAX analysis is also presented. The peaks of Al and reinforcement particles were observed. Small peaks of Fe were also observed, suggesting the abrasion of the steel surface by the reinforcement particles. Particles were homogeneously distributed, and no damage of worn surface was observed for higher load. In few cases, the presence of delamination with intense plastic deformation suggests adhesion wear [5]. The peaks of O in the EDAX analyses indicate the presence of oxidative driven wear. It is well known that Al readily reacts with atmospheric oxygen and forms aluminium oxide at the surface when undergone the counter steel abrasion.

4.2. Radiography Testing of Welds for Quality Assessment

Al-MMCs were placed between the radiation source (Ir-192, Co-60 and Cs-137) and radioactive film. The variation in the image intensities was observed to analyse pores, cracks or discontinuities. Figure 10 shows that the friction stir welded samples (representative) were free from any defect and discontinuity. Pores or pore clusters in the weld samples may weaken the strength in the region but do not alter the weld properties. No pores or pore clusters were observed in the samples. Hence, the samples were free from defects and discontinuities.

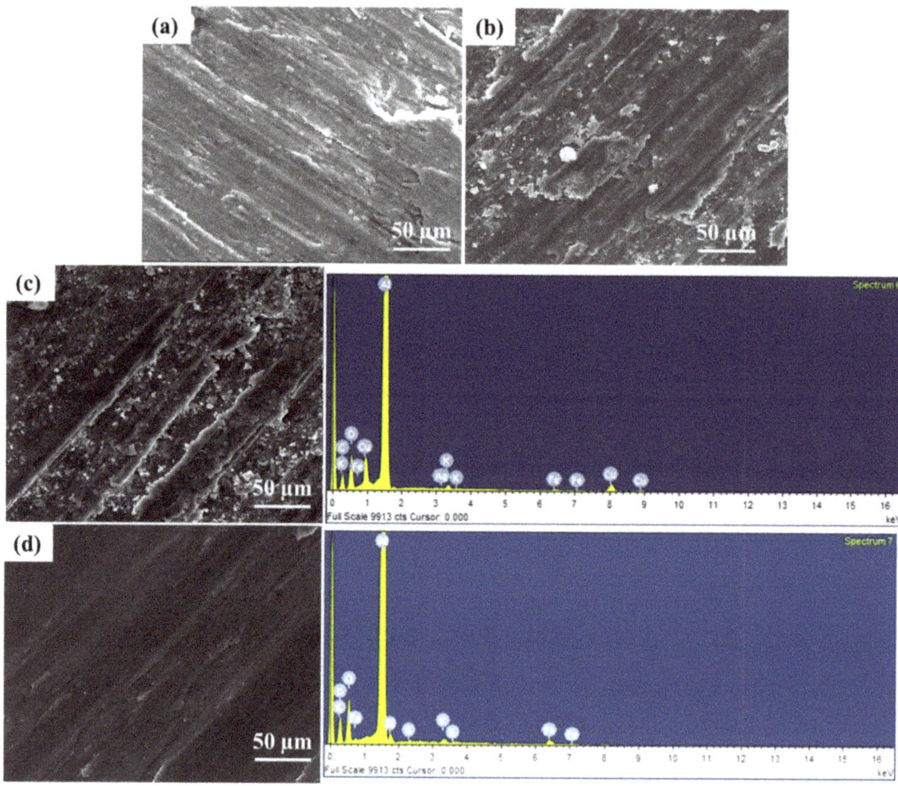

Figure 9. SEM/EDAX for AA 6061 composites with different reinforcement: (**a**) 10%SiC/3%B_4C, (**b**) 8%SiC/5%B_4C, (**c**) 5% SiC/8% B_4C, (**d**) 3% SiC/10% B_4C.

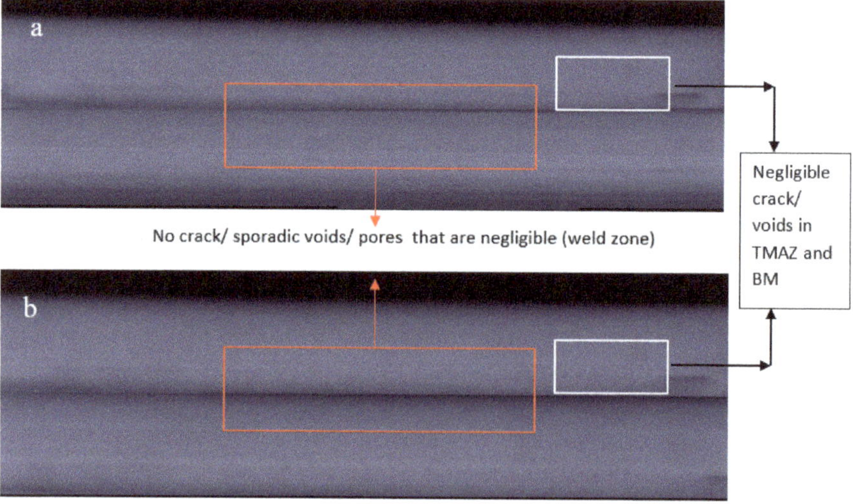

Figure 10. (**a**) Radiographic images of AA6061 welded composite plates reinforced with: (**a**) 5% SiC/8% B_4C, (**b**) 3% SiC/10% B_4C.

4.3. Metallurgical Characterization for Assessment of Weld Microstructures

The distinct macrostructures of different zones (base metal, stir zone, thermomechanically affected zone, heat affected zone) in the friction stir welded Al-MMCs samples are shown in Figure 11. The interfacial boundary was visible on both the advancing and the retreating side. The base material showed fine and enlarged grain particles homogeneously distributed (Figure 11a). At the extremities of the stir zone, the morphology of the microstructure significantly gets altered. (Figure 11b). In the stir zone, dynamic crystallization occurred due to the FSW process. Grain refinement occurred due to heat, mechanical deformation and stirring action of the FSW tool (Figure 11b). The boundary between the thermo-mechanical affected zone and stir zone was not visible, and the grains were distorted, aligned. A sharp transition occurred between these two regions (Figure 11c). Mechanical vibration and heat in the thermomechanical affected zone were less compared to the stir zone.

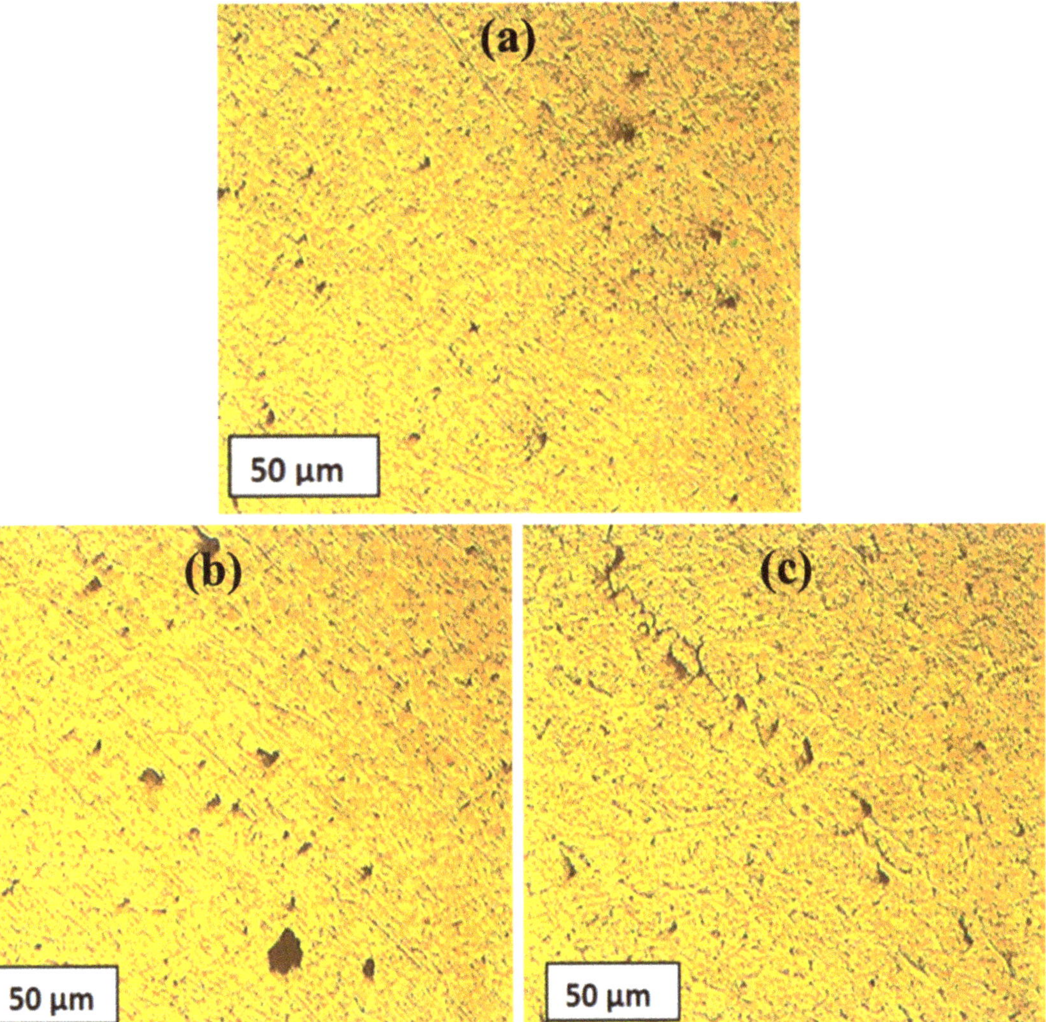

Figure 11. Distinct microstructures of Al-MMC: (**a**) base metal, (**b**) stir zone, (**c**) thermomechanical affected zone/stir zone.

The microstructural examination at the joints reveals the flow of plasticized metal from advancing to the retreating side. The dynamic recrystallized zones of samples obtained from the FSW are shown in Figure 12a,b. It was observed that the reinforcement particles were distributed uniformly, and no undesired higher order reactions in the dynamic recrystallized zones were seen. It is in coherence with other reports [24–26]. The formation of onion rings was observed from Figure 12b due to the composite's cylindrical FSW rotating tool material extrusion. There were no traces of common defects to FSW processes such as tunnels, wormholes and piping due to B_4C dislocation in the Al matrix. The extensive stirring action of FSW resulted in the formation of homogenous and refined grains at the weld zone.

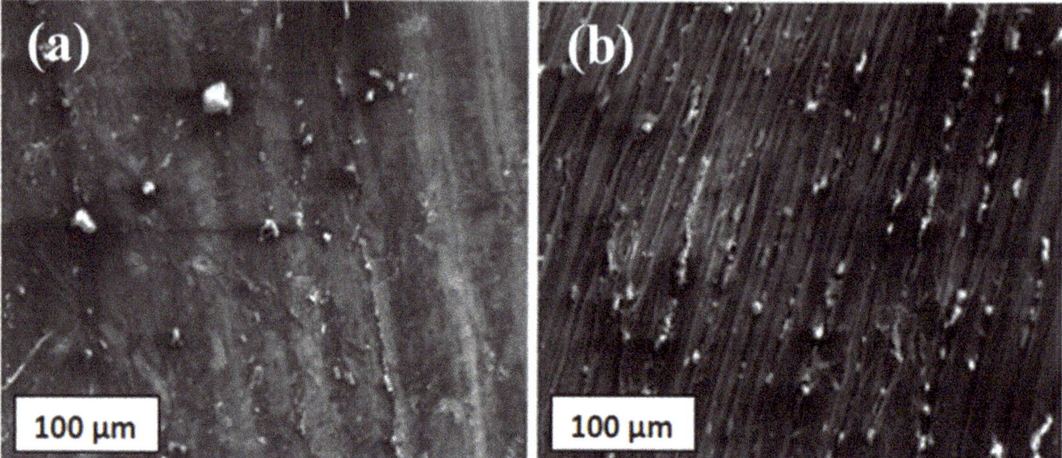

Figure 12. High-resolution SEM images of dynamic recrystallized zones of Al-MMC: (**a**) stir zone, (**b**) onion rings.

Figure 13 reveals the microscopic image of the weld cross-section and different zones in friction stir welded samples. Considerable change was visible in the particle size of the reinforcement phase in different regions of the joint. The reinforcement particles in the weld region revealed fragmentation in the superficial zone (average particle size reduction was about 6 times, whereas in the advancing zone average particle size reduction was about 3 times). This is because the reinforcement particles at the weld are subjected to higher heat flux compared to other regions. However, in the weld nugget region, fragmentation of the particles was significantly less (average particle size reduction was about 2 times). Distinct microstructures were observed at the nugget region of the composite. The homogenous distributions of refined grains were observed in the weld region are mainly attributed to the plastic deformation due to FSW stirring tool. The reinforcement particulate precipitates were fragmented and rearranged by the stirring action. The grains splitting and size reduction lead to the evolution of microstructure that is specific in nature. There is an irregular assortment of low/ high angle boundaries along with considerable quantum of fine equiaxed grains. The microstructure revealed directional realignment indicating close to parallel to the border between the transition zone and the stir zone. This matrix structure formation may be due to the development of deformation-induced boundaries. Besides, the traces of misorientation distribution implies a close relationship between the grain-boundary formation and texture. Henceforth, the underlying microstructural process lays the platform for continuous recrystallization. The refined equiaxed grains originate from grain boundary bulging, which is reflected in concurrent development leading to discontinuous recrystallization [27,28].

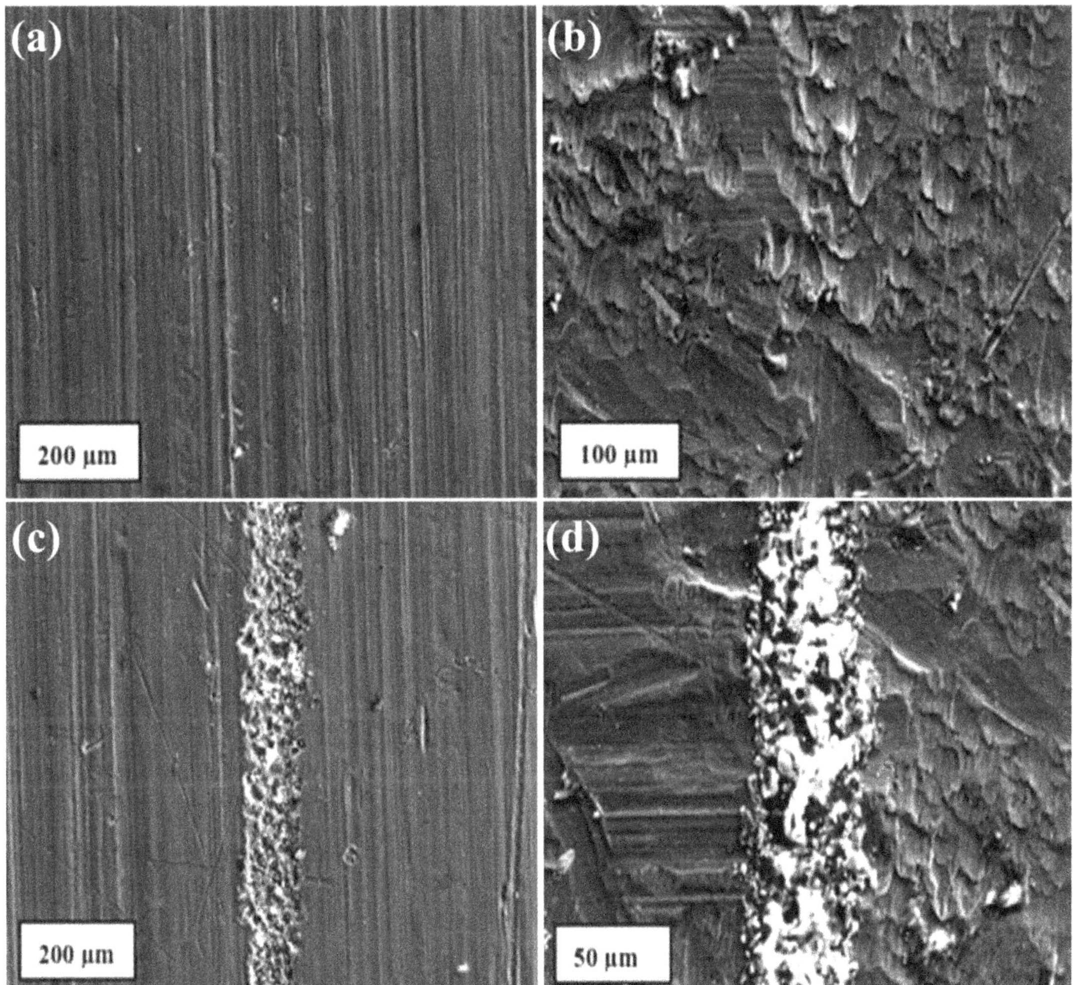

Figure 13. SEM images of weld zone of Al-MMC: (**a**) base material, (**b**) initial stage, (**c**) progressive stage, (**d**) advanced stage.

The EDAX analysis was performed on the weld samples to identify the key elements in the joint and the adjoining regions (Figure 14). The crystalline phase particle distributions were examined by elemental mapping for Al, Si and B. It was observed that small size grains of Si and Al phases are distributed in narrow spaces available between B_4C particle. The Al phase governs the solid solubility of various reinforcement particles. It may be noticed that Al in the SiC pellets is negligible. The uniform distribution of Al, B and Si is observed [29,30].

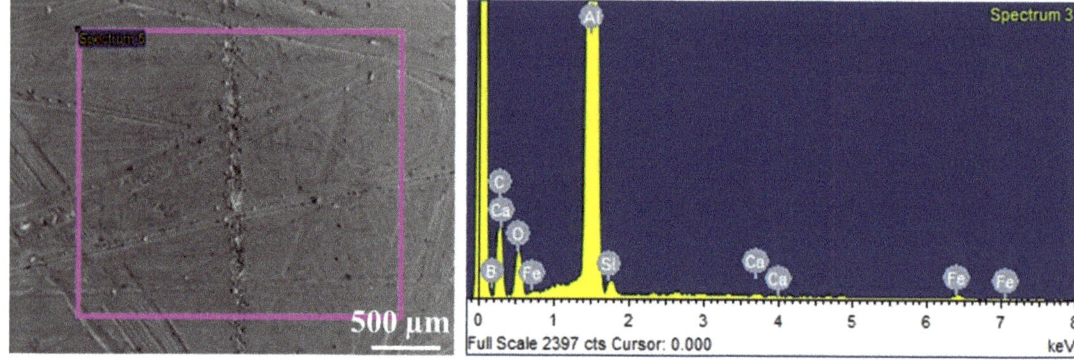

Figure 14. SEM with EDAX of Al-MMC in the weld nugget region.

5. Conclusions

In this study, FSW was employed to join AA6061/SiC/B_4C stir cast composites. The study aimed to identify suitable Al-MMCs with high strength, lightweight and anti-corrosion properties and explore their weldability characteristics and efficiency using FSW. The process parameters of FSW influencing the properties of the weld along with the causes attributed by the reinforcement components were investigated. The conclusions derived from this study are presented as follows:

- The tensile properties of the friction stir welded Al-MMCs improved after reinforcement with SiC and B_4C. The strengthening of the composite joints after the FSW process may be attributed to the dislocation density of SiC and B_4C particles, plastic deformation during joining and the interactions between dislocations. The maximum ultimate tensile stress was found to be around 174 MPa for sample #1 (10% SiC and 3% B_4C). The percentage elongation decreased as the percentage of SiC decreases and B_4C increases.
- The hardness of the Al-MMCs improved considerably by adding reinforcement and subsequent thermal action during the FSW process, indicating an optimal increase as it eliminates brittleness. This is due to hard and tightly bonded SiC and B_4C particles, which constricts dislocation movements in the Al matrix.
- It was seen that wear resistance improved after the reinforcement of Al-MMCs. This is because the reinforcement particles reduced the plastic deformation by constricting the movement of dislocations. The wear rate of composites increased with the increase in load. This is due to the direct metal to metal contact, which leads to wear.
- The microstructural examination at the joints reveals the flow of plasticized metal from advancing to the retreating side. The formation of onion rings in the weld zone was due to the cylindrical FSW rotating tool material impression during the stirring action.

FSW is the feasible and appropriate process for welding Al-MMCs. The results promise to address the requirements of the aerospace and automobile industries. Due to the good weld efficiency, defence sectors like missile launchers and naval sectors with cruise components may adapt these materials along similar lines. The automobile and aerospace industries are the major beneficiaries; nevertheless, their applicability may be extrapolated to marine and defence structures.

Author Contributions: Conceptualization, K.S.A.A., V.M., S.A.V., M.R., A.Y., M.G. and J.W.; methodology, K.S.A.A., V.M., S.A.V., M.R., A.Y., M. G. and J.W.; formal analysis, K.S.A.A., V.M., S.A.V., M.R., A.Y., M.G. and J.W.; resources, K.S.A.A., V.M., S.A.V., M.R., A.Y., M.G. and J.W.; data curation, K.S.A.A., V.M., S.A.V., M.R., A.Y., M.G. and J.W.; writing—original draft preparation, K.S.A.A., V.M., S.A.V., M.R., A.Y., M.G. and J.W.; writing—review and editing, K.S.A.A., V.M., S.A.V., M.R., A.Y., M.G. and J.W.; supervision, K.S.A.A., V.M., S.A.V., M.R., A.Y., M.G. and J.W.; project administration,

K.S.A.A., V.M., S.A.V., M.R., A.Y., M.G. and J.W.; funding, K.S.A.A., V.M., S.A.V., M.R., A.Y., M.G. and J.W. All authors have read and agreed to the published version of the manuscript.

Funding: This research received no external funding.

Institutional Review Board Statement: Not applicable.

Informed Consent Statement: Not applicable.

Data Availability Statement: The authors confirm that the data supporting the findings of this study are available within the article.

Conflicts of Interest: The authors declare no conflict of interest.

References

1. Miracle, D. Metal matrix composites—From science to technological significance. *Compos. Sci. Technol.* **2005**, *65*, 2526–2540. [CrossRef]
2. Rosso, M. Ceramic and metal matrix composites: Routes and properties. *J. Mater. Process. Technol.* **2006**, *175*, 364–375. [CrossRef]
3. Threadgilll, P.L.; Leonard, A.J.; Shercliff, H.R.; Withers, P.J. Friction stir welding of aluminium alloys. *Int. Mater. Rev.* **2009**, *54*, 49–93. [CrossRef]
4. Kumar, S.; Yadav, A.; Patel, V.; Nahak, B.; Kumar, A. Mechanical behaviour of SiC particulate reinforced Cu alloy based metal matrix composite. *Mater. Today: Proc.* **2021**, *41*, 186–190. [CrossRef]
5. Kumar, A.; Kumar, S.; Mukhopadhyay, N.K.; Yadav, A.; Kumar, V.; Winczek, J. Effect of Variation of SiC Reinforcement on Wear Behaviour of AZ91 Alloy Composites. *Materials* **2021**, *14*, 990. [CrossRef]
6. Kumar, A.; Kumar, S.; Mukhopadhyay, N.K.; Yadav, A.; Winczek, J. Effect of SiC Reinforcement and Its Variation on the Mechanical Characteristics of AZ91 Composites. *Materials* **2020**, *13*, 4913. [CrossRef]
7. Sinha, D.K.; Kumar, S.; Kumar, A.; Yadav, A. Optimization of Process Parameters in Continuous Extrusion of Aluminium Alloy. In *Proceedings of the 2018 International Conference on Computational and Characterization Techniques in Engineering & Sciences (CCTES), Lucknow, India, 14–15 September 2018*; Institute of Electrical and Electronics Engineers (IEEE): Manhattan, NY, USA, 2018; pp. 246–251. [CrossRef]
8. Morozova, I.; Obrosov, A.; Naumov, A.; Królicka, A.; Golubev, I.; Bokov, D.O.; Doynov, N.; Weiß, S.; Michailov, V. Impact of Impulses on Microstructural Evolution and Mechanical Performance of Al-Mg-Si Alloy Joined by Impulse Friction Stir Welding. *Materials* **2021**, *14*, 347. [CrossRef]
9. Yadav, A.; Ghosh, A.; Gupta, P.; Kumar, A. Mathematical Modelling of Heat Affected Zone Width in Submerged Arc Welding Process. In *Proceedings of the 2018 International Conference on Computational and Characterization Techniques in Engineering & Sciences (CCTES), Lucknow, India, 14–15 September 2018*; Institute of Electrical and Electronics Engineers (IEEE): Manhattan, NY, USA, 2018; pp. 216–220. [CrossRef]
10. Ramnath, B.V.; Subramanian, S.A.; Rakesh, R.; Krishnan, S.S.; Ramanathan, A.L.A. A review on friction stir welding of aluminium metal matrix composites. *IOP Conf. Ser.: Mater. Sci. Eng.* **2018**, *390*, 012103. [CrossRef]
11. Yadav, A.; Ghosh, A.; Kumar, A. Experimental and numerical study of thermal field and weld bead characteristics in submerged arc welded plate. *J. Mater. Process. Technol.* **2017**, *248*, 262–274. [CrossRef]
12. Ghosh, A.; Yadav, A.; Kumar, A. Modelling and experimental validation of moving tilted volumetric heat source in gas metal arc welding process. *J. Mater. Process. Technol.* **2017**, *239*, 52–65. [CrossRef]
13. Laska, A.; Szkodo, M. Manufacturing Parameters, Materials, and Welds Properties of Butt Friction Stir Welded Joints—Overview. *Materials* **2020**, *13*, 4940. [CrossRef]
14. Milčić, M.; Milčić, D.; Vuherer, T.; Radović, L.; Radisavljević, I.; Đurić, A. Influence of Welding Speed on Fracture Toughness of Friction Stir Welded AA2024-T351 Joints. *Materials* **2021**, *14*, 1561. [CrossRef] [PubMed]
15. Jandaghi, M.; Pouraliakbar, H.; Saboori, A.; Hong, S.; Pavese, M. Comparative Insight into the Interfacial Phase Evolutions during Solution Treatment of Dissimilar Friction Stir Welded AA2198-AA7475 and AA2198-AA6013 Aluminum Sheets. *Materials* **2021**, *14*, 1290. [CrossRef] [PubMed]
16. Derazkola, H.A.; García, E.; Eyvazian, A.; Aberoumand, M. Effects of Rapid Cooling on Properties of Aluminum-Steel Friction Stir Welded Joint. *Materials* **2021**, *14*, 908. [CrossRef] [PubMed]
17. Palanivel, R.; Mathews, P.K., Murugan, N.; Dinaharan, I. Prediction and Optimization of Wear Resistance of Friction Stir Welded Dissimilar Aluminum Alloy. *Procedia Eng.* **2012**, *38*, 578–584. [CrossRef]
18. Suban, A.A.K.; Perumal, M.; Ayyanar, A.; Subbiah, A.V. Microstructural analysis of B4C and SiC reinforced Al alloy metal matrix composite joints. *Int. J. Adv. Manuf. Technol.* **2017**, *93*, 515–525. [CrossRef]
19. Ahn, B.-W.; Choi, D.H.; Kim, Y.-H.; Jung, S.-B. Fabrication of SiCp/AA5083 composite via friction stir welding. *Trans. Nonferrous Met. Soc. China* **2012**, *22*, s634–s638. [CrossRef]
20. Toptan, F.; Kilicarslan, A.; Kerti, I. The Effect of Ti Addition on the Properties of Al-B4C Interface: A Microstructural Study. *Mater. Sci. Forum* **2010**, *636–637*, 192–197. [CrossRef]

21. Xu, W.; Liu, J.; Chen, D.; Luan, G.; Yao, J. Improvements of strength and ductility in aluminum alloy joints via rapid cooling during friction stir welding. *Mater. Sci. Eng. A* **2012**, *548*, 89–98. [CrossRef]
22. Vijay, S.; Murugan, N. Influence of tool pin profile on the metallurgical and mechanical properties of friction stir welded Al–10 wt.% TiB_2 metal matrix composite. *Mater. Des.* **2010**, *31*, 3585–3589. [CrossRef]
23. Wahab, M.N.; Ghazali, M.J.; Daud, A.R. Effect of Aluminum Nitride Addition on Dry Wear Properties of Al-11%Si Alloy Prepared by Stir Casting Process. *Adv. Mater. Res.* **2011**, *264-265*, 614–619. [CrossRef]
24. Kumar, P.; Yadav, A.; Kumar, A. Electron Beam Processing of Sensors Relevant Vacoflux-49 Alloy: Experimental Studies of Thermal Zones and Microstructure. *Arch. Metall. Mater.* **2020**, *65*, 1147–1156. [CrossRef]
25. Kumar, B.A.; Murugan, N. Metallurgical and mechanical characterization of stir cast AA6061-T6–AlNp composite. *Mater. Des.* **2012**, *40*, 52–58. [CrossRef]
26. Singh, R.; Yadav, A. Experimental study of effect of process parameters for heat generation in friction stir welding. *IOP Conf. Series: Mater. Sci. Eng.* **2018**, *402*, 012131. [CrossRef]
27. Zhang, Z.; Du, X.; Li, Z.; Wang, W.; Zhang, J.; Fu, Z. Microstructures and mechanical properties of B_4C–SiC intergranular/intragranular nanocomposite ceramics fabricated from B_4C, Si, and graphite powders. *J. Eur. Ceram. Soc.* **2014**, *34*, 2153–2161. [CrossRef]
28. Shen, Q.; Wu, C.; Luo, G.; Fang, P.; Li, C.; Wang, Y.; Zhang, L. Microstructure and mechanical properties of Al-7075/B4C composites fabricated by plasma activated sintering. *J. Alloys Compd.* **2014**, *588*, 265–270. [CrossRef]
29. Singh, B.B.; Balasubramanian, M. Processing and properties of copper-coated carbon fibre reinforced aluminium alloy composites. *J. Mater. Process. Technol.* **2009**, *209*, 2104–2110. [CrossRef]
30. Prasad, L.; Kumar, N.; Yadav, A.; Kumar, A.; Kumar, V.; Winczek, J. In Situ Formation of ZrB_2 and Its Influence on Wear and Mechanical Properties of ADC12 Alloy Mixed Matrix Composites. *Materials* **2021**, *14*, 2141. [CrossRef]

 *materials* 

Article

Friction Stir Spot Welding of Different Thickness Sheets of Aluminum Alloy AA6082-T6

Mohamed M. Z. Ahmed [1,2,*], Mohamed M. El-Sayed Seleman [2], Essam Ahmed [2], Hagar A. Reyad [2], Kamel Touileb [1] and Ibrahim Albaijan [1]

1 Mechanical Engineering Department, College of Engineering at Al-Kharj, Prince Sattam Bin Abdulaziz University, Al Kharj 16273, Saudi Arabia; k.touileb@psau.edu.sa (K.T.); i.albaijan@psau.edu.sa (I.A.)

2 Department of Metallurgical and Materials Engineering, Faculty of Petroleum and Mining Engineering, Suez University, Suez 43512, Egypt; mohamed.elnagar@suezuniv.edu.eg (M.M.E.-S.S.); essam.ahmed@suezuniv.edu.eg (E.A.); engineering667@yahoo.com (H.A.R.)

* Correspondence: moh.ahmed@psau.edu.sa; Tel.: +966-115-888-273

Abstract: Friction stir spot welding (FSSW) is one of the important variants of the friction stir welding (FSW) process. FSSW has been developed mainly for automotive applications where the different thickness sheets spot welding is essential. In the present work, different thin thickness sheets (1 mm and 2 mm) of AA6082-T6 were welded using FSSW at a constant dwell time of 3 s and different rotation speeds of 400, 600, 800, and 1000 rpm. The FSSW heat input was calculated, and the temperature cycle experience during the FSSW process was recorded. Both starting materials and produced FSSW joints were investigated by macro- and microstructural investigation, a hardness test, and a tensile shear test, and the fractured surfaces were examined using a scanning electron microscope (SEM). The macro examination showed that defect-free spot joints were produced at a wide range of rotation speeds (400–1000 rpm). The microstructural results in terms of grain refining of the stir zone (SZ) of the joints show good support for the mechanical properties of FSSW joints. It was found that the best welding condition was 600 rpm for achieving different thin sheet thicknesses spot joints with the SZ hardness of 95 ± 2 HV0.5 and a tensile shear load of 4300 ± 30 N.

Keywords: aluminum alloys; AA6082; friction stir spot welding; tensile shear load; hardness; microstructure

Citation: Ahmed, M.M.Z.; El-Sayed Seleman, M.M.; Ahmed, E.; Reyad, H.A.; Touileb, K.; Albaijan, I. Friction Stir Spot Welding of Different Thickness Sheets of Aluminum Alloy AA6082-T6. *Materials* **2022**, *15*, 2971. https://doi.org/10.3390/ma15092971

Academic Editors: Józef Iwaszko and Jerzy Winczek

Received: 29 March 2022
Accepted: 14 April 2022
Published: 19 April 2022

Publisher's Note: MDPI stays neutral with regard to jurisdictional claims in published maps and institutional affiliations.

1. Introduction

Nowadays, lightweight materials such as aluminum alloys and/or magnesium alloys and their composites are widely used, especially in automotive and aerospace industries, where saving weight is very important [1–3]. Spot joining technologies for lightweight materials are laser spot welding, resistance spot welding, and riveting. Disadvantages of these technologies include the consumption of tools during joining, large heat distortion, and poor weld strength joints [4–7]. Porosity defects for conventional resistance spot welding and laser spot welding cannot be avoided. Moreover, the riveting process increases the weight of joining parts, besides the cost of drilling needed [8–10].

New welding technology based on the friction stir welding (FSW) principle has been invented and introduced to the aerospace and most transportation industries with the name of friction stir spot welding (FSSW). The FSSW is recommended as a fast and cheap method to be used intensively for joining main structural components of similar [5,11–14] and dissimilar materials [15,16]. Thus, this welding technique is considered a strong competitive method for other spot joining technologies [17,18].

The FSSW is a solid-state joining method whereby a rotating tool is plunged into material sheets in a lap joint setup configuration to a certain depth to weld them via frictional heat generated during the friction stir process [18,19]. Thus, the FSSW is a thermo-mechanical process; the tool is surrounded by plasticized stir zone (SZ) material with high temperature and subjected to severe stress, especially for joining. In the FSSW, the solidification defects are avoided compared to the fusion welding techniques, and the joints are performed in a short time [20,21]. It consists of three stages: the plunge stage with the rotating tool under a certain downward force, stirring at a specific holding time, and the drawing out stage. During the plunging stage, the rotating tool faces high resistance from the workpiece. After that, the material is softened as a result of the plastic deformation and the generated frictional heat. During the dwell period, the frictional heat was raised and allowed the material to flow in the weld zone (WZ). In the last stage, the rotating tool is drawn out from the WZ [15,20].

The friction stir spot welded (FSSWed) joint has a characteristic of a keyhole in the middle. The produced FSSWed joint contains various typical zones, such as SZ, thermo-mechanical affected zone (TMAZ), heat affected zone (HAZ), and base material as found in the FSW joints [18]. In order to achieve spot-welded joints, many FSSW processes have been developed, such as pin-less FSSW, refill FSSW, and swing FSSW [2,20].

The AA 6xxx series (Al, Mg, and Si) alloys are frequently used in transportation and aerospace industries due to good corrosion resistance, excellent extrusion performance, and weldability [22–24]. Yuan et al. [25] used two different tools with the same shoulder feature and different pin geometries, a conventional pin and off-center pin feature, to produce a similar FSSW lap joint of 1 mm thickness AA6016-T4. The obtained results showed that the two tools exhibited a maximum weld separation load of around 3.3 kN at different rotation speeds of 1500 rpm for the conventional pin and 2500 rpm for the off-center pin. In addition, there is no relation between the microhardness results and separation modes of the welded joints. Gao et al. [18] utilized the GDRX model implemented in DEFORM-3D to forecast grain size during the FSSW of AA6082 and observed that grain size refinement owing to dynamic recrystallization could be predicted. Muhayat et al. [26] studied the effect of the pin diameter (3–7 mm) and dwell time (3–9 s) of FSSW similar thickness AA6082-T6 lap joints and reported an increase in dwell time and that the pin diameters decrease in the hardness of the welded joint. Moreover, the excessive heat input at 7 mm pin diameter and 9 s dwell time resulted in a decrease in tensile shear load. Aydin et al. [5] applied high rotation speeds from 1000 to 2500 rpm at a constant dwell time of 7 s, feeding rate of 50 mm/min, and plunge depth of 5 mm to achieve a similar thickness FSSW lap joint of 3 mm AA6082-T6 sheet. The results showed a drop in the hardness of the welded zone compared to the BM. Aydin et al. [27], in other work, recommended a cylindrical pin profile to FSSW of similar AA6082-T6 lap joint of 3 mm sheet thickness among the different pin profiles of conical, hexagonal, triangular, and cylindrical with two grooves used at the applied welding conditions of a relatively high rotation speed of 1500 rpm at different dwell times from 2 to 11 s. Buffaa et al. [28] FSSWed a similar thickness of AA6082-T6 of 1.5 mm sheet alloy at rotation speeds 900, 1500, and 2000 rpm using a cylindrical pin. The obtained results showed that increasing tool rotation causes a decrease in failure load.

In fact, there is a need in the transportation industry to friction stir spot weld different thin thicknesses AA6XXX. Based on the available literature, there is no try to FSSW of dissimilar thin thickness lap joints of AA6082-T6. Moreover, most researchers focused on using high rotational speeds (900–2500 rpm) to achieve the FSSW [5,27,28]. Thus, this study intended to explore the possibility of FSSW dissimilar thin sheet thickness 1 and 2 mm of AA 6082-T6 alloy at relatively lower rotational speeds of 400, 600, and 800. Moreover, the high rotation speed of 1000 rpm was tried for comparison. The microstructure and mechanical properties in terms of hardness and tensile shear load of the produced joints were examined with the light of applied heat input and the FSSW thermal cycle.

2. Methodology

2.1. Starting Materials

The starting material was AA6082-T6 aluminum alloy with two different sheet thicknesses of $1 \times 1000 \times 1000$ mm and $2 \times 1000 \times 1000$ mm were supplied by Future Fond

Company, Milano, Italy. The two sheets have the same chemical composition as listed in Table 1, and Table 2 summarizes the measured mechanical properties.

Table 1. The chemical composition of aluminum alloy AA6082-T6.

Element	Mg	Mn	Si	Fe	Cr	Zn	Cu	Ti	Al
(wt. %)	0.6	0.4	0.75	0.5	0.2	0.2	0.1	0.1	Bal

Table 2. The mechanical properties of the two base material sheets of AA6082-T6.

AA6082-T6	Ultimate Tensile Strength (MPa)	Yield Stress (MPa)	Fracture Strength (MPa)	Hardness (HV)
1 mm sheet	301	296	236	127 ± 3
2 mm sheet	257	252	245	111 ± 2

2.2. Friction Stir Spot Welding Process

The FSSW joints were produced using the friction stir welding/processing machine (EG-FSW-M1) [29]. For spot welding purposes, the two sheets were cut to specimens 100 mm in length and 30 mm in width. The specimens were spot welded in lap joints with a 30 mm overlap for the two different sheet thicknesses at a constant dwell time of 3 s and various rotation speeds of 400, 600, 800, and 1000 rpm. The other spot welding parameters in terms of plunge depth, plunge rate, and tilt angle were kept constant at 2.6 mm, 0.1 mm/s, and 0°, respectively. The tool used was made of steel (AISI H13) with dimensions of 20 mm shoulder diameter, 5 mm pin diameter, and 2.6 mm pin length. The tool has a flat shoulder and a cylindrical pin, as shown in Figure 1. The selection of tool geometry and tool material was based on the literature [5,6,30]. The cylindrical pin profile produces axisymmetric temperature distributions and heat transfer phenomena through the welds. The effect of the circular flow velocity due to the cylindrical pin on axisymmetric temperature distributions is small [6,7,31].

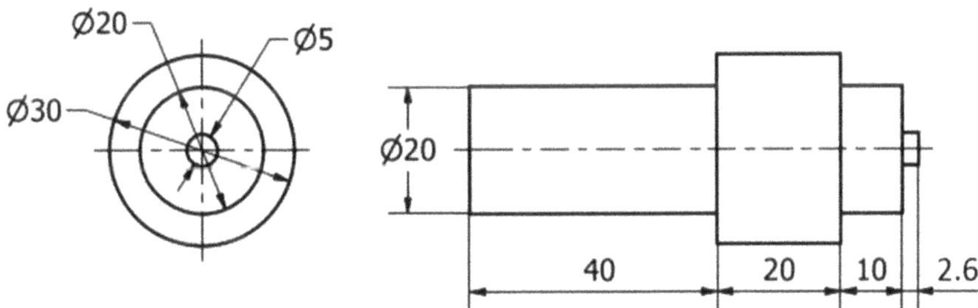

Figure 1. Dimensions of the FSSW tool (all dimensions are in mm).

Figure 2 summarizes the FSSW process carried out, which consists of three steps: (1) lap joint fixing with a clamping system (Figure 2a); (2) plunge stage and stirring action with the rotating tool (Figure 2b); (3) the drawing out stage (Figure 2c). Finally, Figure 2d shows the top view of a representative specimen spot-welded dissimilar thickness AA6082-T6 joint. The modern digital multimeter (type-UT61B, Zhejiang, China) was used to measure the temperature during the FSSW at the applied different rotation speeds (400–1000 rpm) using a thermocouple type "K".

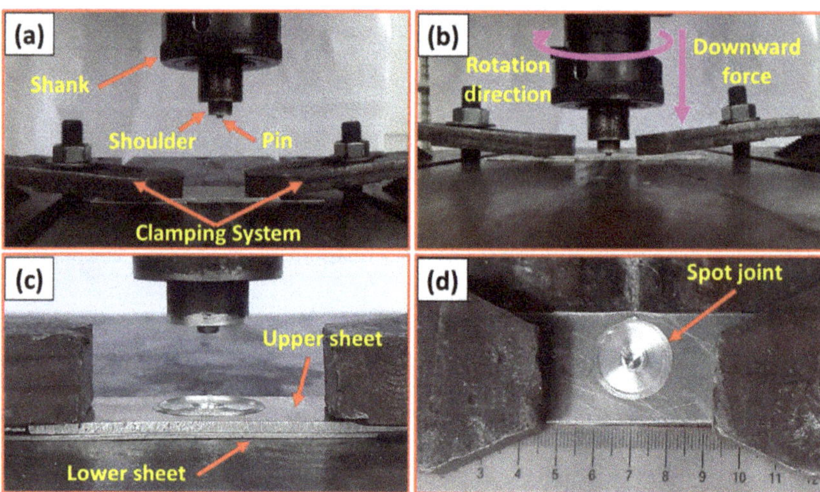

Figure 2. (**a**–**c**) illustrates the steps of the FSSW process of AA6082-T6 lap joints, while (**d**) shows the top view of the produced spot joint.

The produced FSSW joints were sectioned for macrostructure evaluation, microstructure investigation, and hardness test. The cross-sectional specimens were ground with SiC papers up to 2400 grit, then polished using polishing cloth in the presence of alumina paste up to a surface finish of 0.05 µm. Chemical etching of the polished samples was performed using two different compositions of Keller's etcher. The first consists of 25 mL nitric acid (HNO_3), 25 mL hydrochloric acid (HCl), 25 mL methanol (CH_3OH), and one drop of hydrofluoric (HF). The other was 2.5 mL HNO_3, 1.5 mL HCl, 95 mL distilled water, and 1 mL HF. The microstructure investigation was conducted using Olympus optical microscope (OM) (BX41M-LED, Olympus, Tokyo, Japan). The hardness test was performed for obtaining hardness contour maps across the transverse sections of the spot-welded joints. The hardness measurements were carried out using a Vickers hardness tester (model HWDV-75, TTS Unlimited, Osaka, Japan) using an applied load of 5 N with a dwell time of 15 s. The distance between every two indentations was set to 0.75 mm along the cross-section of the welds. The tensile-shear test (the load-carrying capacity) was carried out using a 30-ton universal tensile testing machine (Type-WDW-300D, Guangdong, China) at room temperature. The FSSW joints were tested with a constant loading rate of 0.1 mm/s. The tensile shear test specimens were cut and prepared according to ASTM E 8M-04. Figure 3 illustrates schematic drawings of the tensile-shear test specimen of the FSSWed dissimilar thickness AA6082-T6 joint (Figure 3a) and the tensile test specimen of the BM (Figure 3b). During the tensile-shear test of the joint specimen (Figure 4a), two backing sheets were used to ensure the application of axial load (Figure 4b,c). After tensile testing for the 2 mm thickness BM and the FSSWed fractured joints produced at 400 and 1000 rpm, the fractured surfaces were examined using a scanning electron microscope (SEM-Quanta FEG 250-FEI Company, Hillsboro, OR, USA).

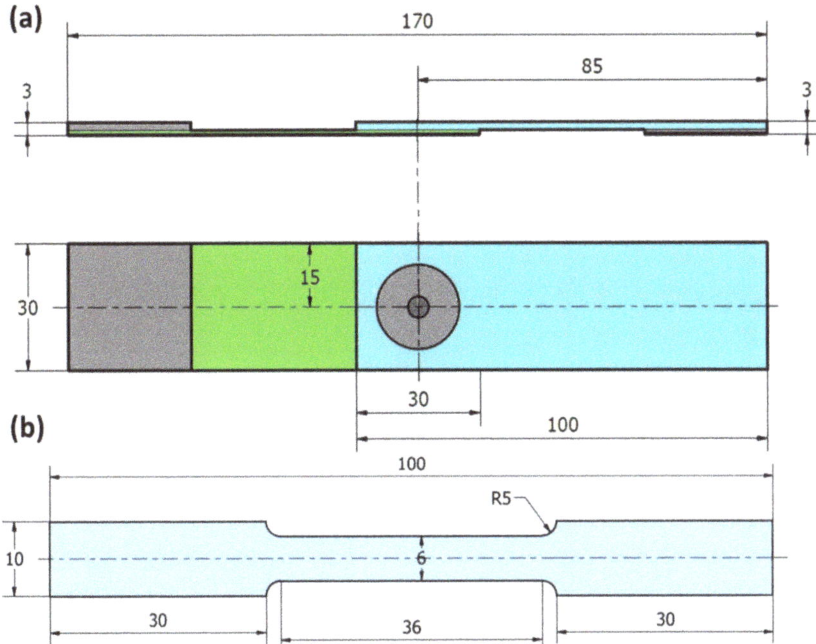

Figure 3. Schematic drawing of (**a**) tensile/shear test samples of FSSW lap joints, (**b**) tensile test sample of AA6082-T6 BM. (All dimensions in mm).

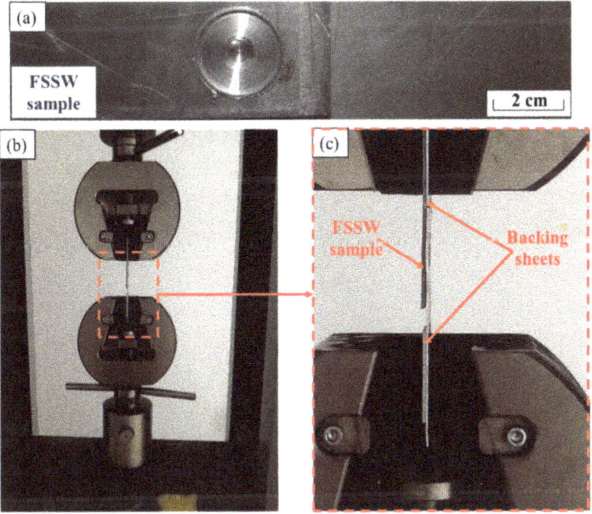

Figure 4. Photo image of (**a**) a dissimilar thickness AA6082-T6 lap joint, (**b**) setup of tensile/shear test of the spot-welded joint, and (**c**) higher magnification of (**b**).

3. Results and Discussion

3.1. FSSWed Joints Surfaces' Appearance

Figure 5 shows the top and bottom typical appearances of the FSSW joints produced at a constant dwell time of 3 s and different rotational speeds of 400, 600, 800, and 1000 rpm.

From a visual inspection, it can be observed that the applied combination parameters were suitable for producing FSSWed joints between different thickness sheets of AA6082-T6. For the top view of the spot-welded joints (Figure 5), it can be seen that circular indentations due to shoulder projection are observed at the different applied parameters and the extruded material flashed to the sides of the shoulder projection is nearly similar. Based on the experimental and published data from our lab [15,32,33], the processing spot welding parameters were carefully selected to eliminate excessive flash during the FSSW. The bottom view of the FSSW joints shows thermal hot spot areas (the affected areas due to the friction stirring process). These thermal hot spot areas get darker with increasing the rotational speed as a result of increasing the generated heat input during the spot dwell time, as can be seen in Figure 5.

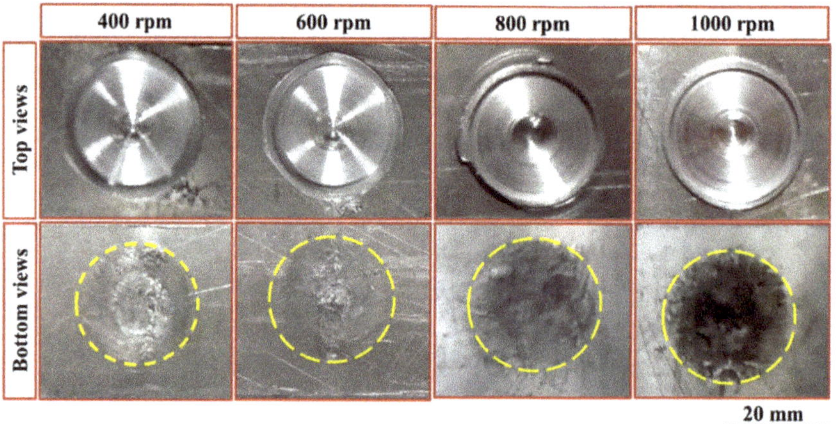

Figure 5. Typical appearances of both sides of dissimilar thin thickness FSSW AA6082-T6 lap joint samples at a constant dwell time of 3 s and different rotation speeds.

3.2. Heat Input Calculation and Temperature Measurement

The heat input generated during the FSSW mainly depends on the tool shoulder profile, pin geometry, rotation speed, friction coefficient, axial downward force, and dwell time [16,33,34]. The thermo-mechanical process during the FSSW converts the produced mechanical energy through tool rotation into heat input in the workpiece. The heat input for the FSSW process (Q) can be calculated using Equation (1) [33,35]:

$$Q = \frac{13}{12}\mu\frac{P}{K_A}\omega r t \ (J), \tag{1}$$

where $P(\mathrm{N})$: applied downward force, K_A: the ratio of the contact area of shoulder profile to tool cross-sectional area, ω (rad/s) equals $2\pi n$ n: the used rotation speed, t (s): the applied dwell time during the FSSW process, and r (m) : tool tip radius. μ (the friction coefficient between tool and aluminum alloy): equals 0.4 [14].

Where:

$$K_A = \frac{shoulder\ radius^2 - pin\ radius^2}{shoulder\ radius^2}\ 0.9375, \tag{2}$$

From Equations (1) and (2), the heat input to produce dissimilar spot lap joints of AA6082-T6 can be calculated using Equation (3).

$$Q = 1.1859 \times 10^{(-3)} \times n \times P \times t, \tag{3}$$

Figure 6 illustrates the generated heat input at the applied different rotational speeds from 400 to 1000 rpm during the FSSW of the dissimilar thin thickness AA6082-T6 spot joints. The generated heat input increases with increasing the rotational speed up to 1000 rpm. The rotation speed of 1000 rpm exhibits the highest heat input energy of 3 kJ.

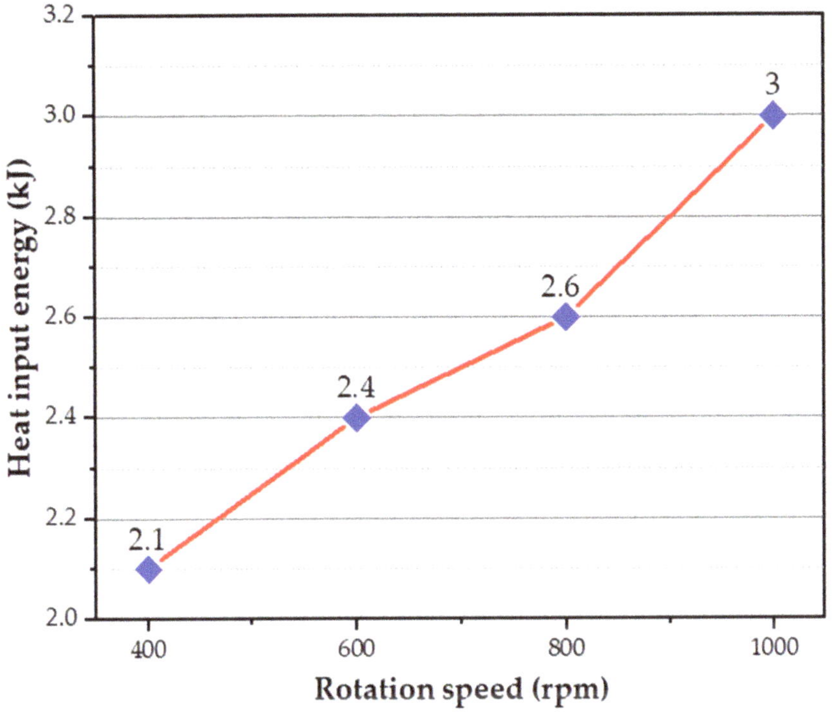

Figure 6. The generated heat input energy during the FSSW against the applied rotational speeds.

The FSSW thermal cycles of the SZ in terms of the measured working temperature as a function of time at the different rotation speeds are shown in Figure 7. It can be seen here that the thermal cycle gives the same trend at the applied different rotation speeds from 400 to 1000 rpm at a constant dwell time of 3 s, with a difference in the measured peak temperatures at each rotation speed. The FSSW of the dissimilar thin thickness AA6082-T6 lap joints process can be divided into three stages. The first stage records the rising temperature gradually by penetrating the tool pin at a constant feed rate of 0.1 mm/s through the two sheets of the lap joint with direct contact of the tool shoulder with the top surface of the upper sheet at the applied rotation speed. The second stage represents the dwell time needed to achieve the spot joint at nearly temperature stability. The third stage (drawing out stage) represents the end of the welding process and air-cooling of the lap joint with gradually temperature loss. Table 3 lists the peak measured temperature in the SZ during the FSSW process. The temperature increases from 236 ± 4 to 367 ± 3 °C with the increasing rotational speed from 400 to 1000 rpm, respectively.

Table 3. The maximum measured temperature during the FSSW process at different rotational speeds.

Rotational Speeds (rpm)	400	600	800	1000
Peak Temperature (°C)	236 ± 4	265 ± 3	308 ± 5	367 ± 3

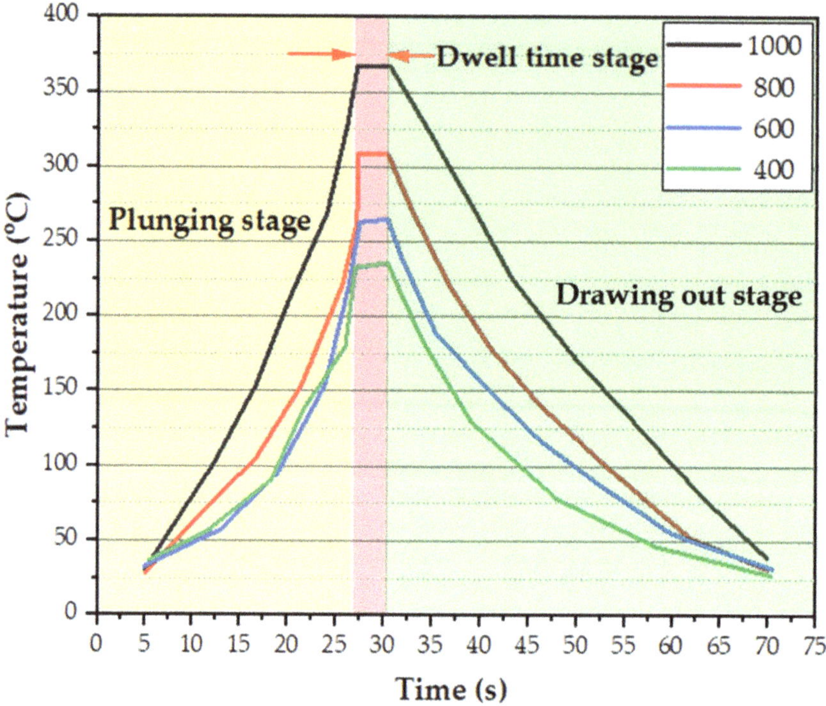

Figure 7. The thermal cycles during the FSSW of the dissimilar thin thickness AA6082-T6 lap joints processed at a constant dwell time of 3 s and the different rotational speeds of 400, 600, 800, and 1000 rpm.

3.3. Transverse Macrographs of the FSSWed Joints

Figure 8 shows the transverse cross-sections macrographs of the FSSWed AA6082-T6 joints. It can be observed that the interface between the overlapped sheets is welded due to plastic deformation and material flow produced by the rotation of the tool pin. In previous studies [15,36], it has been reported that three different regions are established in the spot weld zone. These regions are the flow transition zone under the rotating tool shoulder, the SZ around the tool pin, and the torsion zone under the pin. In fact, the joint efficiency of the FSSW is controlled by various parameters, including welding material, machine parameters (rotational speed, dwell time, and downward force), and tool material and design [1,25,27].

3.4. Microstructure Investigation

The microstructures and grain size histograms of the AA 6082-T6 BMs are shown in Figure 9. It can be seen that the BM microstructures show elongated and larger grain structures in the rolling directions, including very fine dispersed Mg_2Si precipitates and relatively coarse $(Fe, Mn)_3SiAl_{12}$ particles, as shown in Figure 9a,c. Typical microstructure features of the AA 6082-T6 BM are reported by Aydin et al. [27]. It can be remarked that AA 6082-T6 sheets of 1 and 2 mm thickness have mean grain size values of 6.63 $\pm$ 2 µm and 14.14 $\pm$ 1.5 µm, as shown in Figure 9b,d, respectively. The difference in grain size of BMs is related to imposing intense plastic deformation of thickness reduction. More thickness reduction using the rolling process in asymmetric directions causes high internal strain and grain size refining that increases the elongation and the strength [37].

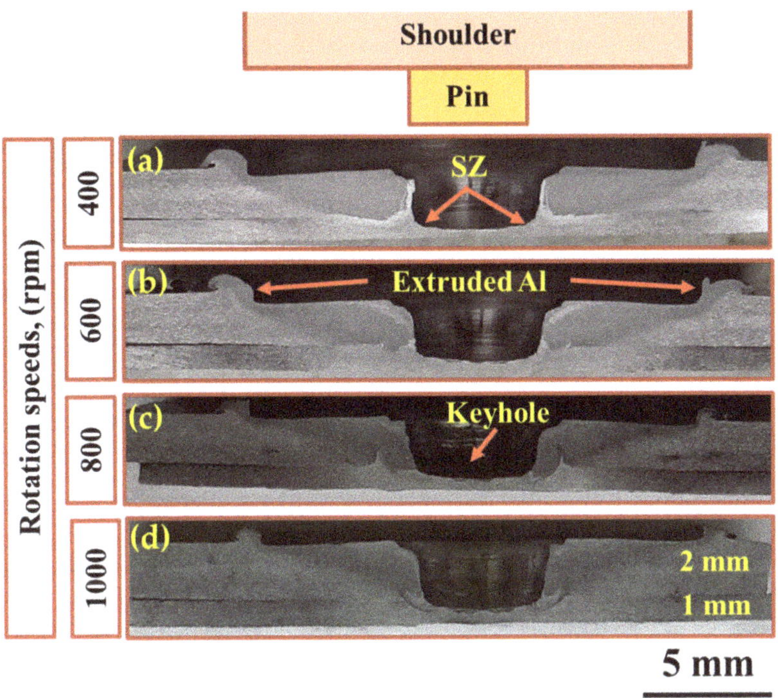

Figure 8. Macrographs of the transverse cross-section of the FSSW lap joints at a constant dwell time of 3 s and rotational speeds of (**a**) 400 rpm, (**b**) 600 rpm, (**c**) 800 rpm, and (**d**) 1000 rpm.

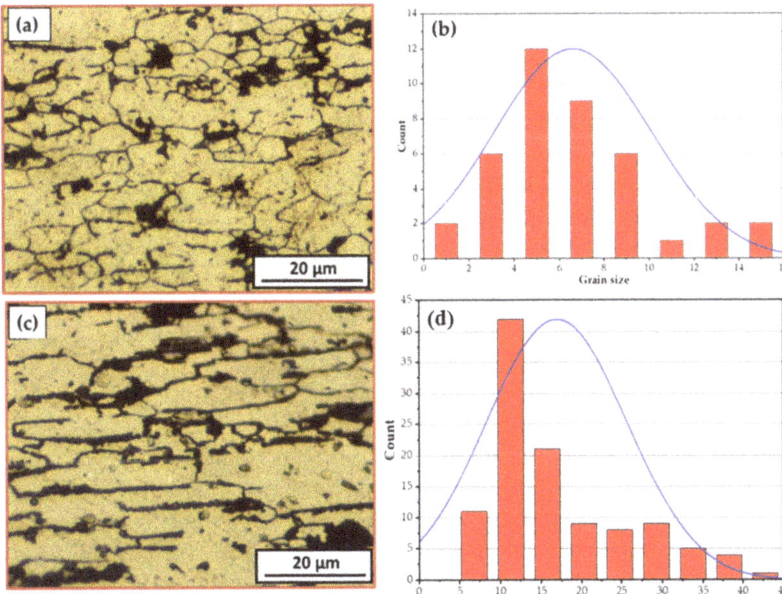

Figure 9. OM images, (**a**,**c**) and grain size histograms, (**b**,**d**) of AA6082-T6 sheet BMs of 1 mm and 2 mm thickness, respectively.

Figure 10 is a representative drawing example for the cross-section of the FSSWed joints to illustrate the locations (P1–P4) examined using an optical microscope to clarify the different zones achieved by the FSSW process at the various parameters. The optical micrographs in Figures 11a, 12a, 13a and 14a illustrate various FSSWed zones (SZ, TMAZ, and HAZ) yielded after the FSSW at a constant dwell time of 3 s using the different rotation speeds of 400, 600, 800, and 1000 rpm. In addition, a keyhole is formed in the center of the FSSW lap joints. The microstructure analysis of the BMs (Figure 9) and S.Z regions confirm distinct modifications in grain morphologies and sizes (Figures 11b, 12b, 13b and 14b). The elongated grains of the AA6082-T6 BM sheets are refined to equiaxed grains in a narrow range distribution at all the applied parameters. This change in the morphology and size of the grains is ascribed to dynamic recrystallization through the SZ [15,26,27,35]. The high temperature experienced and the amount of strain at this high temperature allows the dynamic recrystallization to take place [30,38]. It has been reported that the FSW of aluminum resulted in a high temperature and high strain rate that cause the formation of new fine equiaxed grain structure through dynamic recrystallization [39,40]. During the FSSW of aluminum, as noticed above, the temperature experience ranges between 220 to 350 °C, which is high enough to allow dynamic recrystallization at the high strain rate experienced. Thus, the grain size of the 2 mm AA6082-T6 BM reduced from 14.14 ± 1.5 µm to 1.24, 1.68, 2.37, and 3.52 µm for the stir zones of the joints spot welded at 400, 600, 800, and 1000 rpm, respectively, Figure 15. The average grain size of the SZ is increased with increasing the rotation speed from 400 to 1000 due to the increase in heat input (Figure 6). The coarse particles of (Fe, Mn)$_3$SiAl$_{12}$ in the BMs have fragmented into small particles and redistributed in the aluminum matrix due to the stirring action in the SZ, whereas the fine precipitates (Mg_2Si) are coarse in the SZ, which may be attributed to the dissolution and regrowth during the thermal cycle of welding, followed by air cooling [27,41]. Furthermore, the TMAZ grains (Figures 11c, 12c, 13c and 14c) are noticeably rotated and deformed along with the material flow during the stirring action of the tool material. The grain size in the HAZ is affected by the frictional heat generation, not by plastic deformation (as shown in Figures 11d, 12d, 13d and 14d). Thus, it is expected that a grain growth in HAZ compared to the BM grain size will be observed. The thermal exposure in the HAZ and the TMAZ during the FSSW process resulted in a coarsening of the precipitates. Finally, it can be concluded that the grain size of the different spot welding zones (SZ, TMAZ, HAZ) is directly related to the rotation speed when the other process parameters constant are kept constant. The mean grain size of the three zones increases with increasing tool rotation speed from 400 to 1000 rpm, as given in Figure 15.

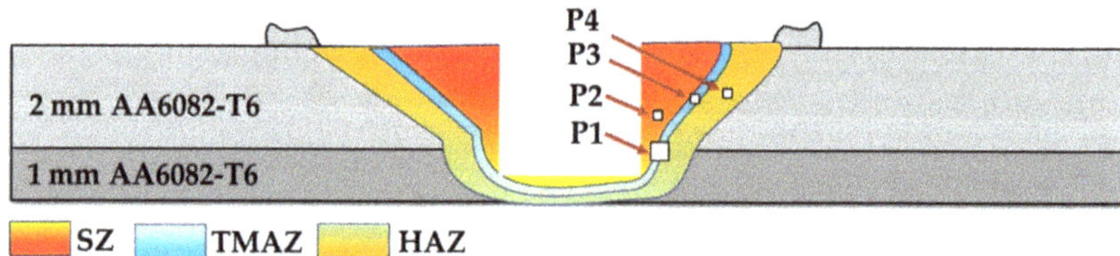

Figure 10. Schematic drawing showing the locations of the selected points (P1–P4) for the microstructure examination of the friction spot weld zone.

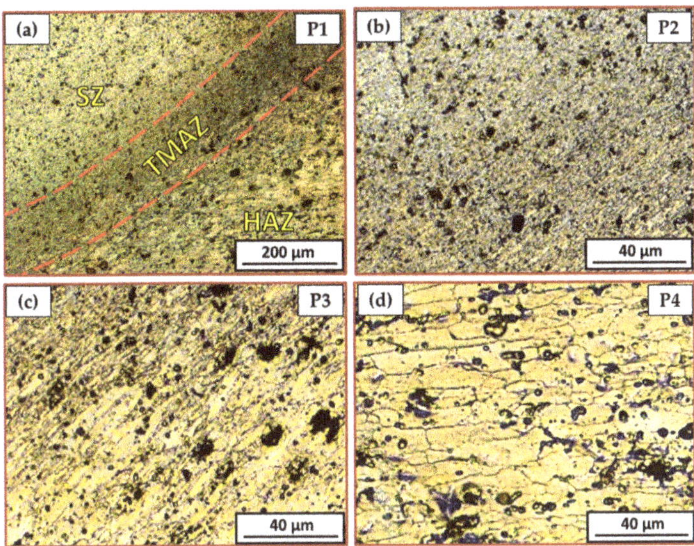

Figure 11. Different microstructure zones of the joint spot welded at 400 rpm where P1 is mixed zone of SZ/TMAZ/HAZ in (**a**), P2 is SZ in (**b**), P3 is TMAZ in (**c**), and P4 is HAZ in (**d**).

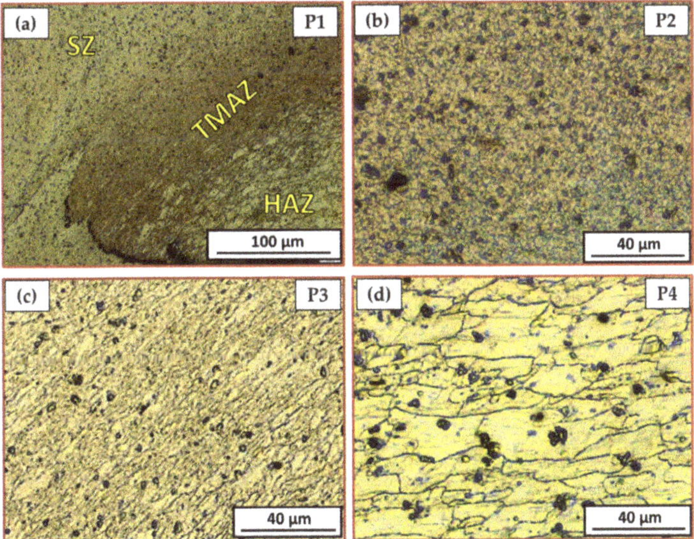

Figure 12. Different microstructure zones of the joint spot welded at 600 rpm where P1 is mixed zone of SZ/TMAZ/HAZ in (**a**), P2 is SZ in (**b**), P3 is TMAZ in (**c**), and P4 is HAZ in (**d**).

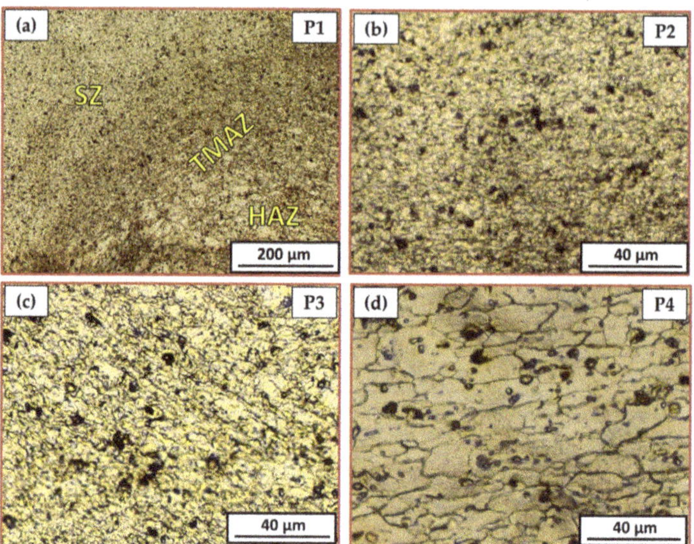

Figure 13. Different microstructure zones of the joint spot welded at 800 rpm where P1 is mixed zone of SZ/TMAZ/HAZ in (**a**), P2 is SZ in (**b**), P3 is TMAZ in (**c**), and P4 is HAZ in (**d**).

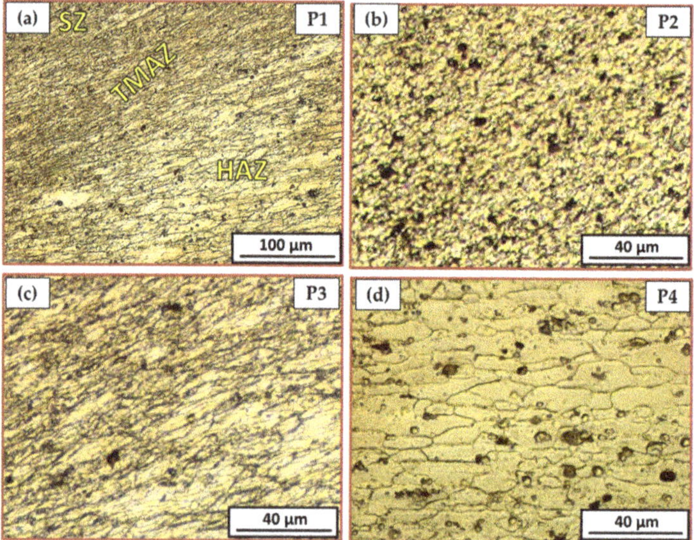

Figure 14. Different microstructure zones of the joint spot welded at 800 rpm where P1 is mixed zone of SZ/TMAZ/HAZ in (**a**), P2 is SZ in (**b**), P3 is TMAZ in (**c**), and P4 is HAZ in (**d**).

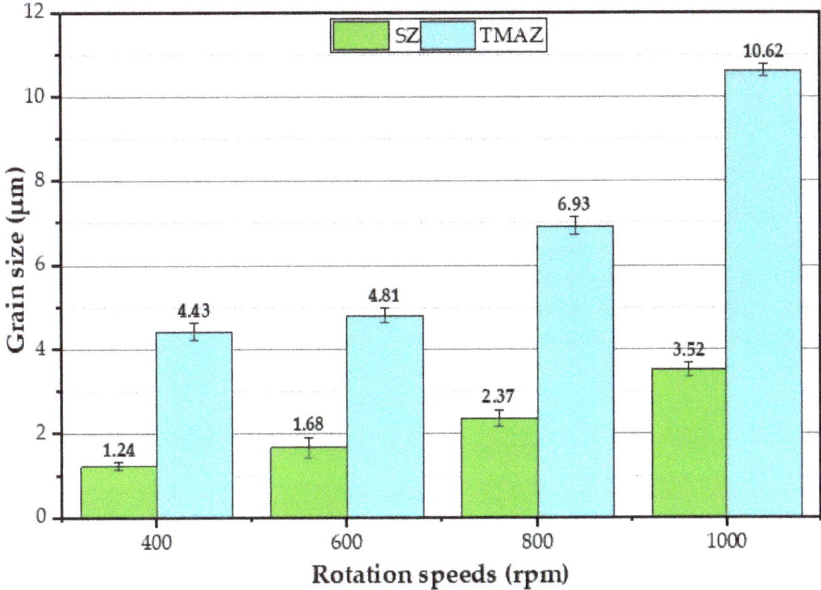

Figure 15. The mean grain size of the SZ and TMAZ of the joints spot welded at 400, 600, 800, and 1000 rpm rotation speed, and 3 s dwell time.

3.5. Hardness Characterization

Hardness was measured on the cross-sections of all the FSSWed joints produced at the rotational speeds of 400, 600, 800, and 1000 rpm and represented in color contoured maps, as shown in Figure 16. It is well known that the hardness through the thickness of the weld zones is controlled by thermal exposure during the FSW process [30]. Moreover, the state of the starting material affects the hardness behavior after FSW. For example, in this study, the starting material is T6 temper condition, which means fully hard through age hardening. Thus, the high thermal exposure is expected to result in hardness reduction in the weld region either due to the coarsening or the dissolution of the hardening precipitates [16,30,40,42–45]. In this study, the hardness contour map across the transverse cross section of the weld zone gives good representation of the hardness behavior. The AA6082-T6 BMs showed the mean hardness value of 127 ± 3 and 111 ± 2 $HV_{0.5}$ for the 1 mm and 2 mm sheet thickness, respectively. Compared to the BM, the hardness of the weld zone of the FSSW joints significantly decreased at all the applied rotation speeds (Figure 16) due to annealing of the AA6082-T6 BMs caused by the frictional heat generated during the FSSW [5,27] and the FSW [8] processes. For each spot-welded joint among the weld zone, the minimum hardness values were observed in the HAZ due to the grain structure and overaging effects. In contrast the higher hardness was recorded in the SZ mainly due to the dynamic recrystallized equiaxed fine grain structure and the reprecipitation process that might take place upon the cooling cycle [30]. The hardness of the TMAZ shows lower values than the SZ and higher values of hardness than the HAZ. The increase of the TMAZ hardness over the HAZ may be ascribed to the high dislocation density promoted by the plastic deformation during the FSSW process [46–48]. The highest hardness values were 95 ± 2, 87 ± 3, and 80 ± 5 $HV_{0.5}$ for the SZ, TMAZ, and HAZ, respectively, of the spot joint welded at 600 rpm (Figure 16b).

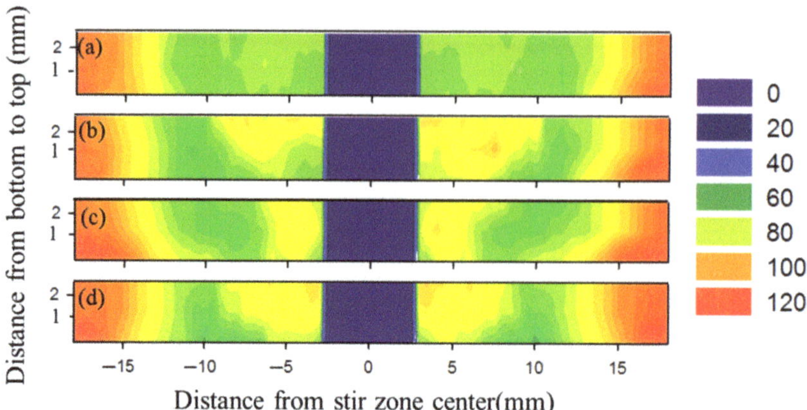

Figure 16. Two-dimensional hardness map of FSSW lap joints (**a**) 400, (**b**) 600, (**c**) 800, and (**d**) 1000 rpm.

3.6. Tensile Shear Testing and Fracture Behavior

Vehicle designers consider the tensile shear performance of the spot-welded joints while developing new models of vehicles. Thus, the tensile shear test for all the produced FSSWed joints was evaluated. It was reported that the tensile shear performance of the FSSWed was significantly affected by the welding process parameters [27,32,35]. The maximum tensile shear load of the FSSWed lap joints produced at the rotation speeds of 400, 600, 800, and 1000 rpm is illustrated in Figure 17. It can be seen here that the FSSW condition of 600 rpm rotation speed at 3 s dwell time shows the highest tensile shear load (4300 $\pm$ 30 N) compared to the load-carrying capacity of the other spot-welded joints. The two sheets of this spot lap joint are still connected after the tensile shear test, indicating high joint efficiency (Figure 18). This increase in tensile shear lap carrying capacity can be attributed to a larger fully bonded section size and lower hook height. In addition, the hardness improvement in the SZ is more than the hardness measured in the SZ of the other FSSWed joints. In contrast, the welded lap sheets produced at 400, 800, and 1000 rpm are completely separated during the tensile shear testing (Figure 19). The failed spot lap joint processed at 400 rpm may be ascribed to insufficient heat input, which affects in mixing during the welding of the dissimilar thickness AA6082-T6 lap joint. The tensile shear loads' carrying capacity of the spot welds processed at higher rotation speeds of 800 and 1000 rpm show a decrease in the maximum tensile shear load. This drop in the tensile shear is likely due to the increased thermal softening in the SZ with rising heat input and lower thickness of the upper sheet underneath the shoulder. Xie et al. [49] and Ohashi et al. [50] concluded that the mechanical properties of the FSSWed lap joints are mainly governed by both the upper sheet underneath the shoulder and the bonded area. It should also be mentioned that the precipitate-free zones around the grain boundaries due to precipitate coarsening might have reduced the strength of SZ.

The fracture surfaces of the tensile-tested BMs are given in Figure 20. It can be seen here that the two fracture modes (ductile and brittle) are detected for the two sheet thicknesses of 1 mm (Figure 20a,c) and 2 mm (Figure 20b,d), that the ductile fracture showed by the aluminum matrix in terms of different dimple sizes in 1 mm sheet and shallow elongated dimples of 2 mm sheet thickness, while the brittle fracture was observed by the presence of precipitates of Mg_2Si and $(Fe, Mn)_3SiAl_{12}$. Microvoids due to the precipitates pullout are also detected on the fracture surface of the two sheets. The observed fracture surface is typically related to the microstructure features of AA6082-T6 BMs. Figure 21 illustrates the SEM fractography of the lower sheet of the FSSWed joint processed at a constant dwell time of 3 s and different rotation speeds of 400 (Figure 20a–c) and 1000 rpm (Figure 20d–f). The two spot lap joints were failed by the tensile shear mechanism. Under the tensile shear loading conditions, the microcracks begin at the partially bonded region at the tip of the hook and propagate preferentially in the horizontal direction at the spot joint interface, shearing the SZ causing failure. The fracture surface of the two joints at the lower sheets shows typically ductile features in terms of very small deep dimples compared to the BMs, indicating the grain refining of the SZ during the FSSW.

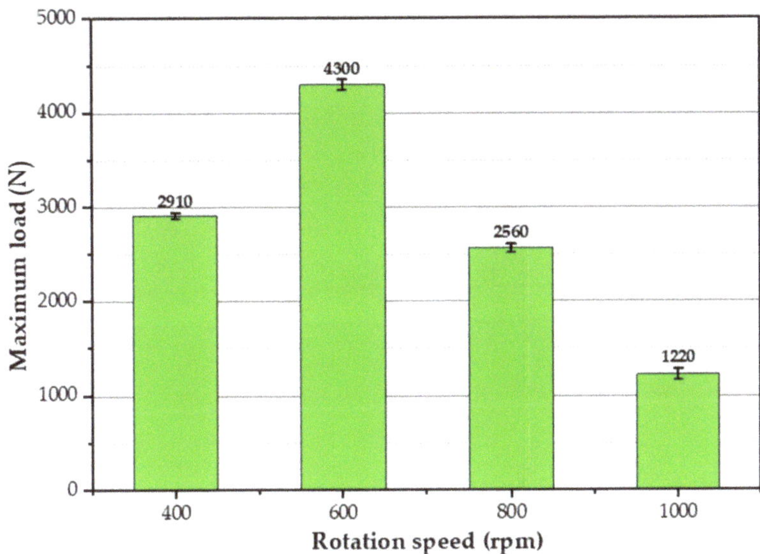

Figure 17. Maximum tensile-shear load of the FSSWed joints processed at a constant dwell time of 3 s and different rotation speeds of 400, 600, 800, and 1000 rpm.

Figure 18. (**a**) top view, (**b**) side view, and (**c**) bottom view photographs of the fracture surface appearance of the FSSWed lap joint processed at 600 rpm and 3 s.

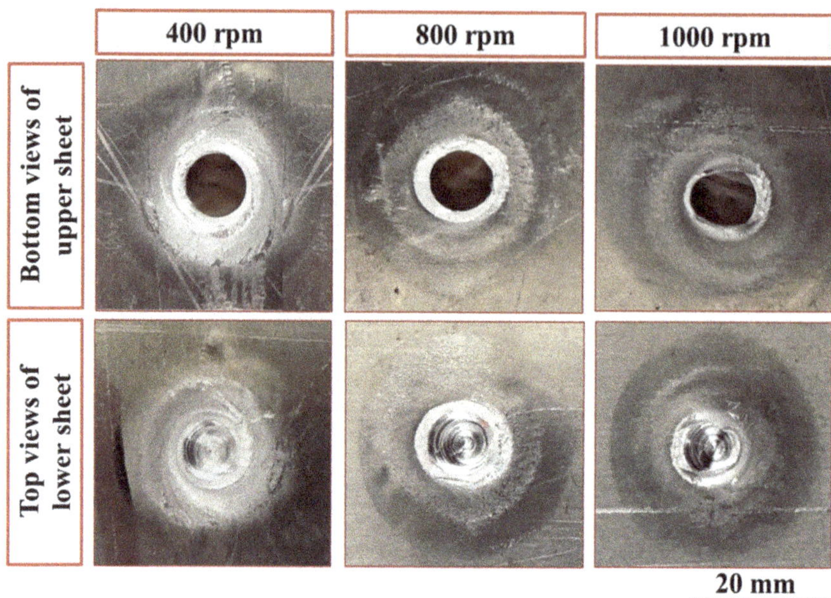

Figure 19. Photographs of fracture surface appearance of the FSSW lap joints processed at 400, 800, and 1000 rpm and 3 s dwell time.

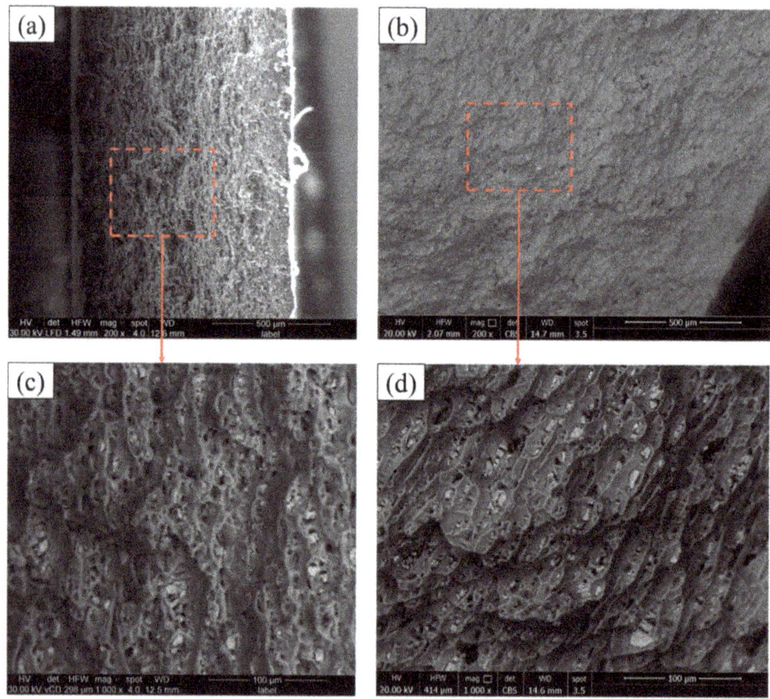

Figure 20. SEM images showing the fracture surface of AA6082-T6 BMs; (**a**,**c**) 1 mm and (**b**,**d**) 2 mm at different magnifications.

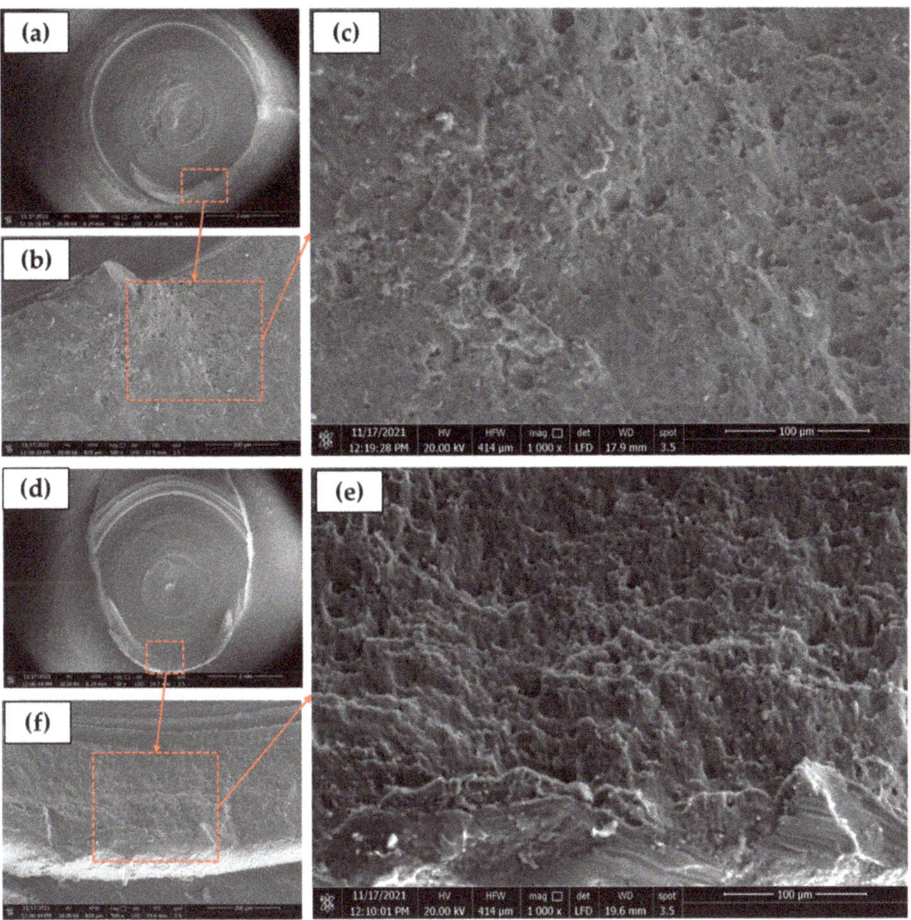

Figure 21. SEM images of FSSW lap joints at (**a**,**b**,**c**) 400 rpm and (**d**,**e**,**f**) 1000 rpm at different magnifications.

4. Conclusions

In the study, different AA6082-T6 thickness sheets of 1 mm and 2 mm were friction stir spot welded at a constant dwell time of 3 s and different rotation speeds of 400, 600, 800, and 1000 rpm. The spot-welded joints were characterized in terms of macro-, microstructure, hardness, and tensile shear testing and fracture surface. Based on the obtained results, the following conclusions can be outlined:

1. The applied FSSW parameters in terms of a constant dwell time of 3 s and different rotation speeds of 400, 600, 800, and 1000 rpm succeeded in welding two different thin sheet thicknesses of AA6082-T6 in spot lap joint configuration.
2. Among the produced joints, the spot-welded joint processed at 600 rpm achieved the highest SZ hardness of 95 ± 2 $HV_{0.5}$ and maximum tensile shear load of 4300 ± 30 N.
3. The FSSW of AA6082-T6 temper condition significantly decreased the hardness in the weld zone (SZ, TMAZ, and HAZ) compared to the BMs, and the SZ showed higher hardness values than the TMAZ and HAZ.
4. The spot-welded joints processed at the lowest rotation speed of 400 rpm and those processed at the highest rotation speed of 1000 rpm failed in the SZ with a ductile fracture mode.

Author Contributions: Conceptualization, M.M.Z.A., K.T., H.A.R. and M.M.E.-S.S.; methodology, H.A.R.; software, H.A.R. and K.T.; validation, M.M.Z.A., K.T., H.A.R. and M.M.E.-S.S.; formal analysis, M.M.Z.A., K.T. and H.A.R.; investigation, H.A.R.; resources, H.A.R., I.A. and E.A.; data curation, H.A.R. and K.T.; writing—original draft preparation, H.A.R., M.M.E.-S.S. and M.M.Z.A.; writing—review and editing, M.M.Z.A., H.A.R., K.T., E.A. and M.M.E.-S.S.; visualization, I.A.; supervision, M.M.Z.A., E.A. and M.M.E.-S.S.; project administration, M.M.Z.A., K.T. and I.A.; funding acquisition, M.M.Z.A., K.T. and M.M.E.-S.S. All authors have read and agreed to the published version of the manuscript.

Funding: The authors extend their appreciation to the Deputyship for Research & Innovation, Ministry of Education in Saudi Arabia for funding this research work through the project number (IF-PSAU-2021/01/18022).

Institutional Review Board Statement: Not applicable.

Informed Consent Statement: Not applicable.

Data Availability Statement: Data will be available upon request through the corresponding author.

Acknowledgments: The authors would like to thank Mohamed I. Habba of Suez University for his support in obtaining some experimental results.

Conflicts of Interest: The authors declare no conflict of interest.

References

1. Suryanarayanan, R.; Sridhar, V.G. Studies on the influence of process parameters in friction stir spot welded joints—A review. *Mater. Today Proc.* **2020**, *37*, 2695–2702. [CrossRef]
2. Yang, X.W.; Fu, T.; Li, W.Y. Friction Stir Spot Welding: A Review on Joint Macro- and Microstructure, Property, and Process Modelling. *Adv. Mater. Sci. Eng.* **2014**, *2014*, 697170. [CrossRef]
3. Tozaki, Y.; Uematsu, Y.; Tokaji, K. A newly developed tool without probe for friction stir spot welding and its performance. *J. Mater. Process. Technol.* **2010**, *210*, 844–851. [CrossRef]
4. Suryanarayanan, R.; Sridhar, V.G. Effect of Process Parameters in Pinless Friction Stir Spot Welding of Al 5754-Al 6061 Alloys. *Metallogr. Microstruct. Anal.* **2020**, *9*, 261–272. [CrossRef]
5. Aydın, H.; Tunçel, O.; Tutar, M.; Bayram, A. Effect of tool pin profile on the hook geometry and mechanical properties of a friction stir spot welded aa6082-t6 aluminum alloy. *Trans. Can. Soc. Mech. Eng.* **2021**, *45*, 233–248. [CrossRef]
6. Shen, Z.; Ding, Y.; Gerlich, A.P. Advances in friction stir spot welding. *Crit. Rev. Solid State Mater. Sci.* **2020**, *45*, 457–534. [CrossRef]
7. Badarinarayan, H.; Shi, Y.; Li, X.; Okamoto, K. Effect of tool geometry on hook formation and static strength of friction stir spot welded aluminum 5754-O sheets. *Int. J. Mach. Tools Manuf.* **2009**, *49*, 814–823. [CrossRef]
8. Liu, Z.; Zhang, H.; Hou, Z.; Feng, H.; Dong, P.; Liaw, P.K. Microstructural origins of mechanical and electrochemical heterogeneities of friction stir welded heat-treatable aluminum alloy. *Mater. Today Commun.* **2020**, *24*, 101229. [CrossRef]
9. Mishra, R.S.; Ma, Z.Y. Friction stir welding and processing. *Mater. Sci. Eng. R Rep.* **2005**, *50*, 1–78. [CrossRef]
10. Hirsch, J. Recent development in aluminium for automotive applications. *Trans. Nonferrous Met. Soc. China* **2014**, *24*, 1995–2002. [CrossRef]
11. Hoziefa, W.; Toschi, S.; Ahmed, M.M.Z.; Morri, A.; Mahdy, A.A.; El-Sayed Seleman, M.M.; El-Mahallawi, I.; Ceschini, L.; Atlam, A. Influence of friction stir processing on the microstructure and mechanical properties of a compocast AA2024-Al_2O_3 nanocomposite. *Mater. Des.* **2016**, *106*, 273–284. [CrossRef]
12. Ahmed, M.M.Z.; Wynne, B.P.; El-Sayed Seleman, M.M.; Rainforth, W.M. A comparison of crystallographic texture and grain structure development in aluminum generated by friction stir welding and high strain torsion. *Mater. Des.* **2016**, *103*, 259–267. [CrossRef]
13. Ahmed, M.M.Z.; Ataya, S.; El-Sayed Seleman, M.M.; Ammar, H.R.; Ahmed, E. Friction stir welding of similar and dissimilar AA7075 and AA5083. *J. Mater. Process. Technol.* **2017**, *242*, 77–91. [CrossRef]
14. Ahmed, M.M.Z.; Habba, M.I.A.; Jouini, N.; Alzahrani, B.; El-Sayed Seleman, M.M.; El-Nikhaily, A. Bobbin tool friction stir welding of aluminum using different tool pin geometries: Mathematical models for the heat generation. *Metals* **2021**, *11*, 438. [CrossRef]
15. Ahmed, M.M.Z.; Ahmed, E.; Hamada, A.S.; Khodir, S.A.; El-Sayed Seleman, M.M.; Wynne, B.P. Microstructure and mechanical properties evolution of friction stir spot welded high-Mn twinning-induced plasticity steel. *Mater. Des.* **2016**, *91*, 378–387. [CrossRef]
16. Ahmed, M.M.Z.; Ataya, S.; El-Sayed Seleman, M.M.; Mahdy, A.M.A.; Alsaleh, N.A.; Ahmed, E. Heat input and mechanical properties investigation of friction stir welded aa5083/aa5754 and aa5083/aa7020. *Metals* **2021**, *11*, 68. [CrossRef]
17. Haghshenas, M.; Gerlich, A.P. Joining of automotive sheet materials by friction-based welding methods: A review. *Eng. Sci. Technol. Int. J.* **2018**, *21*, 130–148. [CrossRef]

18. Gao, Z.; Feng, J.; Wang, Z.; Niu, J.; Sommitsch, C. Dislocation density-based modeling of dynamic recrystallized microstructure and process in friction stir spot welding of AA6082. *Metals* **2019**, *9*, 672. [CrossRef]
19. Hamada, A.S.; Järvenpää, A.; Ahmed, M.M.Z.; Jaskari, M.; Wynne, B.P.; Porter, D.A.; Karjalainen, L.P. The microstructural evolution of friction stir welded AA6082-T6 aluminum alloy during cyclic deformation. *Mater. Sci. Eng. A* **2015**, *642*, 366–376. [CrossRef]
20. Shafee Sab Habeeb, T.K. Analyzing Process of Friction Stir Spot Weld Joint. *Int. J. Sci. Technol. Eng.* **2016**, *3*, 308–313.
21. Pan, T.Y. Friction stir spot welding (FSSW)—A literature review. *SAE Technol. Pap.* **2007**, *2007*, 12. [CrossRef]
22. Majeed, T.; Mehta, Y.; Siddiquee, A.N. Challenges in joining of unequal thickness materials for aerospace applications: A review. *Proc. Inst. Mech. Eng. Part L J. Mater. Des. Appl.* **2021**, *235*, 934–945. [CrossRef]
23. Hoyos, E.; Serna, M.C. Basic tool design guidelines for friction stir welding of aluminum alloys. *Metals* **2021**, *11*, 2042. [CrossRef]
24. Wahid, M.A.; Siddiquee, A.N.; Khan, Z.A. Aluminum alloys in marine construction: Characteristics, application, and problems from a fabrication viewpoint. *Mar. Syst. Ocean Technol.* **2020**, *15*, 70–80. [CrossRef]
25. Yuan, W.; Mishra, R.S.; Webb, S.; Chen, Y.L.; Carlson, B.; Herling, D.R.; Grant, G.J. Effect of tool design and process parameters on properties of Al alloy 6016 friction stir spot welds. *J. Mater. Process. Technol.* **2011**, *211*, 972–977. [CrossRef]
26. Muhayat, N.; Putra, B.P. Triyono Mechanical Properties and Microstructure of Friction Stir Spot Welded 6082-T6 Aluminium Alloy Joint. *MATEC Web Conf.* **2019**, *269*, 01005. [CrossRef]
27. Aydin, H.; Tuncel, O.; Umur, Y.; Tutar, M.; Bayram, A. Effect of welding parameters on microstructure and mechanical properties of aluminum alloy AA6082-T6 friction stir spot welds. *Indian J. Eng. Mater. Sci.* **2017**, *24*, 215–227.
28. Buffa, G.; Fanelli, P.; Fratini, L.; Vivio, F. Influence of joint geometry on micro and macro mechanical properties of friction stir spot welded joints. *Procedia Eng.* **2014**, *81*, 2086–2091. [CrossRef]
29. Ahmed, M.M.Z.; El-Sayed Seleman, M.M.; Shazly, M.; Attallah, M.M.; Ahmed, E. Microstructural Development and Mechanical Properties of Friction Stir Welded Ferritic Stainless Steel AISI 409. *J. Mater. Eng. Perform.* **2019**, *28*, 6391–6406. [CrossRef]
30. Ahmed, M.M.Z.Z.; Wynne, B.P.; Rainforth, W.M.; Addison, A.; Martin, J.P.; Threadgill, P.L. Effect of Tool Geometry and Heat Input on the Hardness, Grain Structure, and Crystallographic Texture of Thick-Section Friction Stir-Welded Aluminium. *Metall. Mater. Trans. A* **2018**, *50*, 271–284. [CrossRef]
31. Hirasawa, S.; Badarinarayan, H.; Okamoto, K.; Tomimura, T. Journal of Materials Processing Technology Analysis of effect of tool geometry on plastic flow during friction stir spot welding using particle method. *J. Mater. Process. Technol.* **2010**, *210*, 1455–1463. [CrossRef]
32. Ahmed, M.M.Z.; Abdul-maksoud, M.A.A.; El-sayed Seleman, M.M.; Mohamed, A.M.A. Effect of dwelling time and plunge depth on the joint properties of the dissimilar friction stir spot welded aluminum and steel. *J. Eng. Res. Res.* **2021**, *9*, 1–20. [CrossRef]
33. Atak, A.; Şık, A.; Özdemir, V. Thermo-Mechanical Modeling of Friction Stir Spot Welding and Numerical Solution with The Finite Element Method. *Int. J. Eng. Appl. Sci.* **2018**, *5*, 70–75.
34. Wiedenhoft, A.G.; de Amorim, H.J.; de Souza Rosendo, T.; Durlo Tier, M.A.; Reguly, A. Effect of heat input on the mechanical behaviour of Al-Cu FSW lap joints. *Mater. Res.* **2018**, *21*, e20170983. [CrossRef]
35. Ataya, S.; Ahmed, M.M.Z.; El-Sayed Seleman, M.M.; Hajlaoui, K.; Latief, F.H.; Soliman, A.M.; Elshaghoul, Y.G.Y.; Habba, M.I.A. Effective Range of FSSW Parameters for High Load-Carrying Capacity of Dissimilar Steel A283M-C/Brass CuZn40 Joints. *Materials* **2022**, *15*, 1394. [CrossRef] [PubMed]
36. Yang, Q.; Mironov, S.; Sato, Y.S.; Okamoto, K. Material flow during friction stir spot welding. *Mater. Sci. Eng. A* **2010**, *527*, 4389–4398. [CrossRef]
37. Ma, C.; Hou, L.; Zhang, J.; Zhuang, L. Influence of thickness reduction per pass on strain, microstructures and mechanical properties of 7050 Al alloy sheet processed by asymmetric rolling. *Mater. Sci. Eng. A* **2016**, *650*, 454–468. [CrossRef]
38. Ahmed, M.Z.; Wynne, B.P.; Rainforth, W.M.; Martin, J.P. Crystallographic Texture Investigation of Thick Section Friction Stir Welded AA6082 and AA5083 Using EBSD. *Key Eng. Mater.* **2018**, *786*, 44–51. [CrossRef]
39. Ahmed, M.M.Z.; Elkady, O.A.; Barakat, W.S.; Mohamed, A.Y.A.; Alsaleh, N.A. The development of wc-based composite tools for friction stir welding of high-softening-temperature materials. *Metals* **2021**, *11*, 285. [CrossRef]
40. Ahmed, M.M.Z.; El-Sayed Seleman, M.M.; Zidan, Z.A.; Ramadan, R.M.; Ataya, S.; Alsaleh, N.A. Microstructure and mechanical properties of dissimilar friction stir welded AA2024-T4/AA7075-T6 T-butt joints. *Metals* **2021**, *11*, 128. [CrossRef]
41. Zhu, R.; Gong, W.-b.; Cui, H. Temperature evolution, microstructure, and properties of friction stir welded ultra-thick 6082 aluminum alloy joints. *Int. J. Adv. Manuf. Technol.* **2020**, *108*, 331–343. [CrossRef]
42. Ahmed, M.M.Z.; El, M.M.; Seleman, S.; Elfishawy, E.; Alzahrani, B.; Touileb, K.; Habba, M.I.A. The Effect of Temper Condition and Feeding Speed on the Additive Manufacturing of AA2011 Parts Using Friction Stir Deposition. *Materials* **2021**, *14*, 6396. [CrossRef] [PubMed]
43. Ahmed, M.M.Z.; Ataya, S.; Seleman, M.M.E.S.; Allam, T.; Alsaleh, N.A.; Ahmed, E. Grain structure, crystallographic texture, and hardening behavior of dissimilar friction stir welded aa5083-o and aa5754-h14. *Metals* **2021**, *11*, 181. [CrossRef]
44. Ahmed, M.M.Z.; Abdelazem, K.A.; El-Sayed Seleman, M.M.; Alzahrani, B.; Touileb, K.; Jouini, N.; El-Batanony, I.G.; Abd El-Aziz, H.M. Friction stir welding of 2205 duplex stainless steel: Feasibility of butt joint groove filling in comparison to gas tungsten arc welding. *Materials* **2021**, *14*, 4597. [CrossRef]
45. Ahmed, M.M.Z.; Wynne, B.P.; Rainforth, W.M.; Threadgill, P.L. Microstructure, crystallographic texture and mechanical properties of friction stir welded AA2017A. *Mater. Charact.* **2012**, *64*, 107–117. [CrossRef]

46. Wan, L.; Huang, Y.; Guo, W.; Lv, S.; Feng, J. Mechanical Properties and Microstructure of 6082-T6 Aluminum Alloy Joints by Self-support Friction Stir Welding. *J. Mater. Sci. Technol.* **2014**, *30*, 1243–1250. [CrossRef]
47. Miller, W.S.; Zhuang, L.; Bottema, J.; Wittebrood, A.J.; De Smet, P.; Haszler, A.; Vieregge, A. Recent development in aluminium alloys for the automotive industry. *Mater. Sci. Eng. A* **2000**, *280*, 37–49. [CrossRef]
48. Dong, P.; Li, H.; Sun, D.; Gong, W.; Liu, J. Effects of welding speed on the microstructure and hardness in friction stir welding joints of 6005A-T6 aluminum alloy. *Mater. Des.* **2013**, *45*, 524–531. [CrossRef]
49. Xie, G.M.; Cui, H.B.; Luo, Z.A.; Yu, W.; Ma, J.; Wang, G.D. Effect of Rotation Rate on Microstructure and Mechanical Properties of Friction Stir Spot Welded DP780 Steel. *J. Mater. Sci. Technol.* **2016**, *32*, 326–332. [CrossRef]
50. Ohashi, R.; Fujimoto, M.; Mironov, S.; Sato, Y.S.; Kokawa, H. Effect of contamination on microstructure in friction stir spot welded DP590 steel. *Sci. Technol. Weld. Join.* **2009**, *14*, 221–227. [CrossRef]

Article

Effect of Microstructure and Tensile Shear Load Characteristics Evaluated by Process Parameters in Friction Stir Lap Welding of Aluminum-Steel with Pipe Shapes

Leejon Choy [1], Myungchang Kang [1,*] and Dongwon Jung [2,*]

[1] Graduate School of Convergence Science, Pusan National University, Busan 46241, Korea; leejonee@hanmail.net

[2] Faculty of Mechanical, Jeju National University, Jeju 63243, Korea

* Correspondence: kangmc@pusan.ac.kr (M.K.); jdwcheju@jejunu.ac.kr (D.J.); Tel.: +82-10-3852-2337 (M.K.); +82-10-3459-2467 (D.J.)

Abstract: In recent years, friction stir welding (FSW) of dissimilar materials has become an important issue in lightweight and eco-friendly bonding technology. Although weight reduction of low-rigidity parts has been achieved, the weight reduction has been minimal because high-rigidity parts such as chassis require the use of iron. Considering the difficulty of welding a pipe shape, it is necessary to understand the effect of process parameters on mechanical performance. As a result of the study by various process parameters affecting the joint between aluminum and steel in the shape of a pipe, it can be seen that the tool penetration depth (TPD) has the most important effect on the tensile shear load (TSL). However, the effect of TPD on intermetallic compound (IMC), which has the most important influence on fracture, has not been well established. In this study, the effect of process parameters on IMC thickness and TSL in FSW of A357 cast aluminum and FB590 high tensile steel was investigated to reduce the weight of the torsion beam shaft of an automobile chassis. After the FSWed experiment, measurements were performed using an optical microscope and scanning electron microscopy (SEM) to investigate the microstructure of the weld. The formation of an IMC layer was observed at the interlayer between aluminum and steel. TPD is a major factor in IMC thickness variation, and there is a direct relationship between IMC thickness reduction and TSL increase, except for certain sections where the welding speed (WS) effect is large. Therefore, in order to improve mechanical properties in friction stir lap welding of aluminum and steel for high-rigidity parts, it is necessary to deepen the TPD at a level where flow is dominant rather than heat input.

Keywords: dissimilar friction stir welding; tool penetration depth (TPD); microstructure; intermetallic compound (IMC) thickness; process parameter

Citation: Choy, L.; Kang, M.; Jung, D. Effect of Microstructure and Tensile Shear Load Characteristics Evaluated by Process Parameters in Friction Stir Lap Welding of Aluminum-Steel with Pipe Shapes. *Materials* **2022**, *15*, 2602. https://doi.org/10.3390/ma15072602

Academic Editors: Józef Iwaszko and Jerzy Winczek

Received: 17 February 2022
Accepted: 17 March 2022
Published: 1 April 2022

1. Introduction

Recently, energy saving for global environment protection has emerged as an important issue in many industries, including the automobile industry [1–4]. Replacing steel with aluminum (Al) alloys to improve vehicle performance and fuel economy is a possible alternative. To replace everything made of steel with aluminum is difficult due to the differences of the mechanical performance of the material [5]. Therefore, a potential alternative is the replacement of a part of the steel with the Al alloy. When doing this, it is important to properly join the steel and Al; however, there are several factors that make dissimilar joining of steel and aluminum difficult. In particular, fusion welding typically results in various types of chemical separation, internal defects, and undesirable IMC in Al/steel welding. Conventional welding methods for dissimilar aluminum alloys and steels can lead to the formation of thick intermetallic (IMC) layers, so an alternative is needed [6]. Friction stir welding (FSW) is a stir welding technology developed by TWI in the UK in 1991 [7]. Because FSW is performed below the melting point of the material, it is possible to solve problems such as solidification defects that occur during melt welding, and it is possible to join lightweight alloys and high-strength alloys. A coupled torsion beam axle using high-strength steel (HSS) FB590 material is mounted on the rear of the car to interlink the tire and the body, support the force received from the tire, adjust the rolling angle when cornering, and the absorb vibrations or impacts of the road side [8]. Hot-rolled FB590 is an HSS that can be used in automotive chassis and suspension applications owing to its excellent crash performance, etc. [9]. In a coupled torsion beam axle steel, A357 material was selected to replace the trailing arm part that is not affected by torsion with Al material. In the aerospace and automotive industries, the alloys A356 and A357 are widely used for casting high-strength parts because they provide a combination of high strength while having good casting properties [10–13]. The FSW of A357 and FB 590 was studied to reduce the weight of automobile parts [14–16]. Park et al. [14] reported that by performing dissimilar FSW of 3 mm thick A357 cast Al and FB590 HSS plate, a tensile shear load (TSL) of 72.8% was achieved compared to the Al base material with TSL of 7912 N. The FSW process is mainly divided into four phases, and the normal force is changed in them [17]. During the FSW process, thermal energy is generated by friction between the tool and the workpiece, and plastic deformation of the workpiece occurs [18]. A study was conducted to determine the effects of different process parameters on the heat input and normal force [19–21]. The helical shape of the scroll tool shoulder improves the vertical force and is especially suitable for curved joints [22,23]. Process parameters that affect the heat input include the tool geometry and tool-related process parameters (offset and tilt), tool rotational speed (TRS), welding speed (WS), plunge speed, dwell time, plunge depth (PD), and tool penetration depth (TPD). To reduce defects in FSW and increase mechanical strength, the influence of process parameters becomes important after the tool design is complete. In a study on tool-related process parameters, it was reported that there is an optimal point of the normal force as the inclination angle of the tool and the tool offset increase [24,25]. In order to reduce the defects of the FSW joint and increase the mechanical strength, a study was conducted on TRS, WS, plunge speed, dwell time, PD, and TPD, which are non-tool process parameters. Process parameters have been shown to affect the mechanical performance and microstructure [26–35]. The joint strength has been shown to increase with increasing TPD [31–34]. In addition, research on various bonding methods between various materials is in progress.

Butts and laps between similar or dissimilar Al alloys [1,17–20,36], butts between Al alloys and steel [14,37–39], and laps [15,16,21,40,41] have been performed. Shamsujjoha et al. [42] studied the effect of tool penetration after friction stir lap welding (FSLW) of AA6061-T6511 and mild steel 1018. Understanding the evolution of microstructures is of considerable interest because the generation of micrograin structures in FSW significantly influences their mechanical properties [24,43,44]. The microstructure is also affected by tool and process parameters. Performing FSW of pipes is hard owing to its complicated geometry, and only a few studies of FSLW have been reported [15,16]. Al and steel materials have near-zero mutual solubility, forming the IMC of Fe_xAl_y [44]. The formation of IMC depends on the solid-state reaction between two substances, Fe and Al [45]. It is proposed that in FSW, the weld strength of aluminum and steel depends on the IMC, which is often formed during the welding operation of dissimilar materials [25,46]. In addition, process parameters directly affect the thickness of IMC [4,26,28,29,39,40,47–49]. Hussein et al. [4] presented a cause-effect diagram for IMC thickness formation. Many studies have been conducted to determine the relationship between process parameters and IMC thickness [25,26,29,33,50–52], while many studies have also been conducted to determine the relationship between the TSL and IMC thickness [53–55]. The latest research trends on IMC thickness and pipes are as follows. Aghajani Derazkola et al. [56] reported that the water-cooled sample with optimal IMC layer thickness showed the highest strength in friction stir joints of AA3003 and A441 AISI steels. A lower cooling rate allows thickening of IMCs and provides more brittleness to decrease the strength of FSW. Mortello et al. [57] obtained best results at feed rates ranging from 1.3 to 1.9 mm/s and rotational speeds of approximately 700 to 800 rpm for friction stir lap joints of AA5083 H111 and S355J2 grades DH36 structural steel. As the IMC thickness increases, the shear force decreases. Pankaj et al. [58] showed that in butt friction stir joints of DH36 steel and AA6061, the IMC thickness decreased with increasing tool offset and increased with increasing tool offset from 0.4 to 1.5 mm. The tool offset was observed to decrease with further increase from 1.5 mm to 2.5 mm. Abd Elnabi et al. [59] reported that the shear tensile strength increased as the tool pin length increased, and there was no change in bonding efficiency at IMC thicknesses of 7.5 μm or more. As a recent research of pipe, Choy et al. [16] used the definitive screening design method in the FSLW of aluminum and steel pipe as a previous study in this paper. PD and TPD are the most important factors influencing TSL, and there is no interaction between PD and TPD for the first time. Sabry et al. [60] reported FSW and submerged friction stir welding (UWFSW) on AA6061 pipe with a diameter of 30 mm and a wall thickness of 2, 3, and 4 mm. Using the full factorial analysis method, it was stated that the higher the TRS, the higher the UTS increased as the WS decreased. Abdullah et al. [61] investigated the rotational and linear velocities of tools in friction stir lap joints of AA5086 and C12200 copper alloy pipes and the microstructure, macrostructure, and mechanical properties of tool joints. Cylindrical pin tools at speed ratios below 10 rev/mm and conical pin tools in the 5–15 rev/mm speed ratio range have been shown to produce tunnel defects in the agitation zone. It is important to study process parameters because the type of IMC, the bonding mechanisms (such as hook joints, micro, and macro structures), and the IMC thickness (related on the axial pressure and heat generation of the weld zone). IMC thickness depends on the relationship between input heat and material flow. In general, increasing TRS increases IMC thickness and increasing WS decreases IMC thickness. In this study, TPD was assumed as the factor that has the greatest influence on material flow among process variables. So far, most of the studies have been mainly conducted on FSW joints to determine the influence of process parameters related to the plate shape. Pipe geometry studies that focus on the influence of process parameters have been mainly conducted on similar butt joints. Even in the study of pipe shape, only a few studies that focus on the effects of microstructures on the FSLW of AL and steel have been reported. In addition, there is little literature describing the effect of the microstructure on TPD, other than TRS and WS. Therefore, in this study, we aim to study the microstructure, as well as the IMC thickness and TSL characteristics according to TPD in pipe-shaped Al and steel FSLW.

2. Experimental Preparation and Methods

2.1. Materials and Tools

The pipes used in this experiment were A357 cast Al and FB590 HSS. The chemical composition of each material is shown in Table 1 [14]. The test specimen, i.e., A357 cast aluminum, was manufactured with an outer diameter of Φ 111 mm, a length of 155 mm, a thickness of 3 mm in the joint, and a thickness of 6 mm in an unbonded part.

Table 1. The chemical composition of base materials [14].

Material	C	Si	Mn	P	S	Cr	Ni
FB590	0.076	0.094	1.472	0.013	0.001	0.019	0.008
Material	**Si**	**Mg**	**Cu**	**Zn**	**Fe**	**Mn**	**Ti**
A357	6.937	0.507	0.034	0.017	0.181	0.007	0.116

FB590 HSS was manufactured with an outer diameter of Φ 105 mm, a length of 110 mm, and a thickness of 3 mm. Figure 1 is a photograph showing the experimental equipment, tool, and TPD. As shown in Figure 1a, the experimental equipment consists of a Winxen milling equipment that supplies the rotational force of the spindle up to 2000 rpm, a chuck fixing both sides of the pipe, and a fixing jig equipped with a bearing to support the pipe. Figure 1b shows the tool used in the FSW processing and the enlarged picture of the scroll

shape of the shoulder. Figure 1c shows the change in TPD according to the change of tool, with pin lengths of 3.0, 3.5, and 4 mm in the boundary layer between 3 mm thick aluminum and steel, in order to investigate the change in IMC thickness according to TPD. To rule out the effect of the PD, the PD was fixed at 0.0 mm. Dotted box arrows indicate increases in TPD. The material of the FSW tool was manufactured using W–Ni–Fe alloy, which is a type of heavy alloy. The tool's pin was processed into a threaded shape with a cylindrical tape, and the tool's shoulder was processed into a parallel scroll shape to increase the z-axis vertical force, improving the frictional heat and stirring during the FSW joining process [14,23].

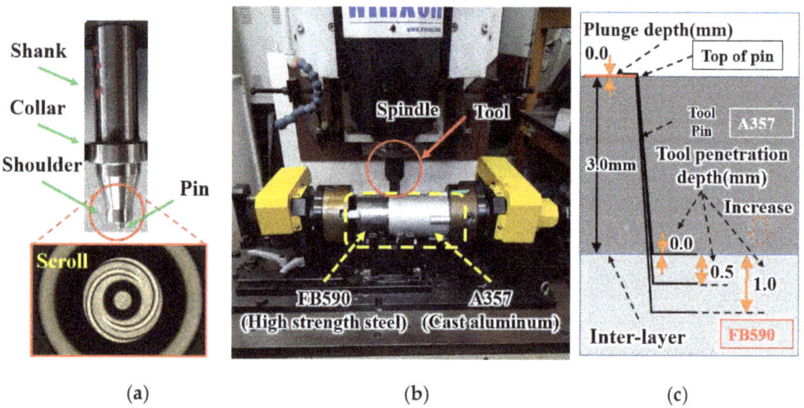

Figure 1. Overall configuration diagram of FSW experiment. (**a**) tool and close view of shoulder; (**b**) photograph of experimental equipment; and (**c**) tool penetration depth (TPD) [16].

The shoulder, pin root, and pin diameter of the tool are 10, 5, and 4 mm, respectively. To determine the effects of TPD, three types of tools with pin lengths of 3.0, 3.5, and 4.0 mm were used.

2.2. Experimental Methods

Figure 2 is a schematic diagram of the FSLW experimental process to investigate the effect of the IMC thickness and TSL according to different process parameters. Figure 2a, b show the definition of process parameters on the experiment. Figure 2a shows the TRS at which the main shaft rotates at a constant speed, the WS at which the tool advances the workpiece for joining the workpiece, the plunge speed descending at a constant speed relative to the TPD, and TPD at a constant plunge speed. After the tool descends to destination point, the descent is stopped, and the definition of dwell time is shown while the tool rotates in a fixed position for a certain period of time. For tools pre-selected by the PD and TPD in Figure 2b, the PD to which the top of the tool pin penetrates the outer diameter of the pipe and the TPD through which the tool pin penetrates the interlayer between steel and Al are shown. Figure 2c shows the thickness of the IMC layer, which is required to investigate the effect of process parameters on the microstructure. Figure 2d is shown to investigate the effect of process parameters on the magnitude of TSL.

According to Figure 2, the experimental method is as follows. After the experiment was performed according to the number of levels of the process parameter based on the set experiment, the change in the thickness of the IMC layer affected by each process parameter was measured. After the measurement, the effects of changes in the TPD and TRS were analyzed.

Shamsujjoha et al. [42] reported a better weld strength when the Al plate was placed along the advancing face on the steel plate. The two materials were washed with ethanol and then installed using the lap method, with A390 cast Al on the top side and FB590 HSS on the bottom side; the lap width was about 30 mm, as illustrated in Figure 1.

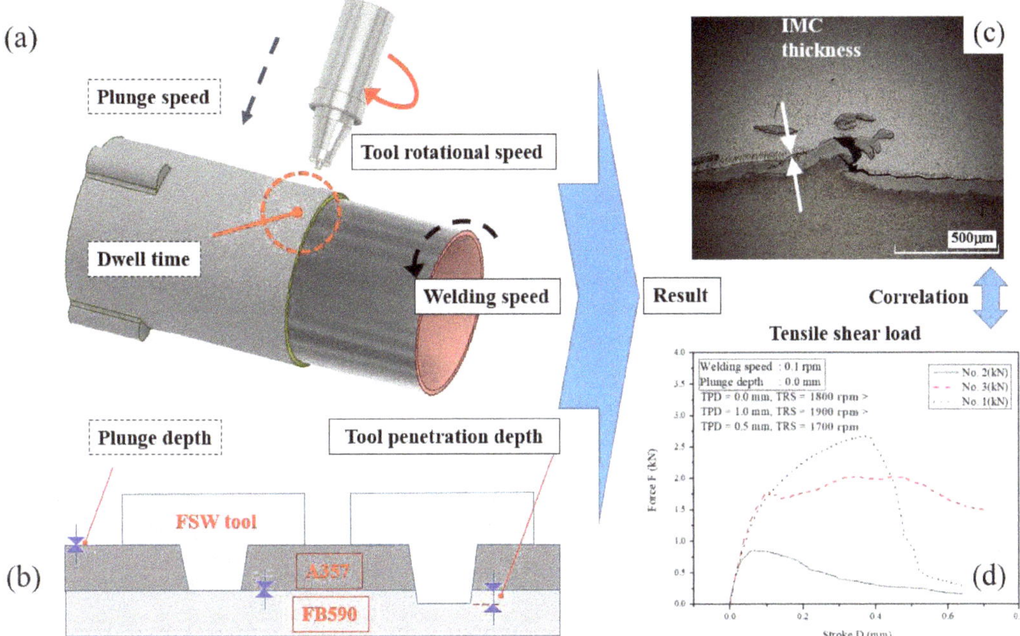

Figure 2. Schematic illustration of FSLW experimental process on IMC thickness and tensile shear load (TSL) by process parameters; (**a**) process parameter; (**b**) plunge depth and TPD; (**c**) IMC thickness; and (**d**) TSL.

Table 2 welding parameters for experiment. Group A is experiment no. 1, no. 2, and no. 3, and group B is experiment no. 4, no. 5, and no. 6.

Table 2. Welding parameters for experiment.

Experiment No.	Welding Speed (rpm)	Tool Penetration Depth (mm)	Tool Rotational Speed (rpm)
1	0.10	0.0	1800
2	0.10	0.5	1700
3	0.10	1.0	1900
4	0.20	0.0	1900
5	0.15	0.5	1800
6	0.20	1.0	1700

The sequence of experiments and the parameter values were selected to investigate the process parameters affecting the IMC thickness and TSL.

Because the PD and the TPD are affected by the pin length of the tool, the PD was fixed at 0.0 mm to prevent mutual interaction and to only investigate the effect of TPD, such that it is not affected by the PD. Tool pin lengths of 3.0, 3.5, and 4.0 mm were selected and used for the TPD experiment. Experimental conditions for the remaining parameters were selected as follows: TRS of 1700–1900 rpm; WS of 0.1–0.2 rpm; and a TPD of 0.0 mm, 0.5 mm, and 1.0 mm. The experiment was divided into two groups to investigate the effect of WS and TPD on the IMC thickness and TSL. The process parameter of one group is 0.1 rpm WS, and the other group has a WS value of 0.15 rpm or more. Figure 1c shows the change according to the TPD and indicates the position of the tool penetrating the steel surface through the boundary layer between A357 and FB590 when PD = 0 and TPD is 0.0, 0.5, and 1.0 mm. In Figure 1c, the box arrows indicate the direction in which the TPD increases. It was concluded that the TPD plays an important role in determining the weld strength [35].

To investigate the effect of the WS and TRS on the IMC thickness and TSL, the same experimental results were divided into two groups based on TRS. One group of reclassified results of experiment was fixed at a WS of 0.1 rpm, and the other group was a WS of 0.15 rpm or more.

Experiments were performed on an FSW machine (Winxen, 22 kN). Pipes are installed on the chuck and fixture, and were lap-bonded, as shown in Figure 1a. Welding is done in quadrants, i.e., 90°, instead of full rotation. To observe the structure of the junction, a rectangular sample was prepared. The tensile test specimen was manufactured according to the ASTM E8 standard, and TSL was measured using a tensile tester (AGX-X, Shimadzu, Kyoto, Japan). The size of the specimen for microstructure measurement was selected to be 20 mm wide and 10 mm long, for hot mounting The specimen was fabricated using EDM wire (NW570II, Doosan, Changwon, Korea) along a line perpendicular to the welding direction. The sample that was used to observe the microstructure of the FSW section was polished with silica sandpaper in a disk grinder and then polished with a diamond cloth. Weld textures and bonding interfaces between Al and steel were studied using optical microscopy (KH-8700, HIROX, Tokyo, Japan) and scanning electron microscopy (JSM-6490, JEOL, Tokyo, Japan).

3. Results and Discussion

3.1. Microstructure Characteristics of Friction Stir Welding (FSW) Joint

Figure 3 shows the optical macrostructure and microstructure of the FSW specimen. Figure 3a shows the macrostructure of the cross section of the FSWed joint according to the experimental conditions. The macrostructure in Figure 3a is enlarged. The TMAZ is shown in Figure 3b and SZ in Figure 3c. The HAZ is shown in Figure 3d, the Al base material (BM) is shown in Figure 3e, and the IMC region is shown in Figure 3f, respectively. In Figure 3a, the broken line of the magnification indicates the shape of the tool pin, and the arrow spacing of the PD indicates the spacing between the outer diameter of A390 cast Al and the shoulder of the tool. The arrow spacing of the TPD indicates the gap between the tip of the tool pin and the interlayer at which the inner diameter of A390 Al and the outer diameter of FB590 HSS meet. The cause of the cavity in Figure 3a is the geometric difference between the pipe and the plate. Akbari et al. [13] found that the contact characteristics of the tool and the workpiece were different due to the small radius of curvature of the pipe. At a PD of 0.1 mm, due to the pipe curvature, the FSW shoulder partially contacts the pipe and only the inner part makes the shoulder part contact the pipe, forming a tunnel cavity. Another cause of large cavity is a lack of material. (1) The upper aluminum material under the influence of process variables is discharged to the outside in the form of flash. (2) The upper aluminum material moves to fill the gap (gap) on the junction interface of aluminum and steel. The lack of material due to the above causes created a large internal cavity.

After FSW, there is an SZ around the center of the joint. There is a TMAZ in which magnitude of grains are increased by plastic flow on the outside of the SZ, and there is an HAZ that is heat-affected but has no plastic deformation on the outside of TMAZ. These zones were observed to have a wider region width than the retreat side on the tool's advance side (AS). To form a bond as the tool is traversed, material flows around the tool in a complex flow pattern depending on the tool geometry and process parameters, such as TRS and WS. Microstructural evolution depends on the change of the welding parameters. A suitable selection of the process parameters results in good material mixing at the joint and a sound weld.

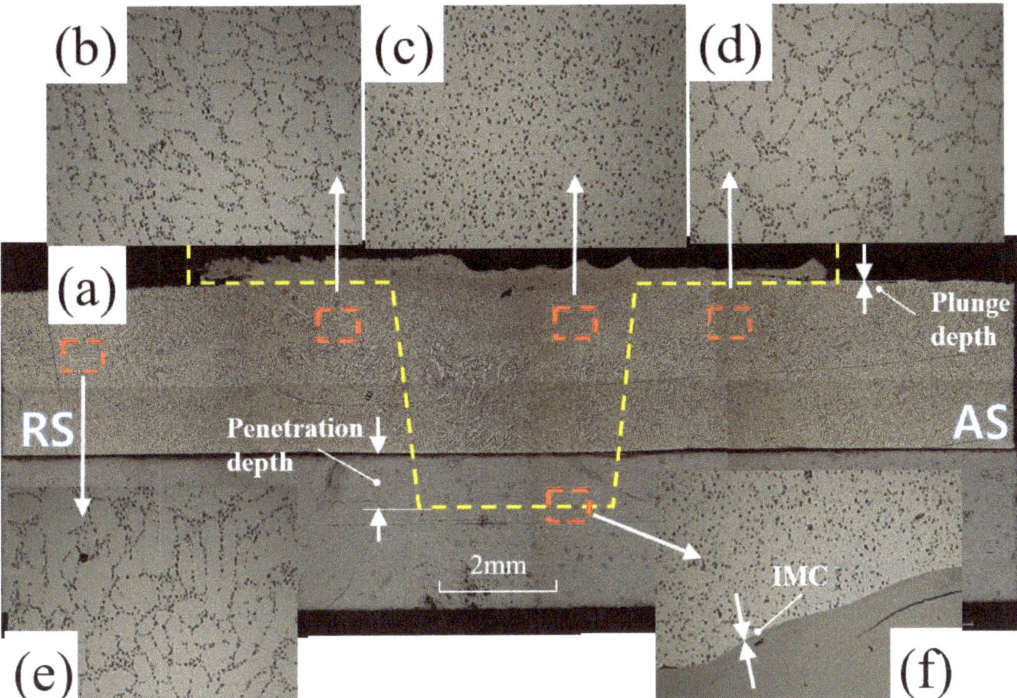

Figure 3. Optical micrographs of the FSW A357 cast alloy and FB590 HSS. (**a**) Overall cross-sectional macrostructure; (**b**) microstructure of the TMAZ; (**c**) microstructure of the SZ; (**d**) microstructure of the HAZ; (**e**) microstructure of the Albase metal; and (**f**) microstructure of the IMC [16].

The SZ in Figure 3c produces a fine grain structure, whereas the TMAZ in Figure 3b has a long grain structure. The SZ region of Figure 3c shows finer particles compared to the TMAZ region and the HAZ region of Figure 3d. In general, the SZ region is also referred to as a "nugget region" and is believed to be formed by dynamic recrystallization. The SZ region has a wider width at the top and decreases in the thickness direction. This is because of the difference in diameter between the shoulder and the pin. In the SZ, a finer grain was formed compared to other regions. Dynamic recrystallization produced skewed and elongated grains in TMAZ and fine grains in SZ. The grain boundary orientation is different in all three regions. The size of the SZ region in Figure 3c is smaller than that of the BM, and the crystalline grains of the BM disappear. This is in agreement with the results of Mahto et al. [21], who reported that new crystals were formed and dynamic recrystallization occurred. The microstructure of TMAZ is shown in Figure 3b. Located between HAZ and SZ, TMAZ is characterized by a highly deformed structure. It shows an elongated non-metallic particle morphology with sub-grain coarsening. The grain structure of the HAZ region, which was not mechanically perturbed by FSW, was generated by static recrystallization and was similar to the base metal in Figure 3e HAZ is much wider at the top surface in contact with the shoulder and is narrow as the pin diameter decreases. In Su et al. [62], the grain size increases in the order of SZ, TMAZ, HAZ, and BM. Therefore, the microstructure was recrystallized in the FSWed region.

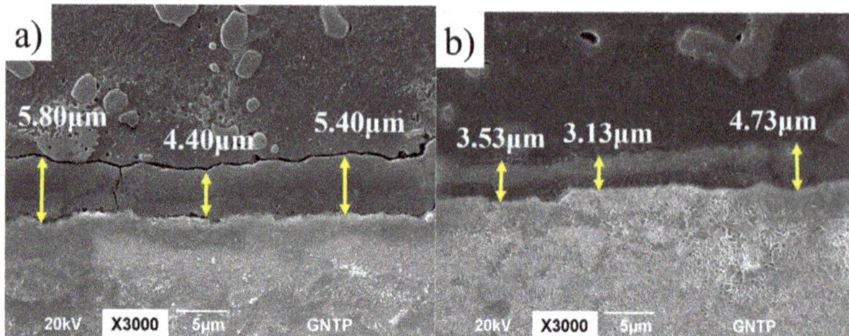

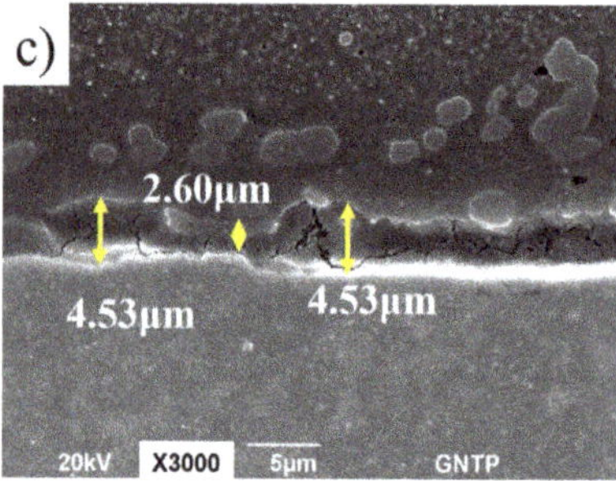

Figure 4. SEM pictures according to TPD of experiment group A. (**a**) TPD = 0.0 mm; (**b**) TPD = 0.5 mm; and (**c**) TPD = 1.0 mm.

3.2. IMC Thickness Characteristics of FSW Joints

In order to investigate the change in the IMC thickness due to the TPD, the PD is fixed at 0.0 mm, and changes in the remaining process parameters are considered. First, the effect on the IMC thickness during FSW according to the TPD of group A with a WS of 0.1 rpm is investigated. Figure 4 shows the change in the IMC thickness during FSW according to the TPD in group A as an SEM photograph. Figure 4a shows the IMC average thickness when the TPD is 0.0 mm in experiment no. 1, Figure 4b shows the IMC average thickness when the TPD is 0.5 mm in experiment no. 2, and Figure 4c shows the IMC average thickness at a TPD of 1.0 mm in experiment no. 3. At a TPD of 0.0 mm, the IMC average thickness is 5.1 μm; at a TPD of 0.5 mm, the IMC average thickness is 3.9 μm; and at a TPD of 1.0 mm, the IMC average thickness is 3.1 μm. It can be seen that the IMC average thickness decreases as the TPD increases. Next, the effect of the TPD on the IMC average thickness was investigated in group B with a WS of 0.15 to 0.2 rpm during FSW. Figure 5 shows the change in the IMC average thickness during FSW according to the TPD in group B as an SEM photograph. Figure 5a–c shows the IMC average thickness of experiment no. 4 with a TPD of 0.0 mm, experiment no. 5 with TPD of 0.5 mm, and an experiment no. 6 with a TPD of 1.0 mm. When the TPD is 0.0 mm, the IMC average thickness is 9.05 μm; at a TPD of 0.5 mm, the IMC average thickness is 6.43 μm; and at a TPD of 1.0 mm, the IMC average thickness is 5.07 μm. As in group A where the WS is 0.1 rpm, it can be seen that the IMC average thickness decreases as the TPD increases.

Mahto et al. [28] reported that the higher the heat input, the thicker the IMC layer was in the FSLW of AA 6061-T6 and AISI 304 stainless steel, each having a thickness of 1 mm. According to Ogura et al. [63], the nature of the IMC formed during welding is determined by the combined influence of base materials stirring and thermal conditions. As the TPD increases, the contact area between the tool and the material increases and the heat input tends to increase. Another trend is that the flow of material will increase due to agitation as the TPD increases. Therefore, it can be deduced that the thickness of the IMC layer decreases because the effect of the material flow due to agitation is greater than the increase in the heat input due to the increase of the TPD.

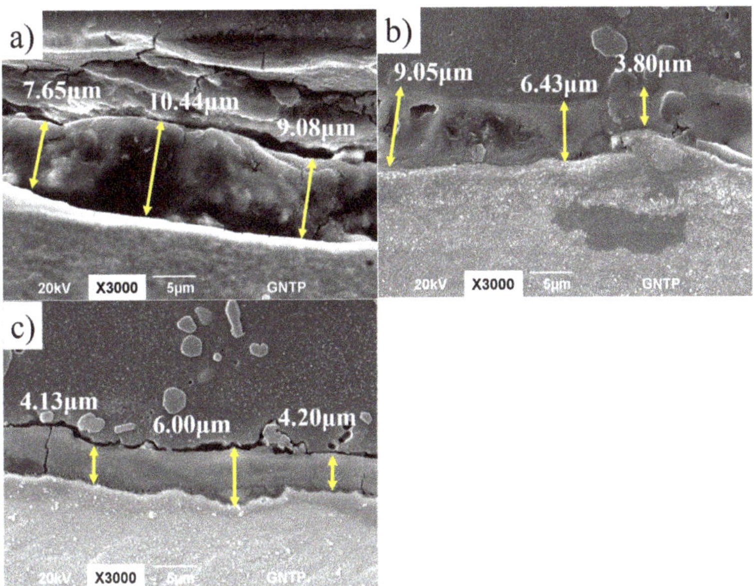

Figure 5. SEM pictures according to TPD of experiment group B. (**a**) TPD = 0.0 mm; (**b**) TPD = 0.5 mm; and (**c**) TPD = 1.0 mm.

3.3. Characteristics of IMC Thickness and Tensile Shear Load (TSL) According to the Tool Penetration Depth (TPD)

Table 3 summarizes the IMC thickness and TSL according to the change in TPD and WSs of 0.1 rpm and 0.15 rpm or more.

Table 3. IMC thickness range and IMC average thickness and TSL on TPD.

Experiment No.	**1**	**2**	**3**	**4**	**5**	**6**
Tool penetration depth (mm)	0.0	0.5	1.0	0.0	0.5	1.0
IMC thickness range (μm)	5.80–4.40	4.73–3.13	4.53–2.60	10.44–7.65	9.05–3.80	6.00–4.13
IMC average thickness (μm)	5.04	3.93	3.56	9.05	6.43	5.07
Tensile shear load (N) [16]	2677.21	860.50	2034.83	807.35	1779.25	1984.72

Figure 6 compares the differences between the two groups with changes in the IMC average thickness according to TPD in group A with a WS of 0.1 rpm and group B with a WS of 0.15 to 0.2 rpm. In Figure 6, group A of experiments no. 1, no. 2, and no. 3 is indicated by a solid line to represent the IMC average thickness change according to the TPD and a WS of 0.1 rpm. It shows a gradual decrease in the IMC average thickness as TPD increases from 0.0 mm to 1.0 mm. In Figure 6, group B of experiments no. 4, no. 5, and no. 6 are indicated by broken lines to represent the IMC average thickness change at WS values of 0.15 and 0.2 rpm. It shows a gradual decrease in the IMC average thickness as the TPD increases from 0.0 mm to 1.0 mm. It can be seen that the IMC average thickness gradually decreased as TPD increased from 0.0 mm to 1.0 mm, regardless of the influence of WS and other process parameters.

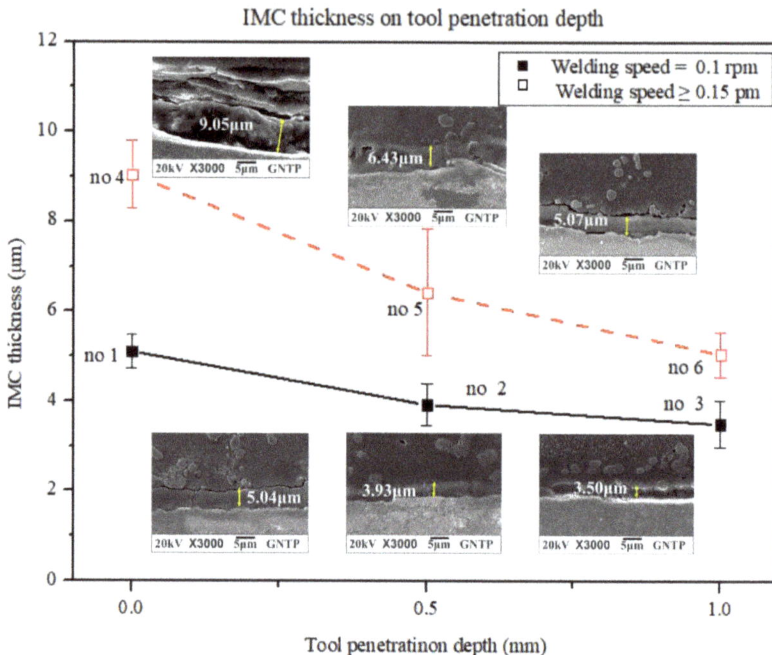

Figure 6. IMC thickness comparison according to TPD for experiment groups A and B.

In addition, it can be seen that the IMC average thickness increases as the WS increases at the same TPD. This is consistent with the experimental results reported by Shen et al. [34]. IMC at the Fe–Al interface decreases with increasing TPD during the FSLW of 2.2-mm-thick AA5754 Al and 2.5-mm-thick DP600 dual-phase steel. However, their results did not provide quantitative results for IMC thickness increase. Mahto et al. [35] performed FSLW between 0.0 mm and 0.3 mm with a pin depth (PLD) of 0.1 mm spacing on 1-mm-thick dissimilar materials AA6061-T6 and AISI304. It is said that the IMC thickness increases as the PLD increases. This is the result of changing the TPD in the narrow range of 0–0.3 mm, and it appears to be different from the current wide region, where the TPD is changed between 0.0 and 1.0 mm. In the study reported by Liu et al. [25], it was proposed that the IMC thickness decreased with the increase of the WS and the increase of the tool offset. In this study, the influence of TPD was dominant, showing the opposite trend. From the experimental results, as the TPD increases, the material fluidity increases and the IMC thickness decreases. This is because the effect of TPD is more dominant on material flow than on heat input.

Figure 7 shows graphs of the variation of the TSL with TPD.

In Figure 7, group A of experiments no. 1, no. 2, and no. 3 is indicated by a dash-dotted line to represent the changes in the TSL with the TSL and a WS value of 0.1 rpm. As TPL increases from 0.0 mm to 1.0 mm, it can be seen that the TSL decreases up to a TPD value of 0.5 mm, and TPD increases above 0.5 mm. In Figure 7, group B of experiments no. 4, no. 5, and no. 6 are indicated by dash-double dotted lines to represent the changes in the TSL for WS values of 0.15 and 0. 2 rpm. It shows a gradual increase in the IMC average thickness as the TPD increases from 0.0 mm to 1.0 mm. It can be seen that the IMC average thickness gradually decreased as TPD increased from 0.0 mm to 1.0 mm, regardless of the influence of WS and other process parameters. As TPD increases, TSL decreases to TPD = 0.5 mm at a WS of 0.1 rpm. At the WS of 0.1 rpm, the TPD increases over 0.5 mm, and at the WS of 0.15 and 0.2 rpm, the TPD increases over the entire region. In the study of TSL according to TPD, Lee et al. [52] reported that the TSL increased as the TPD increased, but this result is due to the difference between FSSW and this study, which has a large initial thermal effect. In order to increase the TSL according to TPD, a WS above a certain speed was not proposed. Shen et al. [34] reported that the TSL decreased and increased with increasing TPD. Contrary to the results of this study, this results from an inability to distinguish between the mutual effects of TPD and PD.

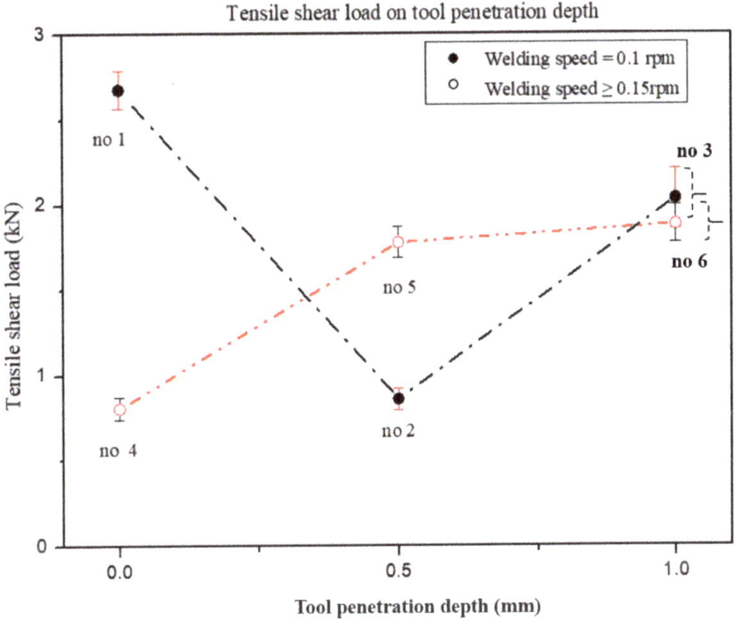

Figure 7. Comparison of TSL with TPD for experiment groups A and B [16].

In Figures 6 and 7, as TPD increases at a WS of 0.1 rpm, the IMC average thickness decreases and TSL decreases until TPD = 0.5 mm. In Figure 8, it can be seen that the IMC average thickness decreases and the TSL increases at TPD = 0.5 mm or more as the TPD increases at the WS of 0.1 rpm. When WS varies from 0.15–0.2 rpm, the IMC average thickness decreases and TSL increases as TPD increases in all regions. Picot et al. [54] used the process parameters of the rotational speed and translational speed of the tool in FSLW of stainless steel 316L and Al alloy 5083 and reported that the TSL decreased with increasing IMC average thickness. Helal et al. [55] used the process parameters of TRS, WS, and tool offset in FSLW of 6061-T6 Al alloy and ultra-low carbon steel and reported that as the IMC average thickness increases, TSL decreases. Kimapong et al. [53] reported that TSL decreased with increasing IMC average thickness using welding parameters such as rotational speed, traversing speed of 0.4 to 1.43 mm/s, and a pin depth in FSLW of A5083 Al alloy and SS400 steel. In previous studies, it was reported that the TSL increased with decreasing IMC average thickness. As WS varies from 0.15–0.2 rpm, it was consistent with the research results of previous researchers. However, at the low WS of 0.1 rpm, for a TPD value up to 0.5 mm, the IMC average thickness decreased as the TPD increased, but the TSL decreased.

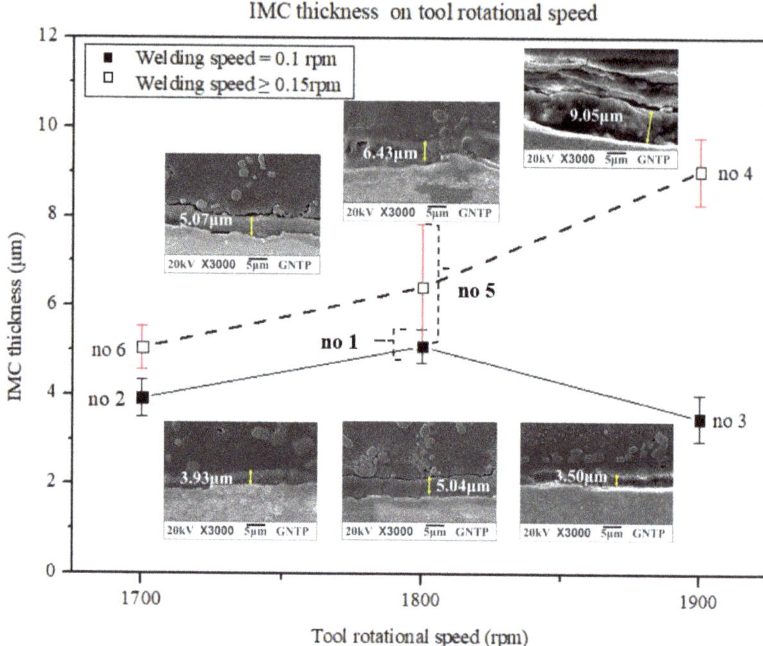

Figure 8. Comparison of IMC thickness for different TRSs of experiment groups A and B.

3.4. Characteristics of IMC Thickness and TSL According to TRS

Table 4 summarizes the IMC thickness and TSL according to the change in the TPD and for WS values of 0.1 rpm and 0.15 rpm or more. Figure 8 shows the relationship between the IMC thickness and TSL according to the change in the TRS.

Table 4. IMC thickness range and thickness average and TSL on TRS.

Experiment No.	2	1	3	6	5	4
Tool rotational speed (rpm)	1700	1800	1900	1700	1800	1900
IMC thickness range (μm)	4.73–3.13	5.80–4.40	4.53–2.60	6.00–4.13	9.05–3.80	10.44–7.65
IMC average thickness (μm)	3.93	5.04	3.56	5.07	6.43	9.05
Tensile shear load (N) [16]	860.50	2677.21	2034.83	1984.72	1779.25	807.35

Figure 8 compares the differences between the two groups in terms of the change in the IMC thickness according to TRS in group A with a WS of 0.1 rpm and group B with a WS of 0.15 to 0.2 rpm. In Figure 8, group A of experiments no. 1, no. 2, and no. 3 is indicated by a solid line to represent the changes in the IMC average thickness change according to TRS and a WS of 0.1 rpm. As TRS increases, it can be seen that the IMC average thickness increases up to TRS of 1800 mm and decreases when TRS exceeds 1800 mm. In group B of experiments no. 4, no. 5, and no. 6 in Figure 8, at WS values of 0.15 and 0.2 rpm, the variation in the IMC average thickness according to TRS is indicated by broken lines. It shows a gradual increase in the IMC average thickness as TRS increases from 1700–1900 rpm. When TRS is increased from 1700–1900 rpm, the tendency of the IMC average thickness change is different according to the effect of WS. As TRS increased, the IMC average thickness decreased for a WS value of 0.1 rpm and when TRS of 1800 rpm or more. When WS was 0.1 rpm, TRS was 1800 rpm or less, and when WS was 0.15 and 0.2 rpm, it showed an increase over the entire range of TRS.

In the study by Kundu et al. [51], the IMC thickness increased with increasing TRS. These results do not indicate conditions for WS above a certain speed. Wan et al. [25] and Das et al. [51] simultaneously studied the increase in the IMC thickness according to the increase of TRS, and the decrease in the IMC thickness according to the increase in WS. However, their study also did not report that the IMC thickness increased with increasing TRS at WS above a certain speed.

Figure 9 is a graph showing the relationship between the TSL and TRS. In the figure, group A of experiments no. 2, no. 1, and no. 3 is indicated by a dash-dotted line to represent the TSL change according to TRS at WS of 0.1 rpm.

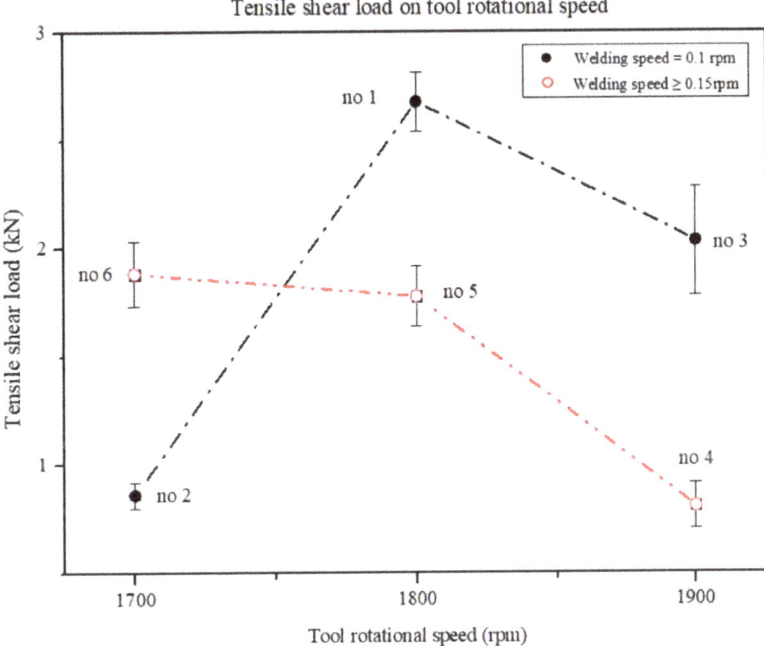

Figure 9. Comparison of TSL with TRS for experiment groups A and B [16].

As TRS increases from 1700 rpm to 1900 rpm, it can be seen that the TSL increases up to a TRS of 1800 rpm, and TSL decreases for a TRS of 1800 or more. In Figure 9, group B of experiments no. 4, no. 5, and no. 6 is indicated by dash-double dotted lines to represent the change in the TSL according to TRS at WS values of 0.15 and 0.2 rpm. As the TRS increases from 1700 rpm to 1900 rpm, the TSL shows a gradual decrease. When TRS is increased from 1700 to 1900 rpm, the tendency for the change in TSL differs according to the effect of WS. As TRS increased at the WS of 0.1 rpm, the TSL increased up to TRS of 1800 rpm. It shows that as TRS increases at the WS of 0.1 rpm, the TSL decreases over the entire range of TRS when TRS is over 1800 rpm and at WS values of 0.15 and 0.2 rpm.

With respect to the TSL and TRS, Kimapong et al. [53] stated that the TSL decreased with increasing TRS. Their study also did not consider the effect of WS on specific regions. In this study, as in the study by Kimapong et al. [53], the TSL decreased as TRS increased at a WS of 0.15 to 0.2 rpm. However, it was found that the TSL increased as TRS increased at the WS of 0.1 rpm and when TRS was 1800 rpm or less.

In Figures 8 and 9, at a WS of 0.1 rpm and below TRS of 1800 rpm, the IMC average thickness increases and the TSL increases as TRS increases. At a WS of 0.1 rpm, if TRS is 1800 rpm or more, the IMC average thickness decreases and the TSL decreases as TRS increases. When WS ranges from 0.15 to 0.2 rpm, the IMC average thickness decreases and TSL decreases as TRS increases in all regions.

Picot et al. [54] used the process parameters of the rotational speed and translational speed of the tool in the FSLW of stainless steel 316L and Al alloy 5083 and reported that the shear strength decreased with increasing IMC average thickness. Helal et al. [55] used the process parameters of TRS, WS, and tool offset in FSLW of 6061-T6 Al alloy and ultra-low carbon steel and reported that as the IMC average thickness increases, TSL decreases. Kimapong et al. [53] reported that the shear strength decreased with increasing IMC thickness using welding parameters such as the rotational speed, traversing speed of 0.4 to 1.43 mm/s, and a pin depth in FSLW of A5083 Al alloy and SS400 steel.

In previous studies, it was reported that the TSL increased with decreasing IMC average thickness. At the WS of 0.15–0.2 rpm, it was consistent with the research results of previous researchers. However, as TRS increased in all regions of TRS at a low WS of 0.1 rpm, the TSL increased as the IMC average thickness increased, and the TSL decreased as the IMC average thickness decreased. Many studies have been done to determine the relationship between the other process parameters and the IMC thickness. A study by Liu et al. [25] indicated that the IMC thickness decreased as the WS increased and the tool offset increased. However, their study results also reported a decrease in the IMC thickness with an increase in the tool offset but did not provide a condition for the WS over a specific speed. Dehghani et al. [29] showed that the IMC thickness increased with increasing PD.

In addition, when the WS was constant, the effect of the TPD was greater than the rotational speed of the tool. This is consistent with the fact that according to the results of this study, the effect of the TPD is dominant. Many studies have also been done to assess the maximum size criterion of the IMC thickness for strength [25,36,46,47]. Jamshidi et al. [36] recommended a critically reactive layer of IMC as thin as 1–2 μm for strong bond strength, as a thicker layer of IMC lowers the TSL and increases the hardness while reducing the bond ductility. However, even if it is thinner than this dimension, the mechanical properties of the joint are said to be improved. Liu et al. [25] reported that the maximum thickness of the IMC at the butt joint of AA6061-T6 and 780/800 trip steel was 1 μm, which was low enough for good weld strength. Movahedi et al. [46] reported that the presence of IMC in FSLW between Al-5083 and St-12 improves the weld strength when the thickness is less than 2 μm. Meanwhile, Sepold et al. [47] showed that excellent mechanical properties could be obtained by minimizing the thickness of the IMC layer to 10 μm or less. In this study, a region with an average IMC thickness of 3.42 to 9.05 μm was obtained over the entire region. In general, the type, size, and amount of IMC formation depends on the input heat, which is controlled by the welding process parameters. An increase in rotational speed and friction time is expected to increase the heat input. Jamshidi et al. [36] also reported that a larger amount of IMC should be formed because the amount of heat input increases with TPD. Dehghani et al. [29] reported that as the PD increased, the interaction between the shoulder and the workpiece increased and the frictional heat was generated more so that the material was sufficiently plasticized, and the material movement around the tool pin was promoted, resulting in higher heat generation.

This study differs from previous studies in that the IMC thickness increases because of the increase in heat input as the PD and TPD increase. It is estimated that the increase in the friction amount with the increase of the TPD leads to an increase in the amount of heat input. However, in the mechanism according to TPD, flow phenomena other than the heat input appear to act as a more dominant factor than the increase in heat input according to TPD. Therefore, it can be estimated that the thickness of the IMC is reduced. Therefore, these results differ from the experimental results of the one factor at a time, in which the effects of PD and TPD were not tested together. Previous studies have shown that the thickness of the IMC increases with increasing TRS. In this study, it was confirmed that the thickness of IMC decreased as TRS increased at a WS value of 0.1 rpm and a TRS of 1800 rpm or higher. In a previous study, the IMC thickness decreased with increasing WS. In this study, the IMC thickness increased with increasing WS. Based on the results obtained, it can be seen that the effect of TPD is a significant factor in the reduction of the IMC thickness.

4. Conclusions

In this study, the following major results were obtained by studying the characteristics of the TSL and IMC thickness according to TPD and TRS in FSLW for dissimilar Al and steel pipe.

1. As a result of observing the structural change of the cross section after FSW, the average thickness of the IMC layer at the interface between aluminum and steel in the microstructure ranged from 3.4 to 9.05 μm, and the formation of the IMC layer was observed.
2. As TPD increased, the average thickness of IMC decreased and TSL increased in the entire experimental area except for the area where TSL decreased at TPD = 0.5 mm or less when WS was 0.1 rpm.
3. When WS was 0.1 rpm, TSL increased with increasing IMC thickness at TRS below 1800 rpm, and TSL decreased with decreasing IMC thickness at TRS above 1800 rpm. TSL decreased with increasing IMC thickness as TRS increased when WS was above 0.15 rpm.
4. TPD is the dominant factor in the change in size of IMC thickness, and IMC thickness decreases with increasing TPD. TSL increases with decreasing IMC thickness. However, at low WS, when TRS is increased, TSL is affected by TPD, and when TPD increased, TSL was affected by TRS.

These results indicate that TPD, IMC thickness, and TSL were directly related, except for in certain areas with low WS. It is assumed that the increase or decrease of TSL in this region is determined by the relative superiority of heat input and material flow.

Therefore, in order to increase the mechanical properties in the FSLW of aluminum and steel for high-rigidity parts, the TPD should be deepened in the weld area where the WS is not high. A suitable TPD should therefore be considered along with tool wear, and continuous material development and the selection of an optimized TPD will play an important role in future dissimilar FSLW.

Author Contributions: L.C. performed the experiments, data analysis, and proofreading and wrote part of paper; M.K. provided basic guidance and supervision of FSW; and D.J. reviewed the paper and supervised the thesis. All authors have read and agreed to the published version of the manuscript.

Funding: This research was supported by Basic Science Research Program through the National Research Foundation of Korea (NRF) funded by the Ministry of Education (2018R1D1A1B0705130214). And The APC was funded by the Korea Institute of Energy Technology Evaluation and Planning(KETEP) grant funded by the Korea government(MOTIE) (20206310100050, A development of remanufacturing technology on the grinding system of complex-function for high-precision).

Institutional Review Board Statement: Not applicable.

Informed Consent Statement: Not applicable.

Data Availability Statement: Not applicable.

Conflicts of Interest: The authors declare no conflict of interest.

References

1. Joo, Y.H.; Park, Y.C.; Lee, Y.M.; Kim, K.H.; Kang, M.C. The Weldability of a Thin Friction Stir Welded Plate of Al5052-H32 using High Frequency Spindle. *J. Korean Soc. Manuf. Process Eng.* **2017**, *16*, 90–95. [CrossRef]
2. Laska, A.; Szkodo, M. Manufacturing Parameters, Materials, and Welds Properties of Butt Friction Stir Welded Joints–Overview. *Materials* **2020**, *13*, 4940. [CrossRef] [PubMed]
3. Park, J.H.; Jeon, S.T.; Lee, T.J.; Kang, J.D.; Kang, M.C. Effect on Drive Point Dynamic Stiffness and Lightweight Chassis Component by using Topology and Topography Optimization. *J. Korean Soc. Manuf. Process Eng.* **2018**, *17*, 141–147. [CrossRef]
4. Hussein, S.A.; Md Tahir, A.S.; Hadzley, A.B. Characteristics of Aluminum-to-Steel Joint Made by Friction Stir Welding: A Review. *Mater. Today Commun.* **2015**, *5*, 32–49. [CrossRef]
5. Ryabov, V.R. *Aluminizing of Steel*; Oxonian Press: New Delhi, India, 1985.

6. Wan, L.; Huang, Y. Friction stir welding of dissimilar aluminum alloys and steels: A review. *Int. J. Adv. Manuf. Technol.* **2018**, *99*, 1781–1811. [CrossRef]
7. Thomas, W.M.; Nicholas, E.D.; Needham, J.C.; Murch, M.G.; Temple-Smith, P.; Dawes, C.J. Friction Welding. U.S. Patent 5,460,317B1, 9 December 1997.
8. Park, J.K.; Kim, Y.S.; Suh, C.H.; Kim, Y.S. Hybrid quenching method of hot stamping for automotive tubular beams. *Proc. Inst. Mech. Eng.* **2015**, *231*, 1–12. [CrossRef]
9. Ducheta, M.; Haouasa, J.; Gibeaub, E.; Pechenota, F.; Honeckera, C.; Muniera, R.; Webera, B. Improvement of the fatigue strength of welds for lightweight chassis application made of Advanced High Strength Steels. *Procedia Struct. Integr.* **2019**, *19*, 585–594. [CrossRef]
10. Sharma, S.R.; Ma, Z.Y.; Mishra, R.S.; Mahoney, M.W. Effect of friction stir processing on fatigue behavior of A356 alloy. *Scr. Mater.* **2004**, *51*, 237–241. [CrossRef]
11. Lee, J.M.; Lee, S.H.; Yoon, J.H.; Kim, K.H. Effects of Melt Treatments on Microstructures and Mechanical Properties of A357 alloy. *J. Korea Foundry Soc.* **2003**, *23*, 69–76.
12. Bouazara, M.; Bouaicha, A.; Ragab, K.A. Fatigue Characteristics and Quality Index of A357 Type Semi-Solid Aluminum Castings Used for Automotive Application. *J. Mater. Eng. Perform.* **2015**, *24*, 3084–3092. [CrossRef]
13. Akbari, M.; Asadi, P. Optimization of microstructural and mechanical properties of friction stir welded A356 pipes using Taguchi method. *Mater. Res. Express* **2019**, *6*, 066545. [CrossRef]
14. Park, J.H.; Park, S.H.; Park, S.H.; Joo, Y.H.; Kang, M.C. Evaluation of Mechanical Properties with Tool Rotational Speed in Dissimilar Cast Aluminum and High-Strength Steel of Lap Jointed Friction Stir Welding. *J. Korean Soc. Manuf. Process Eng.* **2019**, *18*, 90–96. [CrossRef]
15. Choy, L.J.; Park, S.H.; Lee, M.W.; Park, J.H.; Choi, B.J.; Kang, M.C. Tensile Strength Application Using a Definitive Screening Design Method in Friction Stir Welding of Dissimilar Cast Aluminum and High-Strength Steel with Pipe Shape. *J. Korean Soc. Manuf. Process Eng.* **2020**, *19*, 98–104. [CrossRef]
16. Choy, L.; Kim, S.; Park, J.; Kang, M.; Jung, D. Effect of Process Factors on Tensile Shear Load Using the Definitive Screening Design in Friction Stir Lap Welding of Aluminum–Steel with a Pipe Shape. *Materials* **2021**, *14*, 5787. [CrossRef] [PubMed]
17. Forcellese, A.; Simoncini, M.; Casalino, G. Influence of process parameters on the vertical forces generated during friction stir welding of AA6082-T6 and on the mechanical properties of the joints. *Metals* **2017**, *7*, 350. [CrossRef]
18. Wu, C.S.; Zhang, W.B.; Shi, L.; Chen, M.A. Visualization and simulation of the plastic material flow in friction stir welding of aluminium alloy 2024 plates. *Trans. Nonferrous Met. Soc. China* **2012**, *22*, 1445–1451. [CrossRef]
19. Soundararajan, V.; Zekovic, S.; Kovacevic, R. Thermo-mechanical model with adaptive boundary conditions for friction stir welding of Al 6061. *Int. J. Mach. Tools Manuf.* **2005**, *45*, 1577–1587. [CrossRef]
20. Shi, L.; Wu, C.S. Transient model of heat transfer and material flow at different stages of friction stir welding process. *J. Manuf. Processes* **2017**, *25*, 323–339. [CrossRef]
21. Mahto, R.P.; Kumar, R.; Pal, S.K.; Sushanta, K.; Panda, A. comprehensive study on force, temperature, mechanical properties and micro-structural characterizations in friction stir lap welding of dissimilar materials (AA6061-T6 & AISI304). *J. Manuf. Processes* **2018**, *31*, 624–639. [CrossRef]
22. Mishra, R.S.; Ma, Z.Y. Friction stir welding and processing. *Mater. Sci. Eng.* **2005**, *50*, 1–78. [CrossRef]
23. Trimble, D.; O'Donnell, G.E.; Monaghan, J. Characterization of tool shape and rotational speed for increased speed during friction stir welding of AA2024-T3. *J. Manuf. Processes* **2015**, *17*, 141–150. [CrossRef]
24. Rajendran, C.; Srinivasan, K.; Balasubramanian, V.; Balaji, H.; Selvaraj, P. Effect of tool tilt angle on strength and microstructural characteristics of friction stir welded lap joints of AA2014-T6 aluminum alloy. *Trans. Nonferrous Met. Soc. China* **2019**, *29*, 1824–1835. [CrossRef]
25. Liu, X.; Lan, S.; Ni, J. Analysis of process parameters effects on friction stir welding of dissimilar aluminum alloy to advanced high strength steel. *Mater. Des.* **2014**, *59*, 50–62. [CrossRef]
26. Wan, L.; Huang, Y. Microstructure and Mechanical Properties of Al/Steel Friction Stir Lap Weld. *Metals* **2017**, *7*, 542. [CrossRef]
27. Sezhian, M.V.; Giridharan, K.D.; Pushpanathan, P.; Chakravarthi, G.; Stalin, B.; Karthick, A.; Kumar, P.M.; Bharani, M. Microstructural and Mechanical Behaviors of Friction Stir Welded Dissimilar AA6082-AA7075 Joints. *Adv. Mater. Sci. Eng.* **2021**, *13*. [CrossRef]
28. Mahto, R.P.; Bhoje, R.; Pal, S.K.; Joshi, H.S.; Das, S. A study on mechanical properties in friction stir lap welding of AA 6061-T6 and AISI 304. *Mater. Sci. Eng.* **2016**, *652*, 136–144. [CrossRef]
29. Dehghani, M.; Amadeh, A.; Mousavi, S.A. Investigations on the effects of friction stir welding parameters on intermetallic and defect formation in joining aluminum alloy to mild steel. *Mater. Des.* **2013**, *49*, 433–441. [CrossRef]
30. Iqbal, M.P.; Vishwakarma, R.K.; Pal, S.K.; Mandal, P. Influence of plunge depth during friction stir welding of aluminum pipes. *Proc. Inst. Mech. Eng. Part B J. Eng. Manuf.* **2020**, 1–12. [CrossRef]
31. Iqbal, M.P.; Jain, R.; Pal, S.K. Numerical and experimental study on friction stir welding of aluminum alloy pipe. *J. Mater. Process. Technol.* **2019**, *274*, 116258. [CrossRef]
32. El-Kassas, A.M.; Sabry, I.; Mourad, A.-H.I.; Thekkuden, D.T. Characteristics of Potential Sources—Vertical Force, Torque and Current on Penetration Depth for Quality Assessment in Friction Stir Welding of AA 6061 Pipes. *Int. Rev. Aerosp. Eng.* **2019**, *12*, 195–204. [CrossRef]

33. Wei, Y.; Li, J.; Xiong, J.; Zhang, F. Effect of Tool Pin Insertion Depth on Friction Stir Lap Welding of Aluminum to Stainless Steel. *J. Mater. Eng. Perform.* **2013**, *22*, 3005–3013. [CrossRef]
34. Shen, Z.; Chen, Y.; Haghshenas, M.; Gerlich, A.P. Role of welding parameters on interfacial bonding in dissimilar steel/aluminum friction stir welds. *Eng. Sci. Technol. Int. J.* **2015**, *18*, 270–277. [CrossRef]
35. Mahto, R.P.; Pal, S.K. Friction stir lap welding of thin aa6061-t6 and aisi304 sheets at different values of pin penetrations. In *International Manufacturing Science and Engineering Conference*; American Society of Mechanical Engineers: New York, NY, USA, 2018; Volume 51364. [CrossRef]
36. Jamshidi Aval, H.; Falahati Naghibi, M. Orbital friction stir lap welding in tubular parts of aluminium alloy AA5083. *Sci. Technol. Weld. Join.* **2017**, *22*, 562–572. [CrossRef]
37. Herbst, S.; Aengeneyndt, H.; Maier, H.J.; Nürnberger, F. Microstructure and mechanical properties of friction welded steel-aluminum hybrid components after T6 heat treatment. *Mater. Sci. Eng.* **2017**, *696*, 33–41. [CrossRef]
38. Lan, S.; Liu, X.; Ni, J. Microstructural evolution during friction stir welding of dissimilar aluminum alloy to advanced high-strength steel. *Int. J. Adv. Manuf. Technol.* **2016**, *82*, 2183–2193. [CrossRef]
39. Coelho, R.S.; Kostka, A.; Dos Santos, J.F.; Kaysser-Pyzalla, A. Friction-stir dissimilar welding of aluminium alloy to high strength steels: Mechanical properties and their relation to microstructure. *Mater. Sci. Eng.* **2012**, *556*, 175–183. [CrossRef]
40. Haghshenas, M.; Abdel-Gwad, A.; Omran, A.M.; Gökçe, B.; Sahraeinejad, S.; Gerlich, A.P. Friction stir weld assisted diffusion bonding of 5754 aluminum alloy to coated high strength steels. *Mater. Des.* **2014**, *55*, 442–449. [CrossRef]
41. Mahto, R.P.; Anishetty, S.; Sarkar, A.; Mypati, O.; Pal, S.K.; Majumdar, J.D. Interfacial Microstructural and Corrosion Characterizations of Friction Stir Welded AA6061-T6 and AISI304 Materials. *Met. Mater. Int.* **2019**, *25*, 752–767. [CrossRef]
42. Shamsujjoha, M.; Jasthi, B.K.; West, M.; Widener, C. Friction stir lap welding of aluminum to steel using refractory metal pin tools. *J. Eng. Mater. Technol.* **2015**, *137*, 021009. [CrossRef]
43. Ma, Z.Y.; Sharma, S.R.; Mishra, R.S. Microstructural Modification of As-Cast Al-Si-Mg Alloy by Friction Stir Processing. *Metall. Mater. Trans.* **2006**, *37*, 3323–3336. [CrossRef]
44. Springer, H.; Kostka, A.; Payton, E.J.; Raabe, D.; Kaysser-Pyzalla, A.; Eggeler, G. On the formation and growth of intermetallic phases during interdiffusion between low-carbon steel and aluminum alloys. *Acta Mater.* **2011**, *59*, 1586–1600. [CrossRef]
45. Haidara, F.; Record, M.C.; Duployer, B.; Mangelinck, D. Phase formation in Al–Fe thin film systems. *Intermetallics* **2012**, *23*, 143–147. [CrossRef]
46. Movahedi, M.; Kokabi, A.H.; Reihani, S.S.; Najafi, H. Mechanical and microstructural characterization of A1-5083/St-12 lap jointsmade by friction stir welding. *Procedia Eng.* **2011**, *10*, 3297–3303. [CrossRef]
47. Sepold, G.; Kreimeyer, M. Joining of dissimilar materials. In Proceedings of the First International Symposium on High-Power Laser Macroprocessing. International Society for Optics and Photonics, Bellingham, WA, USA, 27–31 May 2003; Volume 4831, pp. 528–533.
48. Jiang, W.H.; Kovacevic, R. Feasibility study of friction stir welding of 6061-T6 aluminium alloy with AISI 1018 steel. *Proc. Inst. Mech. Eng. Part B J. Eng. Manuf.* **2004**, *218*, 1323–1331. [CrossRef]
49. Das, H.; Basak, S.; Das, G. Influence of energy induced from processing parameters on the mechanical properties of friction stir welded lap joint of aluminum to coated steel sheet. *Int. J. Adv. Manuf. Technol.* **2013**, *64*, 1653–1661. [CrossRef]
50. Kundu, S.; Roy, D.; Bhola, R.; Bhattacharjee, D.; Mishra, B.; Chatterjee, S. Microstructure and tensile strength of friction stir welded joints between interstitial free steel and commercially pure aluminium. *Mater. Des.* **2013**, *50*, 370–375. [CrossRef]
51. Das, H.; Jana, S.S.; Pal, T.K.; De, A. Numerical and experimental investigation on friction stir lap welding of aluminium to steel. *Sci. Technol. Weld. Join.* **2014**, *19*, 69–75. [CrossRef]
52. Lee, C.Y.; Choi, D.H.; Yeon, Y.M.; Jung, S.B. Dissimilar friction stir spot welding of low carbon steel and Al–Mg alloy by formation of IMCs. *Sci. Technol. Weld. Join.* **2009**, *14*, 216–220. [CrossRef]
53. Kimapong, K.; Watanabe, T. Lap joint of A5083 aluminum alloy and SS400 steel by friction stir welding. *Mater. Trans.* **2005**, *46*, 835–841. [CrossRef]
54. Picot, F.; Gueydan, A.; Martinez, M.; Moisy, F.; Hug, E. A correlation between the ultimate shear stress and the thickness affected by intermetallic compounds in friction stir welding of dissimilar aluminum alloy–stainless steel joints. *Metals* **2018**, *8*, 179. [CrossRef]
55. Helal, Y.; Boumerzoug, Z. Dissimilar friction stir welding of Al-6061 to steel. In *Proceedings of the AIP Conference Proceedings*; AIP Publishing LLC.: Melville, NY, USA, 2016; Volume 1772, p. 030008.
56. Aghajani Derazkola, H.; García, E.; Eyvazian, A.; Aberoumand, M. Effects of Rapid Cooling on Properties of Aluminum-Steel Friction Stir Welded Joint. *Materials* **2021**, *14*, 908. [CrossRef] [PubMed]
57. Mortello, M.; Pedemonte, M.; Contuzzi, N.; Casalino, G. Experimental Investigation of Material Properties in FSW Dissimilar Aluminum-Steel Lap Joints. *Metals* **2021**, *11*, 1474. [CrossRef]
58. Pankaj, P.; Tiwari, A.; Biswas, P. Impact of varying tool position on the intermetallic compound formation, metallographic/mechanical characteristics of dissimilar DH36 steel, and aluminum alloy friction stir welds. *Weld. World* **2021**, *66*, 239–271. [CrossRef]
59. Abd Elnabi, M.M.; Osman, T.A.; El Mokadem, A.; Elshalakany, A.B. Evaluation of the formation of intermetallic compounds at the intermixing lines and in the nugget of dissimilar steel/aluminum friction stir welds. *J. Mater. Res. Technol.* **2020**, *9*, 10209–10222. [CrossRef]
60. Sabry, I.; Zaafarani, N. *Dry and Underwater Friction Stir Welding of aa6061 Pipes—A Comparative Study*; Materials Science and Engineering Series; IOP Publishing: Bristol, UK, 2021; Volume 1091, p. 012032.

61. Abdullah, M.E.; Elmetwally, H.T.; Yakoub, N.G.; Elsheikh, M.N.; Abd-Eltwab, A.A. Effect of orbital friction crush welding parameters on aluminum tubes. *Int. J. Sci. Technol. Res.* **2020**, *9*, 4483–4486.
62. Su, J.Q.; Nelson, T.W.; Sterling, C.J. Microstructure evolution during FSW/FSP of high strength aluminum alloys. *Mater. Sci. Eng.* **2005**, *405*, 277–286. [CrossRef]
63. Ogura, T.; Nishida, T.; Nishida, H.; Yoshida, T.; Omichi, N.; Hirose, A. Partitioning evaluation of mechanical properties and the interfacial microstructure in a friction stir welded aluminum alloy/stainless steel lap joint. *Scr. Mater.* **2012**, *66*, 531–534. [CrossRef]

Article

The Effect of Temper Condition and Feeding Speed on the Additive Manufacturing of AA2011 Parts Using Friction Stir Deposition

Mohamed M. Z. Ahmed [1,2,*], Mohamed M. El-Sayed Seleman [2], Ebtessam Elfishawy [2,3], Bandar Alzahrani [1], Kamel Touileb [1] and Mohamed I. A. Habba [4]

1 Mechanical Engineering Department, College of Engineering at Al-Kharj, Prince Sattam Bin Abdulaziz University, Al Kharj 16273, Saudi Arabia; ba.alzahrani@psau.edu.sa (B.A.); k.touileb@psau.edu.sa (K.T.)
2 Department of Metallurgical and Materials Engineering, Faculty of Petroleum and Mining Engineering, Suez University, Suez 43512, Egypt; mohamed.elnagar@suezuniv.edu.eg (M.M.E.-S.S.); ebtessam.elfishawy@aucegypt.edu (E.E.)
3 Mechanical Engineering Department, School of Science and Engineering, The American University in Cairo, 11835 New Cairo, Egypt
4 Mechanical Department, Faculty of Technology & Education, Suez University, Suez 43518, Egypt; Mohamed.Atia@suezuniv.edu.eg
* Correspondence: moh.ahmed@psau.edu.sa

Abstract: In the current study, solid-state additive manufacturing (SSAM) of two temper conditions AA2011 was successfully conducted using the friction stir deposition (FSD) process. The AA2011-T6 and AA2011-O consumable bars of 20 mm diameter were used as a feeding material against AA5083 substrate. The effect of the rotation rate and feeding speed of the consumable bars on the macrostructure, microstructure, and hardness of the friction stir deposited (FSD) materials were examined. The AA2011-T6 bars were deposited at a constant rotation rate of 1200 rpm and different feeding speeds of 3, 6, and 9 mm/min, whereas the AA2011-O bars were deposited at a constant rotation rate of 200 mm/min and varied feeding speeds of 1, 2, and 3 mm/min. The obtained microstructure was investigated using an optical microscope and scanning electron microscope equipped with EDS analysis to evaluate microstructural features. Hardness was also assessed as average values and maps. The results showed that this new technique succeeded in producing sound additive manufactured parts at all the applied processing parameters. The microstructures of the additive manufactured parts showed equiaxed refined grains compared to the coarse grain of the starting materials. The detected intermetallics in AA2011 alloy are mainly Al_2Cu and Al_7Cu_2Fe. The improvement in hardness of AA2011-O AMPs reached 163% of the starting material hardness at the applied feeding speed of 1 mm/min. The hardness mapping analysis reveals a homogeneous hardness profile along the building direction. Finally, it can be said that the temper conditions of the starting AA2011 materials govern the selection of the processing parameters in terms of rotation rate and feeding speed and affects the properties of the produced additive manufactured parts in terms of hardness and microstructural features.

Keywords: friction stir deposition; solid-state additive manufacturing; AA2011-T6 and AA2011-O; AA2011 aluminum alloy; microstructure; intermetallics; hardness

Citation: Ahmed, M.M.Z.; El-Sayed Seleman, M.M.; Elfishawy, E.; Alzahrani, B.; Touileb, K.; Habba, M.I.A. The Effect of Temper Condition and Feeding Speed on the Additive Manufacturing of AA2011 Parts Using Friction Stir Deposition. *Materials* **2021**, *14*, 6396. https://doi.org/10.3390/ma14216396

Academic Editor: Józef Iwaszko

Received: 9 September 2021
Accepted: 19 October 2021
Published: 25 October 2021

1. Introduction

Additive manufacturing (AM) is a promising technology in numerous engineering applications. It involves the fabrication of various 3D objects by adding layer by layer material (alloy, plastic, concrete, human tissue, etc.) regardless of any size and shape [1,2]. Fusion-based additive manufacturing (F-BAM) techniques are used for different alloys [3]. Still, they are not suitable for aluminum-based alloys, especially the heat-treatable alloys (2xxx and 7xxx), due to their sensitivity to porosity formation, liquation cracking, segregation, solidification cracking, and anisotropic microstructure [2,4]. In contrast, friction stir deposition (FSD) is solid-state additive manufacturing (S-SAM) technique that can be used to deposit metals and composites [5,6]. The main advantage of the FSD as a solid-state process is that it can eliminate all problems of melting and solidification, and also, the feed material is mainly rods or wire without a need for special specifications of the used feed material. Most of the current works today are carried out on AM of aluminum-based alloys using a S-SAM technique [1,7,8]. In recent years, there has been increasing interest in utilizing FSD in many applications. This technique can be used for many purposes, including additive manufacturing [1,9–11], surface protection [11–13], and repair of defective components [6]. Thus, it can be said that the FSD-based AM is considered a new innovative approach to AM for building 3D parts ultimately in a solid state. The main processing parameters are the rotation rate, feeding speed, downward force, consumable rod material, and substrate material. These parameters govern the heat input and material flow processes. In the FSD process, the final build part's height depends on the layer thickness and the total number of assembly layers. Moreover, the modifications in the geometry of the building design can obtain manufacturing parts with different geometries [1,5]. Thus, the final FSD product is a near-net-shape with enhanced microstructure and isotropic mechanical properties [2,4,14]. Low porosity and low residual stress are the main privileges of the as-deposited part; this will make post-processing heat treatment unnecessary in many cases. However, surface finishing will usually be required [15–17]. Boeing and Airbus companies are considered the first use of additive manufacturing (AM) based on FSW principles [10,18,19]. Meanwhile, Airbus [20] presented the capability of achieving lightweight/low-cost structure parts by manufacturing 2050 Al-Li wing ribs by the FSAM process. Boeing assessed this process as a pre-form fabricating tool for manufacturing energy-efficient structures [10,21,22]. In addition to the capability of the FSD technique to generate material builds of high-performance structures, it is considered an energy and cost-saving process [10,23]. Elfishawy et al. [24] studied the possibility of multi-layers formation of die-cast Al–Si via FSD at the spindle rotation rate of 1200 rpm and different feeding speeds from 3 to 5 mm/min. The results showed sound structure with recrystallized refined grains. Therefore, from a scientific and technological point of view, it is of great importance to study how FSD works for additive manufacturing parts (AMPs) production in heat-treatable aluminum alloys. Although AA2011 is used extensively in aerospace and automotive components, there is a lack of publications discussing the applicability of AA2011 fabrication using FSD. Thus, the current work intends to explore the effect of the initial material conditions of AA2011 alloys on the properties and microstructures of the final produced AMPs. Three levels of feeding speeds of 3, 6, and 9 mm/min were associated with a high rotation rate of 1200 rpm/min to friction stir deposit AA2011-T6, and three other feeding speeds of 1, 2, and 3 mm/min were chosen with a low rotation rate of 200 rpm/min to deposit AA2011-O.

This study aims to study the effect of the consumable rod alloy temper condition on the behavior of the FSD process in terms of the parameters suitable for each temper condition as well as the properties of the AMPs.

2. Materials and Methods

To study the effect of the temper condition of AA2011 alloy on the properties of the produced AMPs, two groups of specimens, AA2011-T6 and AA2011-O, were used as consumable bars against a substrate of AA5083 alloy. The nominal chemical composition of AA2011 is given in Table 1. The annealing process for the as-received was carried out at 415 °C for 2.5 h followed by slow furnace cooling to the room temperature. Figure 1 illustrates the Cu-rich portion of the Al–Cu binary phase diagram with the annealing temperature range indicated [25,26].

Table 1. Chemical composition of AA2011 aluminum alloy (in wt.%).

Cu	Fe	Si	Ti	Zn	Bi	Pb	Ni	Mn	Bal.
5.60	0.72	0.40	0.35	0.30	0.25	0.20	0.02	0.05	Al

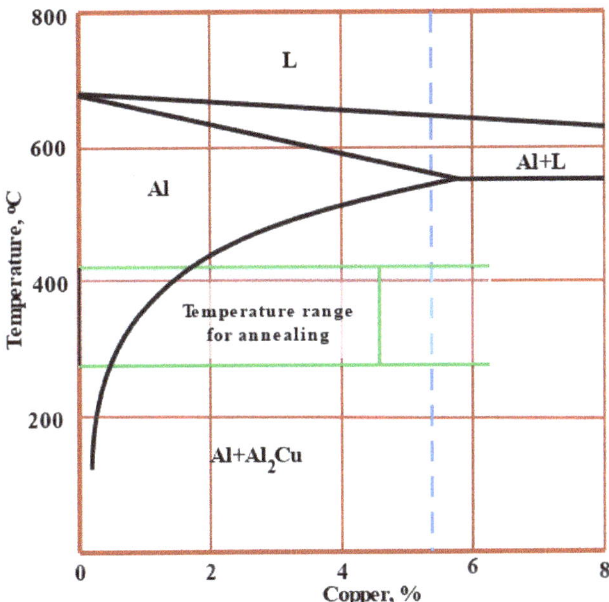

Figure 1. A sketch for the Cu-rich portion of the Al–Cu binary phase diagram with the annealing temperature range indicated.

For comparison, three deposited materials were manufactured from each group of the AA2011-T6 and AA2011-O rods. The FSAM was carried out using the friction stir welding/processing machine (EG-FSW-M1) (Suez University, Suez, Egypt) [27,28]. Table 2 summarizes the deposition process parameters of both Al alloys.

Table 2. Consumable rod dimensions and FSD processing parameters.

Consumable Rod			FSD Parameters			
Material	Initial Length (mm)	Rod Diameter (mm)	Rotation Rate (rpm)	Feeding Speeds (mm/min)		
AA2011-T6	200	20	1200	3	6	9
AA2011-O	110	20	200	1	2	3

The consumable aluminum rods are fixed using the machine shank to ensure the complete fixation of the rods throughout the process; Figure 2 shows a photograph of the actual AM process applied to AA2011. The additive manufacturing (AM) process involves three steps: fixing the consumable Al rod in the spindle shank (Figure 2a) and rotating it at a constant rotation rate while moving downward to reach the substrate material (Figure 2b). Finally, under a continuous feeding speed, the rod plastically deformed due to the high friction and the generated heat between the rod and the substrate that causes the material to transfer from the consumable bar to the substrate to build a material upwards. This process may continue until all the rod length is consumed and became insufficient for more deposition. The shape of the consumed tool tends to form a conical shape, as shown in Figure 2c. For AA2011-T6 group specimens, careful processing parameters were selected based on our experience in the field and the published data [24,29] to produce additive manufacturing parts. The required heat input to friction stir deposit such a hard material limits the process parameters to be 1200 rpm as a spindle rotation rate with 3, 6, and 9 mm/min feeding speeds. For the AA2011-O group specimens, experiments start with shortening each of the three specimens to 110 mm in length. Of this length, 70 mm of the total length was consumed as a fixing base of the rod inside the shank to ensure tight gripping and prevent rod deflection during the deposition process, and 40 mm was functional during the process of friction deposition. Less heat input is needed to deposit this soft material; that is why after many trials, the optimum process parameters obtained were a 200 rpm spindle rotation rate and feeding speeds of 1, 2, and 3 mm/min. Figure 2d,e show schematic drawings of the AMP sections showing hardness measurement points, and the second half of AMP shows the specimens cut for OM and SEM examinations, respectively.

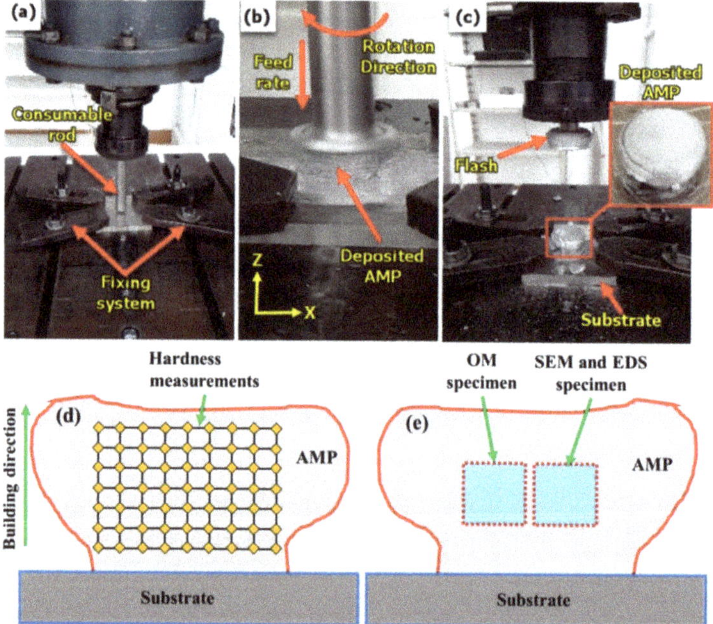

Figure 2. Photographs for the stages of the FSD process: (**a**) Fixing the AA2011 consumable rod and substrate AA5083 on the FSW/FSP machine, (**b**) feeding process during the FSD showing the building up of the part, and (**c**) the end of the deposition process for the additive manufacturing part (AMP). (**d**) and (**e**) are schematic drawings of the AMP sections showing hardness measurement points, and the second half of AMP shows the specimens cut for OM and SEM examinations, respectively.

Additive manufacturing parts (AMPs) have been sectioned vertically along the building direction (z-direction). The deposited layers were oriented perpendicular to the specimen axis/loading direction. The longitudinal sections were prepared according to the standard metallographic procedures by grinding up to 0.05 μm alumina polishing surface finish. The polished sections were investigated using an optical microscope (Olympus, BX41M-LED, Tokyo, Japan) after etching according to ATSM standard E407 using Keller's etchant of the chemical composition of 100 mL distilled water and 3 mL hydrofluoric acid. Microstructural examinations of the AMPs were also carried out using a scanning electron microscope (SEM, FEI, Hillisboro, OR, USA). SEM examination was carried out on the long-transverse sections of the cylindrical friction deposits using secondary electron (SE) imaging modes. Moreover, the grain size of all AM specimens and the base metal have been analyzed by the grain interception method using Olympus Stream Motion Software. A Vickers Hardness Tester (Qness Q10, GmbH, Golling, Austria) with 0.2 kg load and 15 s dwell time was used to evaluate the average hardness of the starting and the AMPs. This test was carried out according to ASTM E92 by measuring twelve readings at least for each AM specimen on the longitudinal sections of the cylindrical friction deposits. The hardness maps were also drawn by collecting four horizontal (perpendicular to building direction) lines and five vertical lines measurements across the AMPs. The free space between any two indentations was 2 mm.

3. Results and Discussions

3.1. Fabrication of AMPs

For conducting the friction stir deposition and forming the AMPs, the axis of the consumable rod is positioned exactly in the center of the square-shaped substrate to ensure the symmetry and homogeneity of heat dissipation through the substrate. Preliminary tests have been carried out to view the behavior of the rod to avoid buckling, physical discontinuities, or other defects of the rod and ensure the build of the part. Based on these preliminary tests, the rotation speeds, feed rate, and length of the consumable rod have been chosen. In addition, the length of the consumable rod out of the shank holder is varied with the temper condition, as the soft alloy tends to buckle easier than the hard alloy that allows more length to be used.

The rubbing between the two surfaces during the rotation and feeding speed of the consumable rod generates frictional heat, which softens the rod's rubbing end, causing plasticized material at the abutting ends. As the process continues, more plasticized material is built up [14,15]. As the required plasticized material thickness is gained, the rotating consumable rod is stopped and withdrawn; this process promotes a deposited layer on the substrate due to torsional shear. Figure 3 illustrates the remains of AA2011-T6 and AA2011-O consumable rods and their AMPs. Figure 3a–c shows the produced AA2011-T6 AMPs fabricated at a constant rotation rate of 1200 rpm at different feeding speeds of 3, 6, and 9 mm/min, respectively. For the AA2011-O specimens, the consumable rods of AA2011-O are softer than the AA2011-T6 rods. Therefore, the energy required to soften the AA2011-O consumable rod is lower than that needed for softening the AA2011-T6 one [30]. Thus, the AM process was conducted after many trials at a constant rotation rate of 200 rpm and various feeding speeds of 1, 2, and 3 mm/min, as shown in Figure 3d–f, respectively. It should be remarked that the higher feeding speeds of 9 mm/min and 3 mm/min at the rotational rates of 1200 and 200 rpm, respectively, are not recommended to fabricated AMPs of AA2011 alloys, where it is not easy to build continuous multi-layers upward to specific height and diameter. The increase in heat input due to an increase in feeding speed over the optimum condition also produces excessive flash around the AMPs, as given in Figure 3c for AA2011-T6 AMP and Figure 3f for AA2011-O AMP. Thus, it was noted that the conical shape at the end of the consumable rods after finishing the FSD process is flattened in a thin thickness, where the other materials are transferred to flash around the fabricated AMPs.

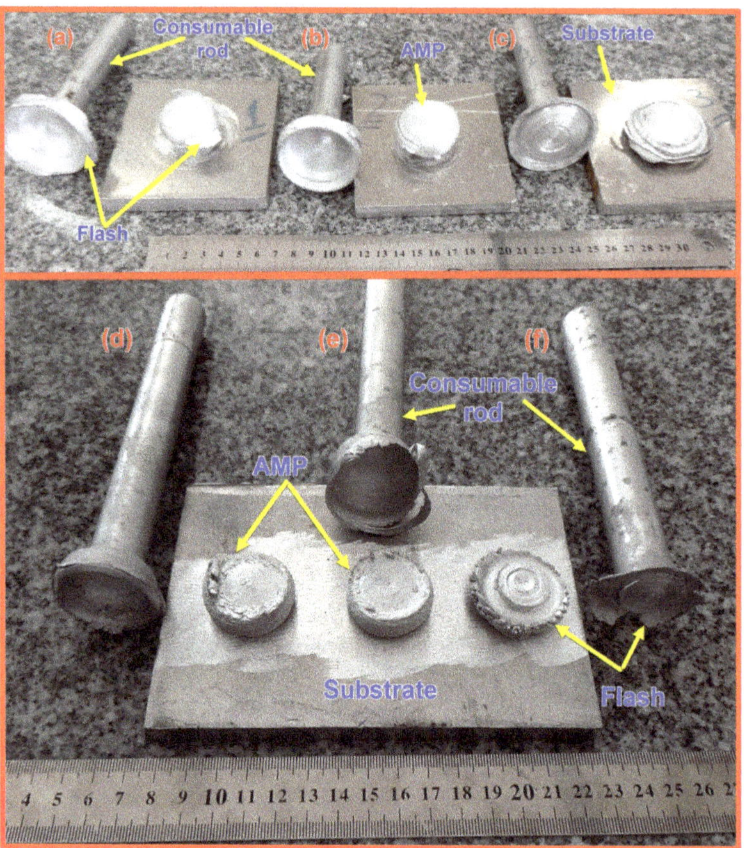

Figure 3. Optical images for the AMPs using FSD and their rods counterparts that remain after obtaining the required part length. AA2011-T6 AMPs processed at 1200 rpm and feeding speeds: (**a**) 3 mm/min, (**b**) 6 mm/min, and (**c**) 9 mm/min. The AA2011-O AMPs processed at 200 rpm and feeding speeds: (**d**) 1 mm/min, (**e**) 2 mm/min, and (**f**) 3 mm/min.

3.2. Macrostructure Examination

Figure 4 illustrates (a) a macrograph of an example of the produced AMP and (b) the AMPs Diameters/Height (D/H) Ratio as a function of the processing feed speed. The visual inspection of the deposit showed that there is significant flash produced from the deposit, which was restacked to the consumable rod. This may be an indication for an overfed condition, in which the feeding speed for the feedstock consumable material is slightly high [16,29]. The possible decrease of the input material feeding speed would mitigate the generation of this produced excess flash. The generation of excessive flash may require post-processing if the geometric accuracy of the final product is sensitive [4]. The macrostructure cross-sections of the produced AMPs show fully continuous dense structures (Figure 3a) without any physical discontinuities or bonding defects at the layer interfaces, indicating the judicious choice of the processing parameters for the AA2011-T6 and AA2011-O aluminum alloys. It can be seen that the D/H of the produced AMPs increases with increasing feeding speed at constant rotation rate for both the AA2011-T6 and the AA2011-O starting materials, as given in Figure 3b. The AA2011 material plasticity during the FSD process is controlled by the amount of heat input introduced in the vortex zone through the AMPs material building from down to up. This phenomenon appears clearly in the AMP geometry based on the applied processing parameters [24].

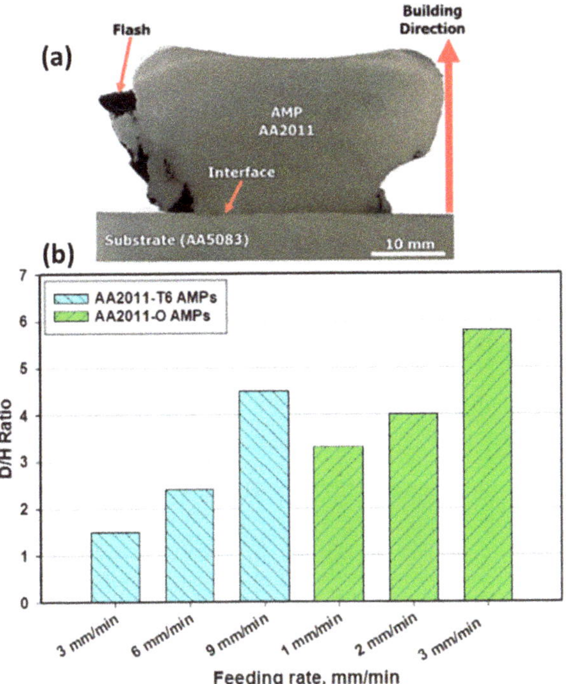

Figure 4. (**a**) The transverse cross-section macrograph for the AMP with the substrate with the building direction and the interface are indicated for the AMP at a constant rotational spindle rate of 1200 rpm and 6 mm/min feeding speeds, (**b**) AMPs Diameters/Height Ratio against feeding speed for all AMPs produced using the different temper conditions and different processing parameters.

3.3. Microstructure Examination

FSD as a thermomechanical process is similar to friction stir welding (FSW) [29–32] and processing (FSP) [28] in heat generation, heat dissipation, and heat transfer mechanisms in the stir zone [33–35]. In the AA2011 AMPs, the heat is generated by dynamic contact friction (DCF) between the consumable tool and AA5083 substrate material. Then, it causes severe plastic deformation of the AA2011 material under the applied downward force and transfers it to continuous build by material flow during the stir deposition process. FSW and FSP generate localized grain refinement in the whole nugget zone (NZ) behind the rotating pin tool. The FSD material is analogous to the NZ in FSW and FSP [17,36]. It was found that the presence of a refined, equiaxed grain structure engaged with the formation of high-angle grain boundaries is an indication of the dynamic recrystallization in FSW of AA2219-T8 [9] and FSP of AA2024/Al_2O_3 nanocomposite [28].

Moreover, Rutherford et al. [9] reported that the reduction in the average volume fraction and size of the intermetallic particles could also be attributed to the severe plastic deformation of the FSD process. Figure 5 shows the optical micrographs of the initial conditions of AA2011 alloys. The microstructure of the AA2011-T6 alloy shows coarse grains as well as the presence of intermetallics in different shapes: rod-like (R), irregular (I), spherical (S), and almost spherical (A-S), as shown in Figure 5a,c. The microstructure grain size in Figure 5c ranges from 30 ± 3 µm to 150 ± 2 µm with an average grain size of 45 ± 8 µm. [37], whereas the microstructure of the AA2011-O rod alloy shows a relatively smaller grain size (Figure 5b,d) than the AA2011-T6 alloy's grain size. The grain size ranged from 8 ± 2 to 75 ± 3 µm with an average grain size of 16 ± 4 µm. Furthermore, the annealing process causes coarsening of the second phase precipitates compared with the

AA2011-T6 material, transferring their shape from the rod-like shape (Figure 5c) to more spheroidal-shaped precipitates (Figure 5d).

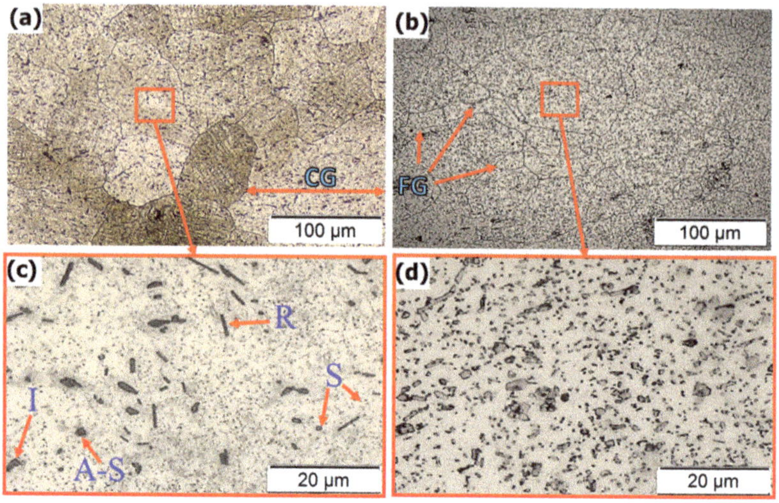

Figure 5. Low and high-magnification optical micrographs of the different temper conditions base material: (**a**,**c**) AA2011-T6 and (**b**,**d**) AA2011-O.

3.3.1. AMPs Parts Produced from AA2011-T6 Alloy

The friction-based processes contribute to the increase of temperature of the material in the stirring zone to the temperature range between 60% and 90% of the melting point of the processed material, which is high enough for the recrystallization during the intensive plastic deformation through the solid-state deposition process [38,39]. Figure 6 represents the microstructures of the AA2011-T6 (Figure 6a) and the AMPs deposited at 1200 rpm spindle rotation rate and different feeding speeds of 3 mm/min (Figure 6b), 6 mm/min (Figure 6c), and 9 mm/min (Figure 6d). It can be seen that the coarse grain structure and precipitates of the AA2011-T6 are refined with the applied FSD process parameters at feed speeds of 3, 6, and 9 m/min. The mean measured grain sizes of AA2011-T6 AMPs were 2.9 ± 0.3, 5.3 ± 0.4, and 11.8 ± 0.5 µm at feeding speeds of 3, 6, and 9 mm/min, respectively. It can be said that the reduction in grain size of AMPs deposited at 3, 6, and 9 mm/min feeding speeds reaches the values of 95.3%, 91.5%, and 82.25%, respectively, compared to the grain size of the AA2011-T6 initial material (62 ± 4 µm). The same results of very fine grains and refined second-phase particles are obtained by Dilip et al. [40] for the multi-layer friction deposits of AA2014-T6 (Al–Cu–Mg–Si alloy system). Consequently, the produced AMPs materials undergo continuous dynamic recrystallization and develop very fine equiaxed grains and refined precipitates [3,41]. In addition, Rutherford et al. [9] reported a significant reduction in the intermetallic particles and grain size after FSD processing of AA6061.

3.3.2. AMPs Parts Produced from AA2011-O

Figure 7 represents the microstructures of the AA2011-O (Figure 7a) and the AMPs deposited at 200 rpm spindle rotation rate and different feeding speeds of 1 mm/min (Figure 7b), 2 mm/min (Figure 7c), and 3 mm/min (Figure 7d). A homogenous fine equiaxed structure has been noticed in all conditions due to the stirring of the grains accompanied with dynamic recrystallization during the additive friction-based process [11,20,42]. The mean grain sizes of AA2011-O AMPs were 0.84 ± 0.05, 0.88 ± 0.06, and 0.94 ± 0.08 µm at feeding speeds 1, 2, and 3 mm/min, respectively. It can be reported that the reduction in grain size of AMPs after FSD reaches not less than the value of ≈98% compared to the grain size of the AA2011-O as-received material (48 ± 4 µm) without any significant difference between the applied feeding speeds.

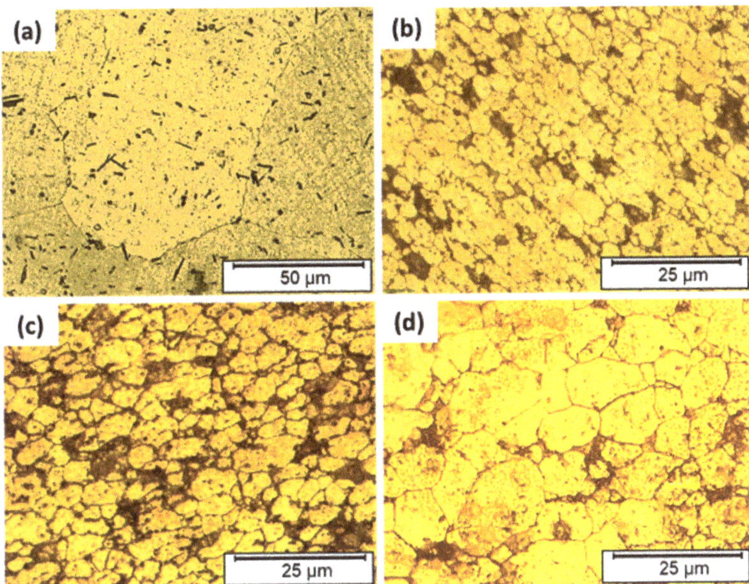

Figure 6. Optical microstructures of the as-received AA2011-T6 (**a**), and AMPs produced t at a rotation rate of 1200 rpm and different feeding speeds of (**b**) 3 mm/min, (**c**) 6 mm/min, and (**d**) 9 mm/min.

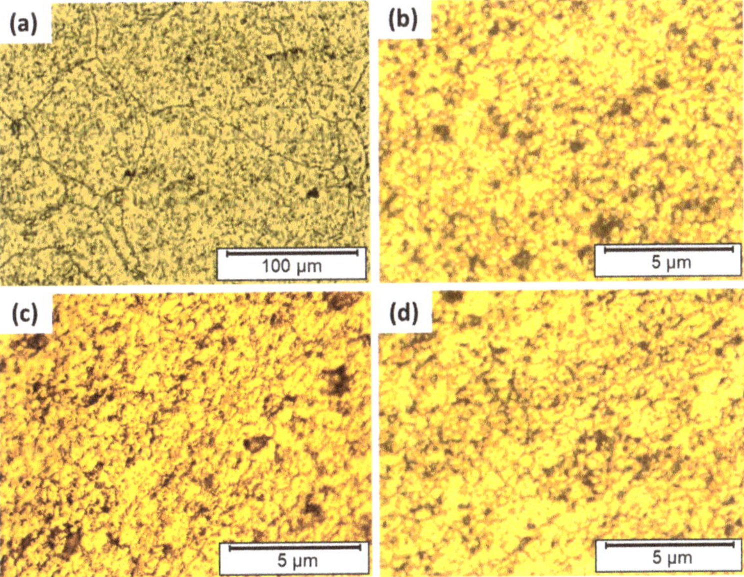

Figure 7. Optical microstructures of the as-received AA2011-O (**a**) and AMPs produced from AA2011-O at a rotation rate of 200 rpm and different feeding speeds of (**b**) 1 mm/min, (**c**) 2 mm/min, and (**d**) 3 mm/min.

SEM was used to examine the present intermetallic precipitates of the as-received AA2011-T6 rod and the friction stir deposited materials at different conditions. Copper is the principal alloying element in AA2011 (Al–Cu alloys). However, other minor alloying elements (Fe, Ti, Zn, and Pb with traces of Ni, Si, and Mn) can also be specified as given in Table 1. During work hardening, an intermetallic phase (Al_2Cu) is precipitated from a supersaturated solid solution. This intermetallic is crystallographically coherent with the Al matrix. Its fine dispersion improves the hardness and strength of the alloy [43]. The non-deformable second-phase precipitates initially present in the base material AA2011 have been fragmented into a smaller size and got uniformly distributed due to the severe plastic deformation involved in the FSD process; see Figure 8. This fragmentation phenomenon is expected in the stir zone of the friction stir welded materials [28,44] and the friction stir deposited materials [6,10,24].

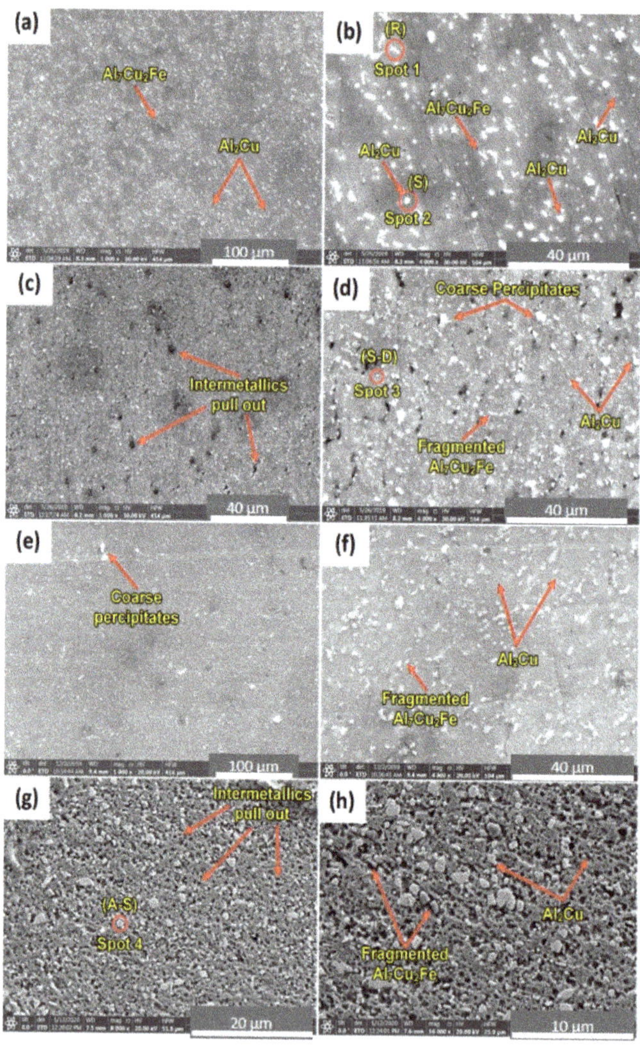

Figure 8. Low and high-magnification SEM micrographs of (**a**,**b**) AA2011-T6 base alloy, (**c**,**d**) FSDed of AA2011-T6 at 1200 rpm—3 mm/min, (**e**,**f**) FSDed AA2011-T6 at 1200 rpm—6 mm/min, and (**g**,**h**) FSDed AA2011-O at 200 rpm—1 mm/min.

The fragmentation of the intermetallics may produce different shapes and sizes. The dispersion of micro and nanoparticles in the aluminum matrix affects the mechanical properties of the AMPs [45–47]. Figure 8 shows low and high magnification SEM micrographs of (a) and (b) AA2011-T6 base alloy (c) and (d) FSDed of AA2011-T6 at 1200 rpm–3 mm/min, (e) and (f) FSDed AA2011-T6 at 1200 rpm—6 mm/min, and (g) and (h) FSDed AA2011-O at 200 rpm—1 mm/min. The effect of the stirring process on the fragmentation and distribution of the intermetallic phases can be seen in (Figure 8c–h) compared with the AA2011-T6 base material (Figure 8a,b). Only two types of precipitates were detected in the base material and AMPs; see Figure 9. The EDS analyses of these precipitates are the rod-like shape (R) Al7Cu2Fe (spot 1 analysis in base material; Figure 8b and represented in Figure 9a) and the Al2Cu phase presents in different shapes given the same EDS analyses Figure 9b. These shapes are spherical (S, spot 2 in Figure 8b), small dots (S-D, spot 3 in AMP produced at 1200 rpm and 3 mm/min; Figure 8d), and almost spherical (A-S, spot 4 in AMP produced at 200 rpm and 1 mm/min; Figure 8g). The detected precipitates for both as-received materials and AMPs are consistent with that reported in the literature [37,43]. It can be remarked that the intermetallics are bonded well with the Al matrix for the as-received materials, as shown in Figure 8a,b. Hence, there is no pull-out detected after the grinding and polishing processes. The pull-out of intermetallics is detected for all the AMPs after grinding and polishing (Figure 8c–f). This indicates the weak bond at the interface between the dispersed intermetallics and the Al matrix as a result of subjecting to the FSD thermomechanical process.

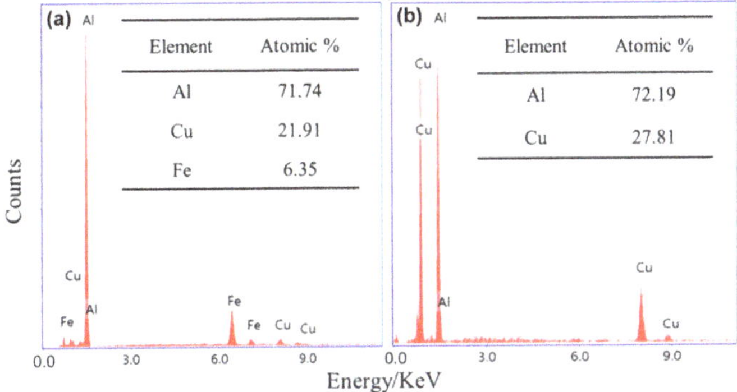

Figure 9. EDS analyses of precipitates show two types of intermetallic precipitates: (**a**) Al_7Cu_2Fe (spot 1 analysis in Figure 8) and (**b**) Al_2Cu (spot 2, 3, and 4 in Figure 8).

Figures 10 and 11 show the EDS elemental mapping of the AMPs produced from AA2011-T6 and AA2011-O at the processing parameters of 1200 rpm spindle rotational rate and 6 mm/min feeding speed and 200 rpm and 1 mm/min, respectively. Figure 10a shows the SEM image obtained from AMP of AA2011-T6 at 1200 rpm and 6 mm/min. It is indicated that the EDS elemental maps confirm the results of the chemical composition of AA2011 in terms of the overall chemical analysis of the alloy as can be seen in Figure 10b. In terms of the elemental maps of the different elements, it can be observed that the Al (c), Fe (e), and Si (f) are homogeneously distributed in their corresponding maps. However, the Cu map in (d) shows a high density of green-colored points where the particles are outlined in the SEM micrographs. Figure 11a shows the SEM image obtained from AMP of AA2011-O at 200 rpm and 1 mm/min. The elemental mapping showed a homogeneous distribution of all elements and phases with no preferences regions along the specimen (no clusters formation), which can be attributed to the severe homogeneous agitation of the material during the FSD process. It is also confirmed the possibility of the Al_7Cu_2Fe and Al_2Cu intermetallics formation and location.

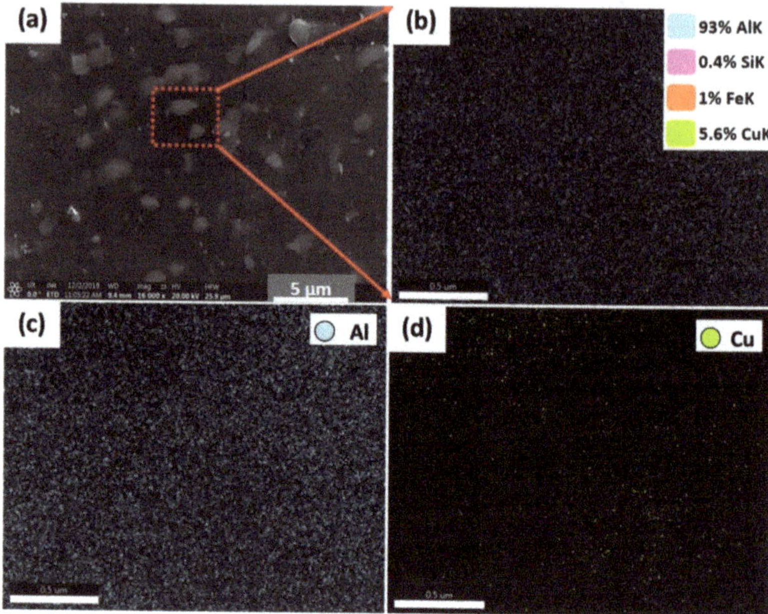

Figure 10. Elemental map distribution of AA2011-T6 AMPs produced at 1200 rpm and 6 mm/min: (**a**) SEM of AMPs, (**b**) map distribution of all alloying elements, and (**c**,**d**) map distribution of Al and Cu, respectively.

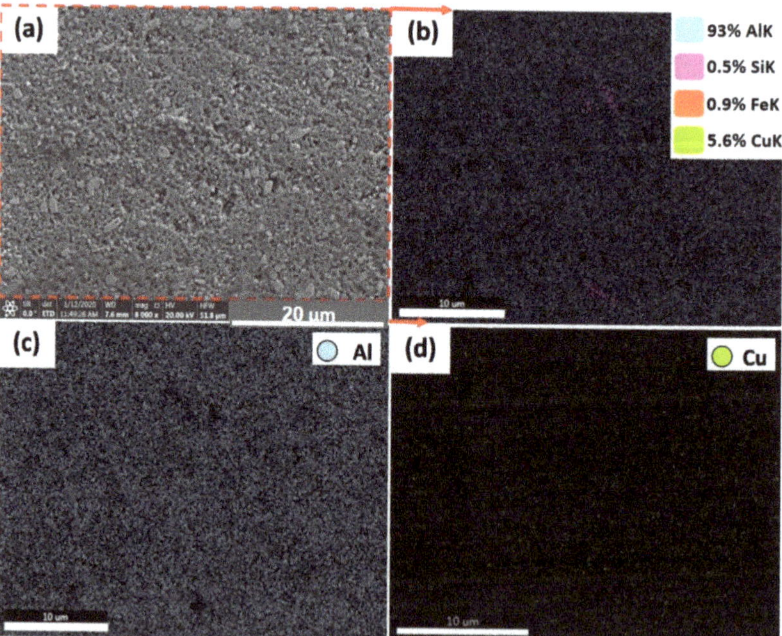

Figure 11. Elemental map distribution of AA2011-O AMPs produced at 200 rpm and 1 mm/min: (**a**) SEM of AMPs, (**b**) map distribution of all alloying elements, and (**c**,**d**) map distribution of Al and Cu, respectively.

3.4. Hardness

Hardness is an important mechanical property to evaluate materials, and its value is controlled by the chemical composition, temper condition, and the FSD process parameters. Figures 12 and 13 represent the average hardness values of the initial materials (AA2011 and AA2011-O) and the produced AMPs at different FSD parameters. The average hardness measurements of the AA2011 AMPs show lower hardness values in comparison to the base metal. Moreover, the hardness decreases as the feeding speed increases, as given in Figure 12. The decrease in hardness percentage (hardness loss %) of AA2011-T6 AMPs reaches the highest value of 49% compared to the AA2011-T6 material at processing parameters of a rotational rate of 1200 rpm and feeding speed of 9 mm/min. At the same rotation rate, the applied feeding speed of 3 mm/min produces an AMP with a hardness loss of 39% compared to the base material. This loss in hardness accompanying the additive manufacturing process was also reported by Dilip et al. [6] for the AMPs of AA2014. They concluded that the friction deposits showed inferior hardness because of their overaged microstructure.

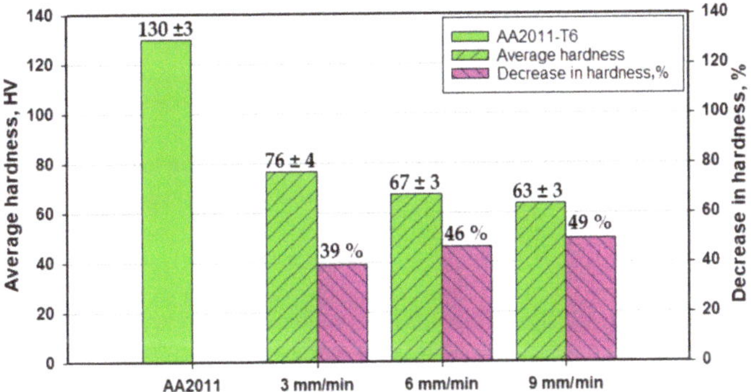

Figure 12. Average hardness and a hardness loss percentage of the AA2011-T6 and its AMPs produced at a rotational rate of 1200 rpm and feed speeds of 3, 6, and 9 mm/min.

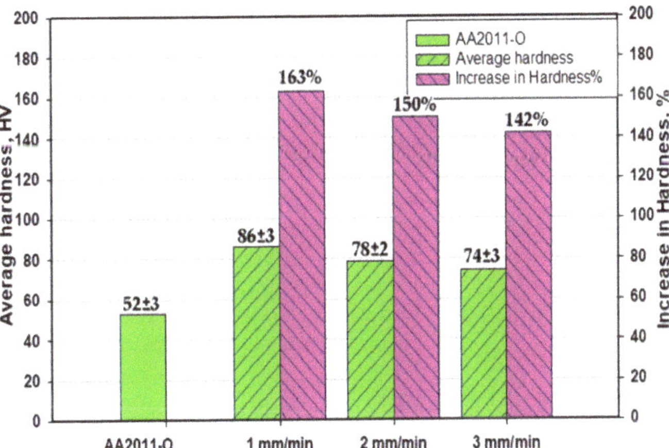

Figure 13. Average hardness and a hardness increase percentage of the AA2011-O and its AMPs produced at a rotational rate of 200 rpm and feed speeds of 1, 2, and 3 mm/min.

The average hardness values of AA2011-O alloy and the AMPs at a rotation rate of 200 rpm and feeding speeds of 1, 2, and 3 mm/min are illustrated in Figure 13. The hardness value of the AMPs produced at the applied feeding speeds is higher than that of the initial condition AA2011-O. This increase in hardness slightly decreases with the increasing feeding speed. The increase in the hardness value of AMP produced at a 1 mm/min feeding speed attains 163% compared with the AA2011-O starting material, whereas the improvement reaches 142% at the applied feeding speed of 3 mm/min.

The high hardness of the as-received AA2011-T6 (130 ± 3 HV) is ascribed to the high internal stresses stored in the material due to the previous production process, such as cold working. Afterward, the annealing process at 415 °C for 2.5 h followed by slow furnace cooling contributed to the stress relief of the alloy, coarsening, and softening of the second phase precipitates. Thus, the hardness of the alloy decreased after annealing, despite the reduction in grain size. It is well known that two mechanisms are controlling the hardness of this alloy. The first one is the reduction in grain size, which is directly proportional to the hardness increase. The second one is the morphology and dispersion of the second phase precipitates. In AMPs fabricated using the hard rods of AA2011-T6, the hardness decreased after deposition because the dominant mechanism that affects the hardness was the fragmentation of the hard precipitates. On the contrary, the hardness of the AMPs manufactured using the soft AA2011-O rods increased after friction deposition as the dominant mechanism, in this case, was the grain size reduction.

4. Conclusions

The current study investigates the effect of alloy temper conditions on the behavior upon friction stir deposition. The deposited AMPs were characterized in terms of macro and microstructural features as well as the hardness distribution. Based on the obtained results, the following conclusions can be outlined:

- The friction stir deposition technique successfully produced sound continuous multi-layered AA2011 AMPs without any physical discontinuities or interfacial defects between layers along the vertical direction.
- The temper condition of the alloy affects the behavior during FSD, and the parameters that worked with the T6 condition does not work with the O condition alloy. The hard condition alloy required high heat input to reach the state of a good deposition, while the soft condition alloy required low heat input to reach the FSD state.
- The optimum condition for FSD of AA2011-T6 was a rotational rate of 1200 rpm at feeding speeds between 3 and 9 mm/min, while the optimum condition for FSD of AA2011-O was a rotational rate of 200 rpm at feeding speeds between 1 and 3 mm/min.
- Microstructural analysis showed that significant grain refining and intermetallic particle fragmentation have been observed in the AMPs from the two temper conditions of AA2011.
- The fine grain formation is mainly dominated by dynamic recrystallization where the average grain size is reduced by decreasing the feeding speed at a constant rotation rate. The use of a lower rotation rate resulted in more refining due to the lower heat input experienced.
- The use of T6 temper alloy has resulted in AMPs with a lower hardness than the starting material that reached about 61% and 51% of the starting material hardness at feeding rates of 3 and 9 mm/min, respectively. However, the use of O temper alloy has resulted in AMPs with higher hardness than the starting material by about 163% hardness enhancement compared to the starting material.

Author Contributions: Conceptualization, M.M.Z.A., E.E. and M.M.E.-S.S.; methodology, M.M.Z.A., E.E., and M.M.E.-S.S.; software, M.M.E.-S.S. and M.I.A.H.; validation, B.A. and K.T.; formal analysis, M.M.E.-S.S. and M.M.Z.A.; investigation, M.M.E.-S.S. and M.M.Z.A.; resources, B.A. and K.T.; data curation, M.M.E.-S.S.; writing—original draft preparation, E.E. and M.M.E.-S.S.; writing—review and editing, M.M.E.-S.S. and M.M.Z.A.; visualization, M.I.A.H.; supervision, M.M.Z.A. and M.M.E.-S.S.; project administration, M.M.Z.A. and M.M.E.-S.S.; funding acquisition, B.A. and K.T. All authors have read and agreed to the published version of the manuscript.

Funding: This research received no external funding.

Institutional Review Board Statement: Not applicable.

Informed Consent Statement: Not applicable.

Data Availability Statement: The data presented in this study are available on request from the corresponding author. The data are not publicly available due to the extremely large size.

Conflicts of Interest: The authors declare no conflict of interest.

References

1. Kumar Srivastava, A.; Kumar, N.; Rai Dixit, A. Friction stir additive manufacturing—An innovative tool to enhance mechanical and microstructural properties. *Mater. Sci. Eng. B Solid-State Mater. Adv. Technol.* **2021**, *263*, 114832. [CrossRef]
2. Khodabakhshi, F.; Gerlich, A.P.P. Potentials and strategies of solid-state additive friction-stir manufacturing technology: A critical review. *J. Manuf. Process.* **2018**, *36*, 77–92. [CrossRef]
3. Zhang, D.; Sun, S.; Qiu, D.; Gibson, M.A.; Dargusch, M.S.; Brandt, M.; Qian, M.; Easton, M. Metal Alloys for Fusion-Based Additive Manufacturing. *Adv. Eng. Mater.* **2018**, *20*, 1–20. [CrossRef]
4. Dilip, J.J.S.; Rafi, H.K.; Ram, G.D.J. A new additive manufacturing process based on friction deposition. *Trans. Indian Inst. Met.* **2011**, *64*, 27–30. [CrossRef]
5. Karthik, G.M.; Ram, G.D.J.; Kottada, R.S. Friction deposition of titanium particle reinforced aluminum matrix composites. *Mater. Sci. Eng. A* **2016**, *653*, 71–83. [CrossRef]
6. Dilip, J.J.S.; Ram, G.D.J. Microstructure evolution in aluminum alloy AA 2014 during multi-layer friction deposition. *Mater. Charact.* **2013**, *86*, 146–151. [CrossRef]
7. Harun, W.S.W.; Kamariah, M.S.I.N.; Muhamad, N.; Ghani, S.A.C.; Ahmad, F.; Mohamed, Z. A review of powder additive manufacturing processes for metallic biomaterials. *Powder Technol.* **2018**, *327*, 128–151. [CrossRef]
8. Singh, S.; Ramakrishna, S.; Singh, R. Material issues in additive manufacturing: A review. *J. Manuf. Process.* **2017**, *25*, 185–200. [CrossRef]
9. Rutherford, B.A.; Avery, D.Z.; Phillips, B.J.; Rao, H.M.; Doherty, K.J.; Allison, P.G.; Brewer, L.N.; Brian Jordon, J. Effect of thermomechanical processing on fatigue behavior in solid-state additive manufacturing of Al-Mg-Si alloy. *Metals* **2020**, *10*, 947. [CrossRef]
10. Griffiths, R.J.; Perry, M.E.J.J.; Sietins, J.M.; Zhu, Y.; Hardwick, N.; Cox, C.D.; Rauch, H.A.; Yu, H.Z. A Perspective on Solid-State Additive Manufacturing of Aluminum Matrix Composites Using MELD. *J. Mater. Eng. Perform.* **2019**, *28*, 648–656. [CrossRef]
11. Dilip, J.J.S.; Babu, S.; Varadha Rajan, S.; Rafi, K.H.; Janaki Ram, G.; Stucker, B.E. Use of friction surfacing for additive manufacturing. *Mater. Manuf. Process.* **2013**, *28*, 189–194. [CrossRef]
12. Su, J.Q.; Nelson, T.W.; Sterling, C.J. Grain refinement of aluminum alloys by friction stir processing. *Philos. Mag.* **2006**, *86*, 1–24. [CrossRef]
13. Gandra, J. Friction stir processing. In *Surface Modification by Solid State Processing*; Woodhead Publishing: Sawston, UK, 2014; pp. 73–111. ISBN 9780857094698.
14. Gupta, G.R.; Parkhurst, J.O.; Ogden, J.A.; Aggleton, P.; Mahal, A. Structural approaches to HIV prevention. *Lancet* **2008**, *372*, 764–775. [CrossRef]
15. Karthik, G.M.; Panikar, S.; Ram, G.D.J.; Kottada, R.S. Additive manufacturing of an aluminum matrix composite reinforced with nanocrystalline high-entropy alloy particles. *Mater. Sci. Eng. A* **2017**, *679*, 193–203. [CrossRef]
16. Andrade, D.G.; Leitão, C.; Rodrigues, D.M. Influence of base material characteristics and process parameters on frictional heat generation during Friction Stir Spot Welding of steels. *J. Manuf. Process.* **2019**, *43*, 98–104. [CrossRef]
17. Ahmed, M.M.Z.; Ataya, S.; Seleman, M.M.E.S.; Allam, T.; Alsaleh, N.A.; Ahmed, E. Grain structure, crystallographic texture, and hardening behavior of dissimilar friction stir welded aa5083-o and aa5754-h14. *Metals* **2021**, *11*, 181. [CrossRef]
18. Guru, P.R.; Khan MD, F.; Panigrahi, S.K.; Ram, G.D.J. Enhancing strength, ductility and machinability of a Al-Si cast alloy by friction stir processing. *J. Manuf. Process.* **2015**, *18*, 67–74. [CrossRef]
19. Jahangir, M.N.; Mamun, M.A.H.; Sealy, M.P. A review of additive manufacturing of magnesium alloys. In *AIP Conference Proceedings*; American Institute of Physics Inc.: College Park, MD, USA, 2018; Volume 1980, p. 030026.
20. Mishra, R.S.; Palanivel, S. Building without melting: A short review of friction-based additive manufacturing techniques. *Int. J. Addit. Subtractive Mater. Manuf.* **2017**, *1*, 82. [CrossRef]
21. Bose, S.; Ke, D.; Sahasrabudhe, H.; Bandyopadhyay, A. Additive manufacturing of biomaterials. *Prog. Mater. Sci.* **2018**, *93*, 45–111. [CrossRef]
22. Palanivel, S.; Nelaturu, P.; Glass, B.; Mishra, R.S.S. Friction stir additive manufacturing for high structural performance through microstructural control in an Mg based WE43 alloy. *Mater. Des.* **2015**, *65*, 934–952. [CrossRef]
23. Padhy, G.K.; Wu, C.S.; Gao, S. Friction stir based welding and processing technologies—Processes, parameters, microstructures and applications: A review. *J. Mater. Sci. Technol.* **2018**, *34*, 1–38. [CrossRef]
24. Elfishawy, E.; Ahmed, M.M.Z.; El-Sayed Seleman, M.M. Additive manufacturing of aluminum using friction stir deposition. In *TMS 2020 149th Annual Meeting & Exhibition Supplemental Proceedings*; The Minerals, Metals & Materials Series; Springer: Cham, Switzerland, 2020. [CrossRef]
25. Boyer, H.E. Heat Treating of Nonferrous Alloys. *Metallogr. Microstruct. Anal.* **2013**, *2*, 190–195. [CrossRef]

26. Committee, A.H. Heat Treating of Aluminum Alloys. In *ASM Handbook, Volume 4: Heat Treating*; ASM Handbook Committee: Russell Township, OH, USA, 1991; pp. 841–879. [CrossRef]
27. Ahmed, M.M.Z.; Ataya, S.; El-Sayed Seleman, M.M.; Ammar, H.R.; Ahmed, E. Friction stir welding of similar and dissimilar AA7075 and AA5083. *J. Mater. Process. Technol.* **2017**, *242*, 77–91. [CrossRef]
28. Hoziefa, W.; Toschi, S.; Ahmed, M.M.Z.; Morri, A.; Mahdy, A.A.; El-Sayed Seleman, M.M.; El-Mahallawi, I.; Ceschini, L.; Atlam, A. Influence of friction stir processing on the microstructure and mechanical properties of a compocast AA2024-Al2O3 nanocomposite. *Mater. Des.* **2016**, *106*, 273–284. [CrossRef]
29. Alzahrani, B.; El-Sayed Seleman, M.M.; Ahmed, M.M.Z.; Elfishawy, E.; Ahmed, A.M.Z.; Touileb, K.; Jouini, N.; Habba, M.I.A. The Applicability of Die Cast A356 Alloy to Additive Friction Stir Deposition at Various Feeding Speeds. *Materials* **2021**, *14*, 6018. [CrossRef] [PubMed]
30. Lukács, J.; Meilinger, Á.; Pósalaky, D. High cycle fatigue and fatigue crack propagation design curves for 5754-H22 and 6082-T6 aluminium alloys and their friction stir welded joints. *Weld. World* **2018**, *62*, 737–749. [CrossRef]
31. Ahmed, M.M.Z.; Ataya, S.; El-Sayed Seleman, M.M.; Mahdy, A.M.A.; Alsaleh, N.A.; Ahmed, E. Heat input and mechanical properties investigation of friction stir welded aa5083/aa5754 and aa5083/aa7020. *Metals* **2021**, *11*, 1–20.
32. Ahmed, M.M.Z.; Jouini, N.; Alzahrani, B.; Seleman, M.M.E.-S.; Jhaheen, M. Dissimilar Friction Stir Welding of AA2024 and AISI 1018: Microstructure and Mechanical Properties. *Metals* **2021**, *11*, 330. [CrossRef]
33. Zayed, E.M.; El-Tayeb, N.S.M.; Ahmed, M.M.; Rashad, R.M. Development and Characterization of AA5083 Reinforced with SiC and Al2O3 Particles by Friction Stir Processing. In *Engineering Design Applications*; Öchsner, A., Altenbach, H., Eds.; Springer International Publishing: Cham, Switzerland, 2019; pp. 11–26. ISBN 978-3-319-79005-3.
34. Hamada, A.S.; Järvenpää, A.; Ahmed, M.M.Z.; Jaskari, M.; Wynne, B.P.; Porter, D.A.; Karjalainen, L.P. The microstructural evolution of friction stir welded AA6082-T6 aluminum alloy during cyclic deformation. *Mater. Sci. Eng. A* **2015**, *642*. [CrossRef]
35. Tonelli, L.; Morri, A.; Toschi, S.; Shaaban, M.; Ammar, H.R.; Ahmed, M.M.Z.; Ramadan, R.M.; El-Mahallawi, I.; Ceschini, L. Effect of FSP parameters and tool geometry on microstructure, hardness, and wear properties of AA7075 with and without reinforcing B4C ceramic particles. *Int. J. Adv. Manuf. Technol.* **2019**, *102*, 366–376. [CrossRef]
36. Kang, J.; Feng, Z.C.; Frankel, G.S.; Huang, I.W.; Wang, G.Q.; Wu, A.P. Friction Stir Welding of Al Alloy 2219-T8: Part I-Evolution of Precipitates and Formation of Abnormal Al2Cu Agglomerates. *Metall. Mater. Trans. A Phys. Metall. Mater. Sci.* **2016**, *47*, 4553–4565. [CrossRef]
37. Proni, C.T.W.; D'Ávila, M.A.; Zoqui, E.J. Thixoformability evaluation of AA2011 and AA2014 alloys. *Int. J. Mater. Res.* **2013**, *104*, 1182–1196. [CrossRef]
38. Mishra, R.S.; Ma, Z.Y. Friction stir welding and processing. *Mater. Sci. Eng.* **2005**, *50*, 1–78. [CrossRef]
39. Rivera, O.G.; Allison, P.G.; Brewer, L.N.; Rodriguez, O.L.; Jordon, J.B.; Liu, T.; Whittington, W.R.; Martens, R.L.; McClelland, Z.; Mason, C.J.T.; et al. Influence of texture and grain refinement on the mechanical behavior of AA2219 fabricated by high shear solid state material deposition. *Mater. Sci. Eng. A* **2018**, *724*, 547–558. [CrossRef]
40. Fouad, D.M.; El-Garaihy, W.H.; Ahmed, M.M.Z.; Albaijan, I.; Seleman, M.M.E.; Salem, H.G. Grain Structure Evolution and Mechanical Properties of Multi-Channel Spiral Twist Extruded AA5083. *Metals* **2021**, *11*, 1276. [CrossRef]
41. Rathee, S.; Srivastava, M.; Maheshwari, S.; Kundra, T.K.; Siddiquee, A.N. *Friction Based Additive Manufacturing Technologies*; CRC Press: Boca Raton, FL, USA, 2018; ISBN 9780815392361.
42. Mishra, R.S.; Mahoney, M.W.; McFadden, S.X.; Mara, N.A.; Mukherjee, A.K. 32-High strain rate superplasticity in a friction stir processed 7075 Al alloy. *Scr. Mater.* **1999**, *42*, 163–168. [CrossRef]
43. Niclas, Å. A calorimetric analysis and solid-solubility examination of aluminium alloys containing low-melting-point elements. Master's Thesis, KTH Royal Institute of Technology, Stockholm, Sweden, 2012.
44. Ahmed, M.M.Z.; El-Sayed Seleman, M.M.; Zidan, Z.A.; Ramadan, R.M.; Ataya, S.; Alsaleh, N.A. Microstructure and mechanical properties of dissimilar friction stir welded AA2024-T4/AA7075-T6 T-butt joints. *Metals* **2021**, *11*, 128. [CrossRef]
45. Zhang, W.W.; Lin, B.; Zhang, D.T.; Li, Y.Y. Microstructures and mechanical properties of squeeze cast Al-5.0Cu-0.6Mn alloys with different Fe content. *Mater. Des.* **2013**, *52*, 225–233. [CrossRef]
46. Xu, D.; Zhu, C.; Xu, C.; Chen, K. Microstructures and tensile fracture behavior of 2219 wrought al–cu alloys with different impurity of fe. *Metals* **2021**, *11*, 174. [CrossRef]
47. Zhang, W.; Lin, B.; Fan, J.; Zhang, D.; Li, Y. Microstructures and mechanical properties of heat-treated Al-5.0Cu-0.5Fe squeeze cast alloys with different Mn/Fe ratio. *Mater. Sci. Eng. A* **2013**, *588*, 366–375. [CrossRef]

Article

Effective Range of FSSW Parameters for High Load-Carrying Capacity of Dissimilar Steel A283M-C/Brass CuZn40 Joints

Sabbah Ataya [1,2], Mohamed M. Z. Ahmed [3,*], Mohamed M. El-Sayed Seleman [2], Khalil Hajlaoui [1], Fahamsyah H. Latief [1], Ahmed M. Soliman [4], Yousef G. Y. Elshaghoul [5] and Mohamed I. A. Habba [6]

1 Department of Mechanical Engineering, College of Engineering, Imam Mohammad Ibn Saud Islamic University, Riyadh 11432, Saudi Arabia; smataya@imamu.edu.sa (S.A.); kmhajlaoui@imamu.edu.sa (K.H.); fhlatief@imamu.edu.sa (F.H.L.)
2 Department of Metallurgical and Materials Engineering, Faculty of Petroleum and Mining Engineering, Suez University, Suez 43512, Egypt; mohamed.elnagar@suezuniv.edu.eg
3 Mechanical Engineering Department, College of Engineering at Al Kharj, Prince Sattam Bin Abdulaziz University, Al Kharj 16273, Saudi Arabia
4 Department of Mechanical Engineering, College of Engineering, Jouf University, Sakaka 72388, Saudi Arabia; amsoliman@ju.edu.sa
5 Mechanical Engineering Department, Faculty of Engineering, Suez University, Suez 43518, Egypt; yousef.gamal@eng.suezuniv.edu.eg
6 Mechanical Department, Faculty of Technology & Education, Suez University, Suez 43518, Egypt; mohamed.atia@suezuniv.edu.eg
* Correspondence: moh.ahmed@psau.edu.sa

Abstract: In the current study, a 2 mm thick low-carbon steel sheet (A283M—Grade C) was joined with a brass sheet (CuZn40) of 1 mm thickness using friction stir spot welding (FSSW). Different welding parameters including rotational speeds of 1000, 1250, and 1500 rpm, and dwell times of 5, 10, 20, and 30 s were applied to explore the effective range of parameters to have FSSW joints with high load-carrying capacity. The joint quality of the friction stir spot-welded (FSSWed) dissimilar materials was evaluated via visual examination, tensile lap shear test, hardness test, and macro- and microstructural investigation using SEM. Moreover, EDS analysis was applied to examine the mixing at the interfaces of the dissimilar materials. Heat input calculation for the FSSW of steel–brass was found to be linearly proportional with the number of revolutions per spot joint, with maximum heat input obtained of 11 kJ at the number of revolutions of 500. The temperature measurement during FSSW showed agreement with the heat input dependence on the number of revolution. However, at the same revolutions of 500, it was found that the higher rotation speed of 1500 rpm resulted in higher temperature of 583 °C compared to 535 °C at rotation speed of 1000 rpm. This implies the significant effect for the rotation speed in the increase of temperature. The macro investigations of the friction stir spot-welded joints transverse sections showed sound joints at the different investigated parameters with significant joint ligament between the steel and brass. FSSW of steel/brass joints with a number of revolutions ranging between 250 to 500 revolutions per spot at appropriate tool speed range (1000–1500 rpm) produces joints with high load-carrying capacity from 4 kN to 7.5 kN. The hardness showed an increase in the carbon steel (lower sheet) with maximum of 248 HV and an increase of brass hardness at mixed interface between brass and steel with significant reduction in the stir zone hardness. Microstructural investigation of the joint zone showed mechanical mixing between steel and brass with the steel extruded from the lower sheet into the upper brass sheet.

Keywords: Friction stir spot welding; low-carbon steel; brass; load-carrying capacity; hardness; microstructure

Citation: Ataya, S.; Ahmed, M.M.Z.; El-Sayed Seleman, M.M.; Hajlaoui, K.; Latief, F.H.; Soliman, A.M.; Elshaghoul, Y.G.Y.; Habba, M.I.A. Effective Range of FSSW Parameters for High Load-Carrying Capacity of Dissimilar Steel A283M-C/Brass CuZn40 Joints. *Materials* **2022**, *15*, 1394. https://doi.org/10.3390/ma15041394

Academic Editors: Józef Iwaszko, Jerzy Winczek and Andrey Belyakov

Received: 29 January 2022
Accepted: 11 February 2022
Published: 14 February 2022

1. Introduction

In 1991, friction stir welding (FSW) was initially introduced by TWI in the United Kingdom as a newly developed solid-state welding process, particularly for joining aluminum alloys with specific requirements that should be fulfilled [1,2]. Then friction stir spot welding (FSSW) was developed as one of its variant for local joining of similar and dissimilar sheets [3–9] as a promising technique with similarities to the basic concepts of linear FSW with a specific requirement where no lateral movement of the tool is required [10]. More interestingly, the rotating tool pierces the sheets that are being welded and then produces a stir zone. The stir zone or nugget in the friction processed materials characterized by fine, dynamic, recrystallized microstructures due the severe plastic deformation experienced at high temperatures [11–14]. Indeed, the FSSW can be operated under the bulk melting point, which allows this technique to prevent the formation of defects attributed to porosity, solidification, thermal distortion, etc. [15,16]. The FSSW technique has been applied for joining dissimilar materials [17] such as aluminum–magnesium [18], aluminum–copper [19], and aluminum–steel [10,20]. Moreover, previous investigations stated that FSSW had been employed for various steels such as high-strength [21], advanced high-strength [22], and ultrahigh-strength steels [23]. Moreover, the steels comprising mixture structures of ferrite and martensite are termed dual-phase (DP) steel. These DP steels are actually a kind of low-carbon steel consisting of ferrite and martensite microstructures. DP steels have captured particular interest due to their good combination of high strength, ductility, and weldability of these alloys and have been used widely for over the past few years in the automotive industry [24]. Some methods are used to improve the mechanical properties of DP steels: one of them is adding alloying elements into the DP steels to modify their microstructure. However, as a result of the addition of the alloying elements at a high percentage, it may cause the DP steel to become more prone to cracking during the welding process [25]. Many attempts have been carried out to join steel sheets of interstitial-free (IF) and dual-phase (DP) steel sheets by FSSW with a convex shoulder tool. Two different combinations were used: one with IF as the top sheet (IF/DP) and another with DP as the top sheet (DP/IF) [26]. Turning to copper [27] and its alloys [28], they have been widely used in many engineering applications due copper's excellent electrical and thermal conductivities, superior corrosion resistance, and good strength and ductility. More importantly, conventional fusion welding processes cannot be undertaken to manufacture copper joints since they demonstrate high thermal diffusivity, which is approximately 10–100 times higher compared with nickel alloys and most steels [29]. Meran [30] reported that copper and copper alloys have been successfully manufactured via the FSW method. Gao et al. [31] investigated the properties of dissimilar lap joints of commercial brass (CuZn40) to plain carbon steel (S25C) by considering the welding speed effect using the FSW method. The results showed that the grain size, hardness at the stirred zone, and tensile shear fracture load of the joints were significantly changed when the welding speed were varied. With respect to brass (Cu–Zn alloy), it demonstrates high strength, plasticity, high hardness, and good corrosion resistance, which makes brass suitable to be used as structural materials in many industrial applications [32]. Nevertheless, it is very difficult to clad brass to steel by fusion welding because of the strength loss in the fusion zone due to the evaporation of Zn, as well as the large differences in the thermal physical properties between brass and steel, such as melting point, thermal conductivity, and thermal expansion coefficient [31]. To handle this problem, friction welding as a solid-state welding process has been employed in recent years. Luo et al. [33] used the CT-130-type special inertia friction welding machine to finish the H90 brass/high carbon steel dissimilar metals radial friction welding process, and Kimura et al. [34] achieved the brass/low-carbon steel dissimilar metal welding by friction welding. However, the friction welding has a large limitation in the shape of the joint, which should be a body of rotation such as a pipe and a rod-type joint. In addition, yellow brass (Cu/Zn 63/37) has an excellent capacity for cold working and tin coating by hot dipping; moreover, brass CuZn37 has excellent cold working properties [35]. Most brasses are not normally ranked as heat-treatable; some brasses are cast or hot-worked in the duplex α/β state and then annealed at about 450 °C to convert the microstructure into a single phase of better resistance to corrosion [35]. Due to the relative high temperature associated with FSSW of brass, recrystallization accompanied with microsegregation leading to separation of β-phase takes place [31], which leads to some softening. This behavior could be avoided either by applying proper stirring speed [31] or by postweld heat treatment [34].

There are some research works on the friction stir spot welding of similar and dissimilar metals and alloys [36–39]. However, limited reports are found on the friction stir spot welding of low-carbon steel and brass, which is important for the use of joints. Based on the above reasons, it is interesting to study the manufacturing of low-carbon steel and brass as a lap joint by FSSW method. The aim of this study is to investigate the effective range of FSSW combinations of parameters of rotational speed and dwell time to obtain a high load-carrying capacity of dissimilar steel A283M-C/brass CuZn40 sheet FSSW joints. The examined rotational speed are 1000, 1250, and 1500 rpm, and the applied dwell times are 5, 10, 20, and 30 s. The resulting load-carrying capacity of the joints are measured and judged by the resulting heat generated and temperature raise at the joint during the FSSW process. Moreover, the cohesion quality of the joints is examined by the metallographic examination and EDS analysis. Normalized process parameters will be determined to function in the selection of the effective range of conditions for further FSSW processes.

2. Methodology

2.1. Materials

The FSSW was carried out on lap joints of the brass (CuZn40) and low-carbon steel. CuZn40 brass samples were cut from cold-rolled sheet of 1 mm thickness produced by Wieland Company, Ulm, Germany. The low-carbon steel (St 44-2) sheet was produced and delivered by Ezz Dekheila Steel (EZDK) Company, Alexandria, Egypt, in the form of 2 mm thick sheets. The chemical composition of the used low-carbon steel and brass have been analyzed using Foundry-Master pro, Oxford Instruments, Abingdon, UK. The steel grade is equivalent to A283M grade C according to ASTM A283M-18 [40]. Preparing for the FSSW lap joints, the starting copper and steel sheets were cut into plates of the width of 100 mm and length of 200 mm. The chemical composition of both starting materials is listed in Table 1. The mechanical properties that were obtained according to ASTM E8/E8M-21 [41] are also included in Table 2.

Table 1. The chemical composition of low-carbon steel (St 44-2) and brass (CuZn40) sheets (in Wt.%).

Element (Wt. %)	C	Si	Mn	P	S	Cr	Ni	V	Nb	Cu	Zn	Pb
St 44-2	0.052	0.034	0.875	0.01	0.003	0.017	0.05	0.036	0.026	0.03	-	0.005
CuZn40	-	0.002	0.005	0.007	-	-	-	-	-	61.3	38.6	0.025

Table 2. The mechanical properties of the starting sheet materials: low-carbon steel (steel 44-2) and brass (CuZn40).

Material	Properties			
	Yield Stress σ_y (MPa)	Tensile Strength σ_{uts} (MPa)	Fracture Strain ε_{Fr} (%)	Hardness HV
Steel 44-2	461	525	19.4	223 ± 2.4
CuZn40	250	434	40	128 ± 2.3

For microstructure investigation, the brass and steel samples were separately mechanically ground up to the final stage of SiC grit of 2400. (i) Brass was primarily polished using a solution of the following mixture: 260 mL oxide (Al_2O_3) suspension with file alumina particle (0.05 μm), 0.5 mL HF, 1 mL HNO_3, and 40 mL H_2O_2. (ii) The samples were finally etched during the final polishing stage using Klemm III and potassium disulfide. The specimens were polished and etched as in steps (i) and (ii) for two or three times, until a scratch-free surface and the microstructure were revealed. Steel was etched after polishing in Nital 2% for 5 s. The microstructure of the starting brass and low-carbon steel is shown in Figure 1a,b, respectively. From the binary Cu–Zn system, brass with copper content less than about 63 wt.% is composed of a single α-phase [42]. With increasing the zinc content, more than 37 wt.% β-phase starts to appear. In the current study, brass with little dispersed β-phase can be distinguished due to the cold deformation of the alloy, as shown Figure 2a. Figure 2b shows the microstructure of the low-carbon steel used in this study. As can be seen from this micrograph, the microstructures are composed mainly of relatively large ferritic grains, while some smaller pearlite grains or dispersed pearlitic islands at the ferritic triple points of ferrite grains in small portions are spread across the whole area.

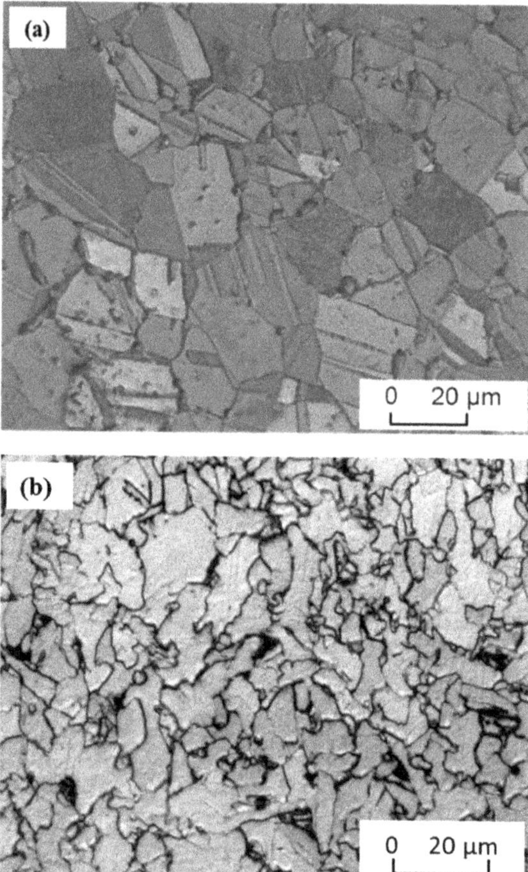

Figure 1. Optical microstructure of base materials used in this study. (**a**) Microstructure of brass (CuZn40) and (**b**) microstructure of low-carbon steel (steel 44).

2.2. Friction Stir Spot Welding (FSSW)

The difficulty in fusion welding of the dissimilar metallic alloys, especially those are far different in their physical and mechanical properties, comes from the selection of the filler material that can match both alloys and also the high heat input that can severely deteriorate the properties of the welded area. Thus, the use of the solid state welding process can the best alternative to overcome these limitations.

In this study, steel and brass sheets were cut into the dimensions of 100 mm × 200 mm. For the FSSW lap joint, the steel sheet was used as the lower sheet, and the brass sheet was used as the top sheet with an overlap of 40 mm. A fixture plate with rectangular slots of 30 mm × 30 mm slots was used to clamp the sheet and to support and stabilize the sheets during FSSW. Figure 2a shows the arrangement of the FSSW process, and the fixture plate is shown in Figure 2b. FSSW joints were conducted using a home-manufactured FSW machine with main specifications of load of 100 kN, Torque of 140 Nm, and rotation speed up to 3000 rpm existing in the FSW lab at Suez University, Egypt. FSSW tools made from tungsten carbide (WC) with a cylindrical probe (length: 0.9 mm, diameter: 8.2 mm) and shoulder diameter of 20 mm were used. There are many design considerations for manufacturing of FSW and FSSW tools [43–45]. However, the tool material was selected based on the tool material required for welding high-melting-point materials, as the work deals with both steel and copper alloys. Furthermore, a simple cylindrical tapered geometry design was used, as the preparation of complex features in the WC is quite difficult, and it will wear away very quickly.

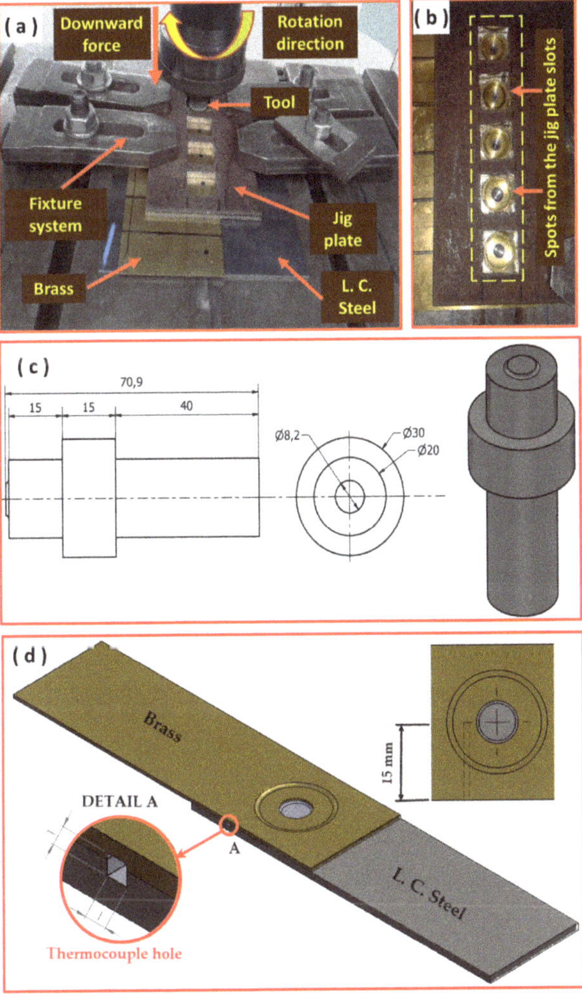

Figure 2. (**a**) FSSW process arrangement, (**b**) fixture plate supporting the sheets of spot joints, (**c**) graphical representation of the used FSSW tool arrangement, and (**d**) temperature measurement during FSSW (all dimensions in mm).

Figure 2.c shows a graphical representation of the applied FSSW tool. The WC tool was fixed in a tool steel holder that made from H13 tool steel. After the welding process, the welded sheets were cut for further characterization.

The FSSW joints were conducted at different rotation speeds of 1000, 1250, and 1500 rpm at a constant plunge rate of 0.1 mm/s, zero tilt angle, and distance controlled mode. The applied vertical load during welding was ranging between 11 to 14 kN. The temperature near to the stir zone was measured for FSSW experiments covering the applied number of rotations (83, 167, 250, 333, 500 R). A Modern Digital Multimeter (MDM) model UT61B, Zhejiang, China with thermocouple type "K" was used to measure the temperature. The thermocouple was inserted between the brass and steel sheets (as shown in Figure 2d), and the temperature values were recorded and plotted as a function of the time.

2.3. *FSSW lap Joints Evaluation*

The starting sheets of copper and steel were cut into 100 mm × 200 mm. Copper was placed over steel with an overlap area of 40 mm × 200 mm. After welding, the whole dimensions of welded pair of materials were 160 mm × 200 mm, and it was cut into individual samples for investigations perpendicular to the length (200 mm); each is 30 mm in width and containing one spot, as shown in Figure 3.

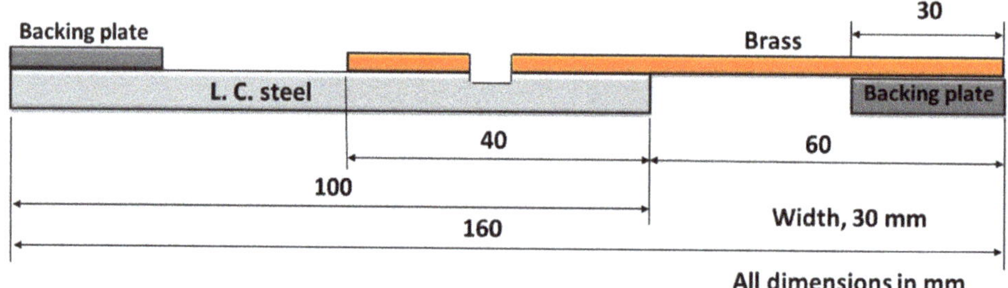

Figure 3. FSSW tensile shear test specimen with two backing plates to ensure axial loading of the specimens during the tensile shear test.

Lap shear tensile test was carried out on the FSSW joints with configuration of the sample, as shown in Figure 3, to measure the load-carrying capacity of the joints. The tensile test samples for the FSSW joints are of the width of 30 mm and a total length of 160 mm, with an overlap length of 40 mm, which are the same dimensions of the original welded plates. To ensure the axial loading of the test specimen, two packing pieces (30 mm × 30 mm) were adhesively joined on the weld specimens, as shown in Figure 3 [46]. A universal tensile testing machine, type Instron model 4210, (Norwood, MA, USA was used at crosshead speed of 0.05 mm/s. The load cells of the FSW and the tensile testing machine are calibrated by a specialized company; the same was performed periodically for the displacement measuring units.

The transverse cross section of the spot joints was investigated through macrographs and hardness measurements on a diagonal line through the joined materials. Low-load Vickers hardness tester HWDV-75, TTS Unlimited, Osaka, Japan was used with an indentation interspacing distance of 2 mm using a test load of 10 N and a dwell time of 15 s.

FEI Quanta FEG 250 Field Emission Gun Scanning Electron Microscope (FEGSEM), FEI company (Hillsboro, OR, USA), equipped with EDAX-OIM7 and controlled by TEAM software was used for microstructure investigation and elemental analysis of the joints. EDS line and point elemental analysis was carried out using a scan step size of 0.5 μm.

3. Results and Discussion

Many FSSW conditions were examined to explore the effect of the different parameters on developing high-quality dissimilar lap joints between brass (CuZn40) and low-carbon steel (St 44-2).

3.1. Heat Input in the FSSW Process

To determine the effect of the different welding parameters, the number of parameters has been reduced by combining the rotational speed (rpm) and the dwell time (s) into the number revolutions per joint as follows:

$$\text{Total number of revolutions (R)} = \text{Tool rotation speed per second} \times \text{dwell time (s)}$$

Now, the rotational speed and the dwell time are reduced into one parameter, namely, the number of revolutions per spot (R). Moreover, the vertical load which represents the downward force (applied pressure) by the tool will be used to calculate the heat input in the welding process. Considering the friction force between the tool (pin surfaces and the pin shoulder) and considering an average friction coefficient (μ) equal to 0.5 [47–49] and using the applied vertical force (F), the rotational speed (ω) and the heat input (Q) are calculated as follows [50]:

$$Q = \frac{13}{12}\,\mu\,\frac{F}{K_A}\,\omega\,r_p \tag{1}$$

where r_p is the tool pin radius which equal to 4.1 mm and K_A is the value of shoulder contact surface area (A_C) divided by the total shoulder surface area (A) [50], which can be calculated as follows:

$$K_A = \frac{A_C}{A} = \frac{\pi\left(r_s^2 - r_p^2\right)}{\pi r_s^2} = \frac{\left(10^2 - 4.1^2\right)}{10^2} = 0.8319 \tag{2}$$

where r_s is the tool shoulder radius and r_p is the tool pin radius.

Basically, the angular speed $\omega(\text{rad/min}) = 2\pi N\ (\text{rpm})$ can be expressed in (rad/s) as follows:

$$\omega(\text{rad/s}) = \frac{2\pi}{60} N(\text{rps}) \tag{3}$$

We can substitute the values of Equations (2) and (3) into Equation (1) and rewrite Equation (1) as follows:

$$Q(\text{J/s}) = \frac{13}{12} * 0.5 * \frac{F}{0.8319}\,\frac{2\pi}{60} * 0.0041(\text{Nm/s}) \tag{4}$$

Simply, heat input in the current case can be calculated as:

$$Q\ (\text{kJ/s}) = 2.794 * 10^{-7}\ \omega\ (\text{rad/s}) * F\ (\text{N}) \tag{5}$$

Equation (5) is valid and used for friction spot welding when multiplied by the linear travel speed. However, in FSSW, there is no travel speed but rotation in the same position for a dwell time, so that multiplying Q with dwell time D_t (in s) gives the total generated heat (E) per spot weld, as described by Equation (6).

$$E = Q * D_t\ \ (\text{kJ}) \tag{6}$$

Figure 4 shows the effect of the total number of revolutions on the heat generated per spot weld. This figure shows the effect of the number of rotations as the key parameter in the heat generation upon FSSW process. Clearly it can be observed that increasing the number of rotations increases the heat input per spot. This will significantly affect the joints quality and microstructure. The heat input ranges from 11 kJ to 1.5 kJ based on the number of revolutions per spot.

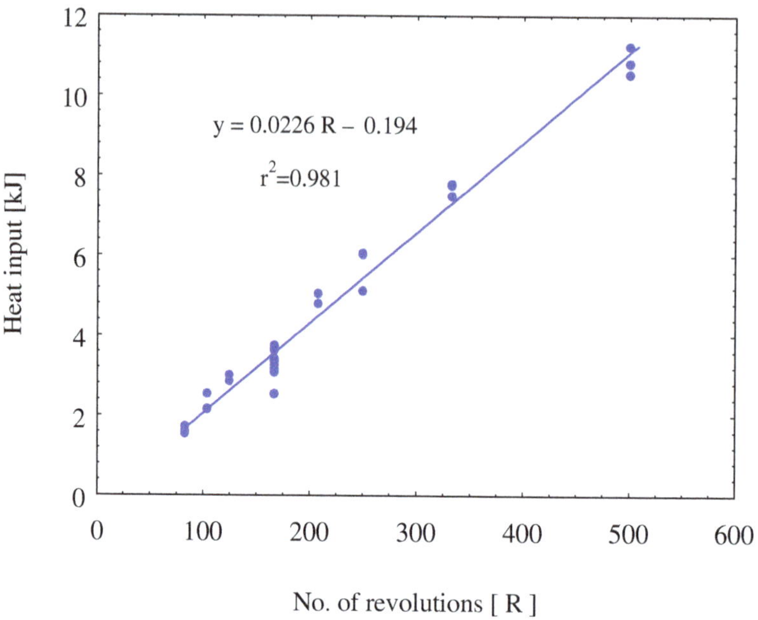

Figure 4. Linear fitting of the total heat input per spot weld as a function of the number of rotations on FSSW of low-carbon steel/brass (CuZn40) joints.

Table 3 gives the calculated heat input at the different FSSW conditions, and Figure 4 shows the heat input against the number of revolutions. In Figure 4, the average total heat input has been well linear fitted with the total number of revolutions (R) per spot weld with a high coefficient of determination (r^2 = 0.981). The linear equation describing the relation of the total generated heat per spot and the number of rotations is:

$$\text{Heat (kJ)} = 0.0226\ \text{R} - 0.194 \quad (7)$$

Temperature was measured during the FSSW process for different experiments covering the applied number of revolutions range (83 to 500R) using different combinations of revolutions per minutes and dwell times. Figure 5 includes the curves of the measured temperature during the FSSW for some experiments. It is clear that the controlling parameter for heat input or temperature rise is not only the number of revolutions per spot but also the speed of the revolution. Practicing 500 revolutions within 20 s using a speed 1500 rpm is more effective than practicing the same number of revolutions in a time of 30 s using 1000 rpm. The first condition (1500 rpm, 20 s) raised the temperature to 583 °C, and a long time was needed to cool again, while a temperature of 535 °C was reached at the second condition (1000 rpm, 30 s). Comparing the experiments carried at the same speed of 1000 rpm for different dwell times of 5, 10, 20, and 30 s, it is obvious that the short dwell time is effective to raise the FSSW temperature, whereas the long dwell time is needed to generate heat at a rate higher than the heat dissipation rate through the different metallic units adjacent to the FSSW nuggets: base materials, jig plate, clamping units, FSSW tool, and the machine table. At 1000 rpm, the maximum attained temperature is 535 °C after 30 s; the maximum temperature falls to 323 °C at the shortest dwell time of 5 s (see Figure 5 and Table 4).

In Table 4, the maximum measured temperatures are listed for some FSSW experiments. Considering the melting of low-carbon steel (St 44-2) as 1539 °C and that for brass (CuZn40) as 900 °C [42], the relative absolute temperatures (T/T_m) were calculated to show how far apart are the different welding temperatures from the recrystallization temperatures of the welded materials. It is obvious that the attained temperature during FSSW lies within the recrystallization temperature range of brass, while it is only at the boundaries or lower than the recrystallization temperature of steel.

Table 3. Welding conditions, the generated heat input, and the tensile lap shear maximum force carried by the joint.

Max. Tensile Load	Energy	Power	No. of Rotations	Machine Load	Dwell Time	Rotation Speed	Joint Numbering
(kN)	(kJ)	(kJ/s)	(R)	(kN)	(s)	(rpm)	
2.7	1.7	0.34	83	11.7	5	1000	1
3	1.6	0.32	83	10.96	5	1000	2
– *	1.6	0.31	83	10.7	5	1000	3
–	1.5	0.31	83	10.5	5	1000	4
–	2.5	0.51	104	13.8	5	1250	5
–	2.1	0.43	104	11.7	5	1250	6
1.3 **	3	0.6	125	13.6	5	1500	7
–	2.9	0.6	125	13	5	1500	8
3.6	3.3	0.33	167	11.4	10	1000	9
3.3	3.7	0.37	167	12.8	10	1000	10
–	3.2	0.32	167	11.1	10	1000	11
6	2.5	0.25	167	8.6	10	1000	12
4.9	3.2	0.32	167	10.7	10	1000	13
–	3.1	0.31	167	10.5	10	1000	14
4	3.4	0.34	167	11.6	10	1000	15
1.5	3.6	0.36	167	12.4	10	1000	16
–	3.7	0.37	167	12.5	10	1000	17
2.4	4.8	0.48	208	13.1	10	1250	18
–	5	0.5	208	13.8	10	1250	19
3.2	5.1	0.51	250	11.6	10	1500	20
5.2	6	0.6	250	13.7	10	1500	21
2	6	0.6	250	13.8	10	1500	22
3.8	NA	NA	250	NA	10	1500	23
1.6	NA	NA	250	NA	10	1500	24
7.5	NA	NA	250	NA	10	1500	25
7.1	NA	NA	250	NA	10	1500	26
6.3	NA	NA	250	NA	10	1500	27
7.5	7.8	0.39	333	1.3	20	1000	28
7.5	7.5	0.37	333	12.8	20	1000	29
–	7.8	0.4	333	13.3	20	1000	30
6.4	11.2	0.37	500	12.8	30	1000	31
5.4	10.8	0.36	500	12.3	30	1000	32
–	10.5	0.35	500	12	30	1000	33

* The joints that show "–" in the "Max. tensile load" have been consumed in investigations other than tensile test such as hardness testing and macro- and microstructure investigation. ** FSSW joints with a maximum load lower than 2 kN have been considered as failed joints.

Table 4. Measured temperature during brass/steel FSSW experiments related to the absolute melting temperature of the base materials.

Rotation Speed, (rpm)	Time (s)	No. of Rotations, (R)	Maximum Temperature, (°C)	T/T_m (Steel)	T/T_m (Brass)
1500	20	500	583	0.47	0.73
1000	30	500	535	0.45	0.69
1000	20	333	472	0.41	0.64
1500	10	250	420	0.38	0.59
1000	10	167	363	0.35	0.54
1000	5	83	323	0.33	0.51

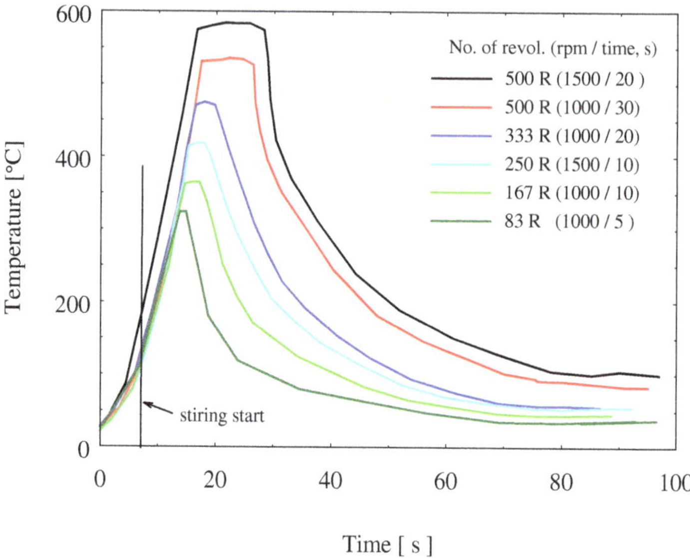

Figure 5. Temperature measured during some FSSW of low-carbon steel/brass joints.

3.2. *Joint Quality and Macroscopic Investigation*

Some samples were macroscopically investigated by a cut through the spot zone. The cut samples were molded to be easily handled during the metallographic preparation process. Due to the stirring occurred by the welding tool pin, a so-called keyhole is generated. Usually, this keyhole reaches the lower sheet of the joint, and the materials of the upper sheet are softened, pressed, and extruded to flow under the tool shoulder. With the applied force from the machine, the tool shoulder continues pressing the material removed by the pin, causing further extrusion of the softened brass to make it flow around the tool shoulder, producing a flash-like extruded brass, as shown in Figure 6.

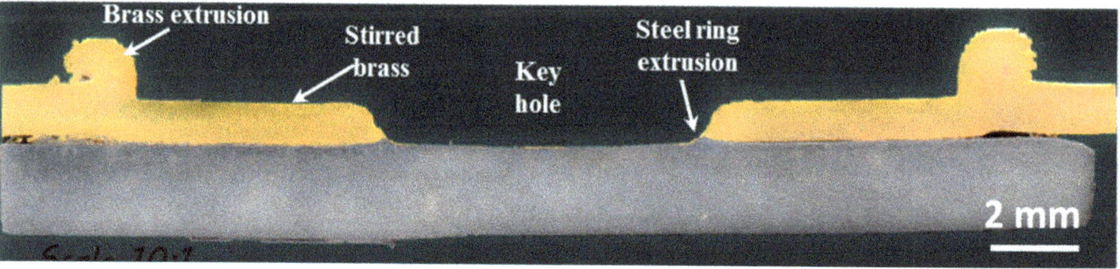

Figure 6. Macro image of an FSSW steel/brass joint showing the different zones using a total of 167 revolutions (rpm = 1000 and dwell time = 10 s).

Depending on the extent of the dwell time and the rotational speed, the tool pin may further press the underlying sheet (steel), causing little extrusion and producing a slight steel ring. Figure 6 also indicates that there is good cohesion between the brass and the undelaying steel sheet at the stirred region beneath the tool shoulder. This cohesion region is generated mainly by the heat generated from the friction process and the pressure practiced by the tool shoulder on the brass.

Figure 7 includes macroscopic images showing a comparison of the FSSW joint form at different welding conditions. Figure 7a shows the joint form at a low number of rotations (83 R) where the keyhole is still not well-formed. With increasing the number of rotations to 167 R, the weld nugget seems uniform around the keyhole (as shown in Figure 7b). Repeating the previous conditions while increasing the nominal downward displacement from 1.4 to 1.8 mm makes the keyhole deeper (Figure 7c) with more steel extrusion to the sides of the keyhole, and the cohesion of brass to steel appears with a lower area around the keyhole. Increasing the number of revolutions per spot to 333 and 500 increases the quality and the area of cohesion of brass to steel, as shown in Figure 7d,e, respectively. This is mainly due to the increased heat input for these two conditions, as shown in Figure 4 and Table 3.

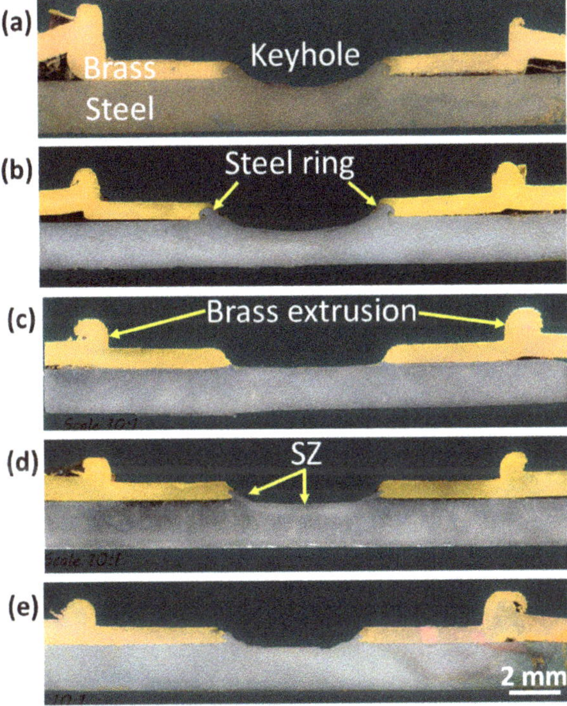

Figure 7. Macro images of FSSW steel/brass joints at different revolutions (R): (**a**) 83 R; (**b**) 167 R (penetration = 1.4 mm); (**c**) 167 R (1.8 mm); (**d**) 333 R; and (**e**) 500 R.

3.3. Tensile Lap Shear Test Results

Tensile lap shear tests (TLST) were conducted on the FSSWed samples to evaluate the load-carrying capacity of the joints. The TLST results are presented in the load-displacement curves as shown in Figure 8. Figure 9 illustrates the fracture surface of lower surface of upper brass sheet and the upper surface of the lower steel sheet after the tensile test of the spot joints given in Figure 8. The lap shear tensile curves of the joints welded with the same rotation speed (1000 rpm) and different dwell times (5, 10, 20, and 30 s), which produced a different number of rotations per spot (from 83 to 500 revolutions), are shown in Figure 8a. This figure shows a variation of the tensile strength of the joint with the variation of the number of revolutions per spot. At the lower number of revolutions (83–167 R; Figure 9a–h), the displacement and the maximum force seem lower than that at the higher number of revolutions (333 and 500; Figure 9i–l) per spot. This is due to the improved joint quality in terms of good cohesion of brass on steel with the effect of increased total heat input per spot. The highest reached maximum tensile force at a rotational speed of 1000 rpm and the total number of rotations of 333 was 7.51 kN.

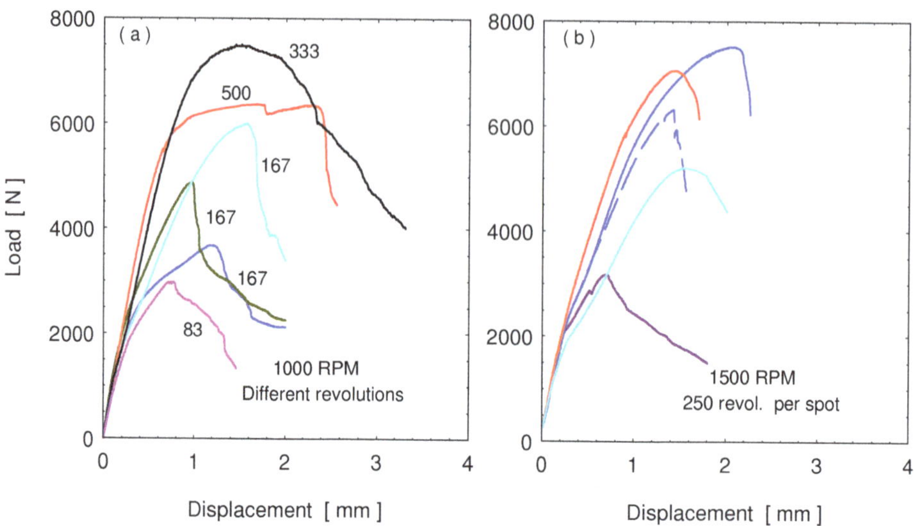

Figure 8. Load-displacement curves of joined samples at rotational speed of (**a**) at 1000 rpm and number of revolutions per spot (83, 167, 333, and 500 R) and (**b**) at 1500 rpm and 250 revolutions per spot.

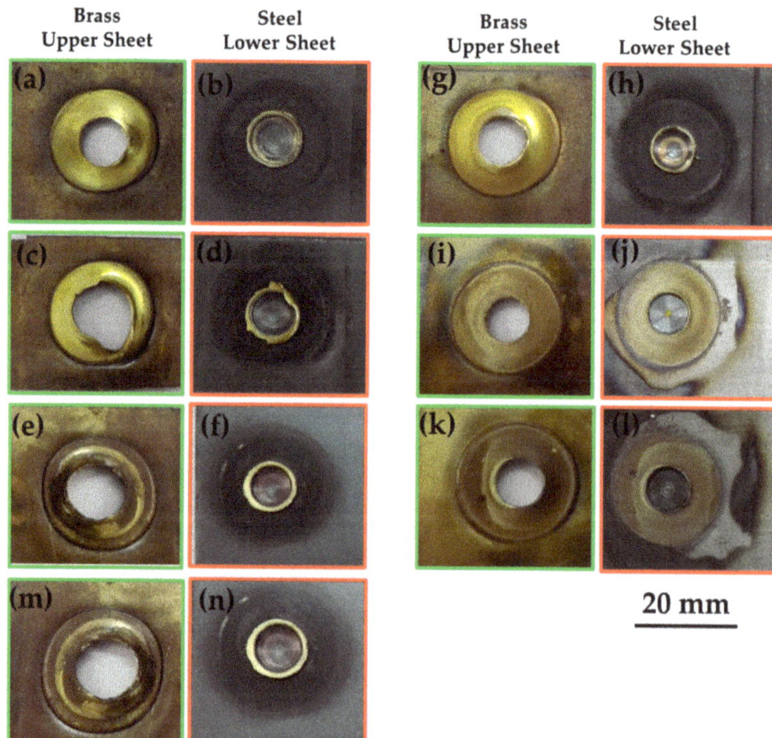

Figure 9. Photo images of the dissimilar fracture surfaces of the lap shear tensile tested FSSW steel/brass joints produced at different numbers of revolutions (R); (**a**,**b**) at 83 R, (**c**–**h**) at 167 R, (**i**,**j**) at 333 R, (**k**,**l**) at 500 R, and (**m**,**n**) at 250 R.

The FSSW of brass/steel welded at 250 R (1500 rpm and 10 s) is not reproducible as detected and represented in Figure 8b, where all displayed curves are for the same welding condition, and the resulting tensile data are very different. However, the maximum tensile force of 7.52 kN was attained at a rotation speed of 1500 rpm, or the total number of rotations per spot of 250. It can also be noticed from Figure 8a,b, that, with increased heat input (at the higher number of revolutions; 250, 333, and 500), the extent of displacement before fracture is higher than other conditions.

Table 3 includes the maximum tensile load for the different tested samples. The average of the maximum load carried by the joint is presented against the number of rotations per spot weld, as shown in Figure 10. It can be indicated that the maximum tensile increases with increasing the number of stirring rotations, which coincides with the highest total heat input per spot (Table 3). It can be noticed that there is a considerable error bar, which is calculated by the standard deviation of the obtained results for the same conditions. This means that the quality described by the maximum carried load of the joints produced by this manufacturing process is still not identical for the same condition, but there is a general trend showing increased load-carrying capacity of the joint with increasing the total generated heat during the process, which can be achieved by increasing the number of rotations per spot at considerable revolution speed in the range from 1000 to 1500 rpm.

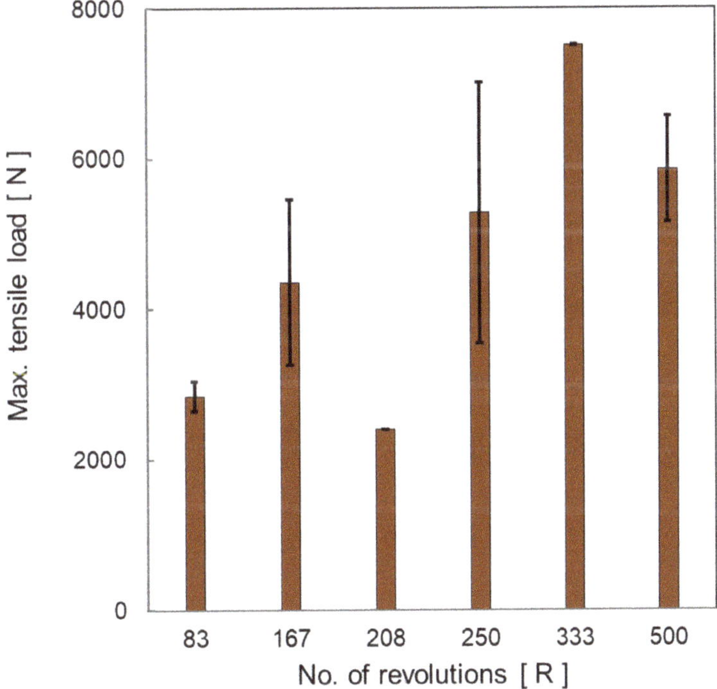

Figure 10. Average maximum tensile load of the tensile lap shear test of FSSW steel/brass joints manufactured by different revolutions per spot weld.

To evaluate the success of such FSSW joints, the tensile shear results are compared with other works. According to the reported results by Abdullah and Hussein [51] for the FSSW of copper with carbon steel at plunge depths of 0.2 and 0.4 mm and different rotation speeds of 1120, 1400, and 1800 rpm, the tensile shear force of the welded joints was increased with an increase of the rotation speed and plunge depth, and the tensile shear force ranged between 510 N and 4.56 kN. The higher load-carrying capacities are attained mostly at a higher plunge depth using 1400 and 1800 rpm. The results from Figner et al. [52] show that FSSW of aluminum alloy AA5754 with galvanized steel using rotational speed ranging between 800 and 3200 rpm, dwell time range of 0 to 0.8 s, and plunge depth of 2.2 mm resulted in a maximum tensile lap shear force of 8.30 kN. However, the base materials of the joints of these results, or at least the softer component (aluminum alloy AA5754), have a higher strength than brass of the current work. The same can be mentioned for the joints [53] of AA5083, which were welded with steel alloy at various FSSW parameters. The shear forces of the joints with a maximum value of 4.02 kN was attained. Moreover, much lower shear force (3.20 kN) has been recorded by Sun et al. [20] for FSSW joints of AA6061-T6 with mild steel, which was achieved at a rotational speed of 700 rpm and a dwell time of 2 s. Mubiayi et al. [54] studied FSSW of pure copper (C11000) with the soft aluminum alloy AA1060; the higher load-carrying capacities of 5.23 kN and 4.84 kN were obtained at rotational speeds of 800 and 1200 rpm, respectively. A plunge depth of 1 mm was used for these successful cases. The base materials used by Mubiayi et al. [54] were softer than the current study materials. Based on the previously mentioned results and conditions, it could be concluded that the current attained results are comparable or even better among the achieved results from FSSW works.

3.4. Hardness Results of FSSW Joints

Hardness measurements were carried out on the FSSW joint to explore the effect of stirring on the development of joint material strength. Figure 11 shows the hardness profiles measured on the joint sheet at different stirring dwell times, which produced different numbers of revolutions per joint.

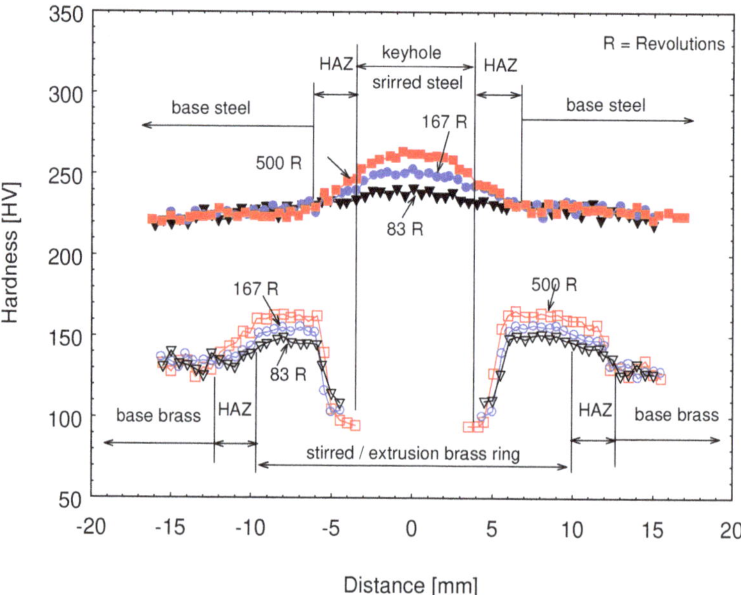

Figure 11. Vickers hardness profile measured on stirred brass and steel at the rotation speed of 1000 rpm for different dwell times, i.e., different revolutions (R) per spot (5 s: 83 R, 10 s: 167 R, and 30 s: 500 R).

FSW or FSSW of materials with melting points very far apart such as steel/brass does not produce completely mixed dissimilar materials at the joint regions. Even if cohesion takes place, no wide range mixing occurs: materials of the top and bottom sheet stay distinguishable. Thus, hardness measurements were conducted on the welding regions of the two joined sheets individually. It seems that a small area of the steel sheet was affected, which is slightly bigger than the tool pin area. Beyond this area, the steel was not affected, and the hardness values of steel were scattered within a band around the base steel sheet hardness value (223 HV). At longer dwell time and higher number of rotations (500 R), the heat input was at a maximum, which provided a better condition for stirring the steel beneath the tool pin, leading to a clear increase of the hardness values reaching its maximum (262 HV), as shown in Figure 11 for steel in the middle zone. At the lower number of rotations (167 R), there was a little increase in the hardness value under the keyhole, reaching 248 HV, while a slight increase of the hardness above the hardness value of the base steel sheet was recorded a value of 238 HV on stirring for 83 R.

Brass containing 40 Zn has a melting temperature of around 900 °C. This means a temperature above 315 °C can lead to recrystallization of cold-worked brass. However, a much higher temperature (583 °C) was recorded on the FSSW of the brass/steel sheet (see Figure 5 and Table 4). Along with this high temperature, we also must keep in mind the mechanical stirring pressing and extrusion of brass, which is performed by the tool pin and shoulder. This gave us two oppositely working effects. The hardness measurements reflect the dominant effect on the strength of brass. Starting from the base brass sheet, the hardness increased with increasing the number of revolutions per spot. The highest hardness value of 163 HV was reached using 500 R at the stirred zone, while brass hardness using a number of rotations 83 R per spot showed a hardness value of about 149 HV. Both are located within the hardness ranges of the highest strength temper conditions ($\geq$140 HV). At the highest temperature zone near the keyhole, brass began to recrystallize and became softer than the base cold-worked bass sheet. The lowest hardness value around 92 HV was recorded for the spot-welded test using 500 R, where the highest temperature was reached at the keyhole.

3.5. Microstructure, SEM, and EDS Analysis

Figure 12 shows the microstructure of the right side of the cross-section through the keyhole of the brass/steel joint produced by 333 rotations. Figure 12a shows good cohesion of brass to steel by the pressing force practiced by the tool shoulder at the generated heat and temperature, which could be around 472 °C (as indicated in Table 4). Frictional stirring of steel sheet by the bottom surface of the tool pin extruded some of the steel into brass. Penetration of the extruded steel into brass could increase the joint strength. The extruded steel appears in the form of a ring, as shown in the macro images shown in Figure 7, appears in the form of double rings, as shown in Figure 12a,b shows higher magnification of the region around the extruded steel where the brass is highly deformed and highly heat-affected by stirring and extrusion of both steel and brass.

Microstructure investigation of the left side of the cross-section of the joint carried out at 1000 rpm and dwell time of 20 s (333 R) is shown in Figure 13. The general features explained in the right side of the joint in Figure 12 can also be seen in the microstructure of the left side of the joint shown in Figure 13, except the fracture of the steel chip extruded in the brass sheet. The upper steel extrusion chip seems thinner and especially longer than the lower one. This thinner chip faced resistance to flow in brass by the pressure from the tool shoulder; thus, this thin chip is bent and fractured. Near the keyhole, the temperature is at maximum due to the friction of the tool and the steel sheet. This high temperature facilitates the flow of steel extrusion into softened brass. Away from the keyhole, the heat sink increases, and the temperature steeply falls. This can be an additional resistance to the flow of the steel chip into brass. Although the fracture of steel extrusion can be considered as a discontinuity in the joint, it does not represent a welding defect as long as the area of cohesion of brass to steel is large enough. From the tensile lap shear test results (Figure 10 and Table 3), this welding condition (1000 rpm, 20 s, 333 R) produced a joint with the highest load-carrying capacity, where the average maximum load was 7505 N.

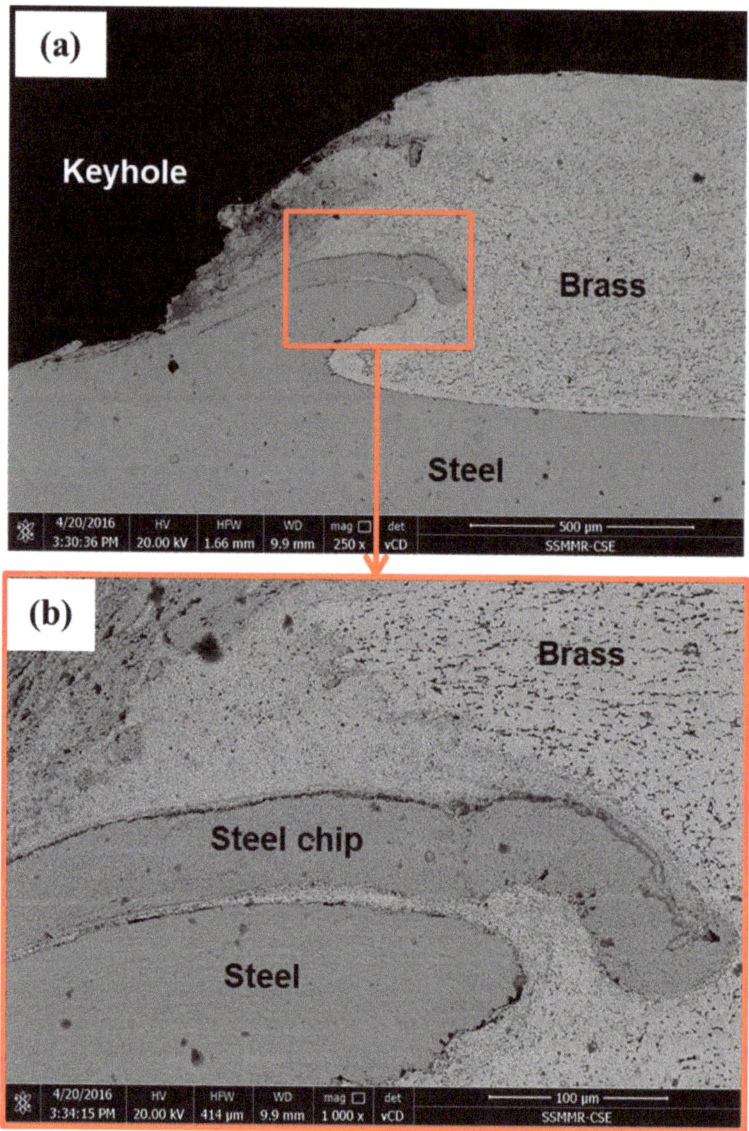

Figure 12. BSE SEM microstructure, right side of a cross-section through the keyhole of the joint produced by FSSW at 1000 rpm, dwell time of 20 s (333 R) (**a**) extrusion and flow of stirred steel into brass, and (**b**) higher magnification of the extruded steel region in an image (**a**).

EDS line scan was carried out on the sample welded at 1000 rpm for a dwell time of 20 s at the region containing steel and brass; see Figure 14. This image is a small area from the image shown in Figure 12. Normally, the analysis of steel is rich in iron, and the brass region is rich in copper and zinc. Let us concentrate on the transition length between steel and brass. There could be mutual atomic diffusion between steel and brass, which can be shown from light concentration gradient on the sound steel/brass interface. In friction stir welding of the brass/steel spot joint using a tool rotation of 1000 rpm and linear travel speeds of 250, 500, and 600 mm/min, Gao et al. [31,55] found that there is a mutual diffusion between iron and brass (Cu and Zn) over a length ranging between 70 and 120 nm. In the present work, the dwell time of 20 s using the same rotational speed (1000 rpm) offers a higher chance to introduce higher heat input in a localized spot. This increases the opportunity for mutual diffusion than that attained in FSW of brass/steel conducted by Gao et al. [31,55].

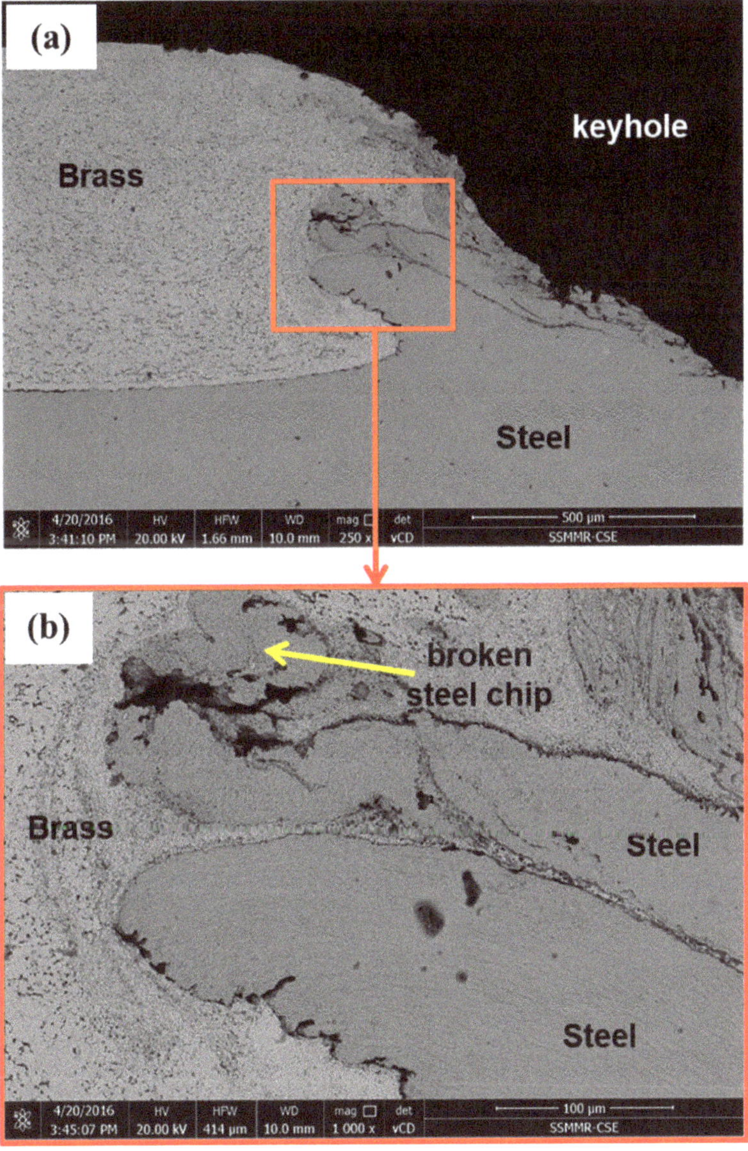

Figure 13. BSE SEM microstructure images, left side of a cross-section through the keyhole of the joint produced by FSSW at 1000 rpm, dwell time of 20 s (333 R). (**a**) Flow of steel extrusion into brass, and (**b**) higher magnification of the extruded steel in an image (**a**).

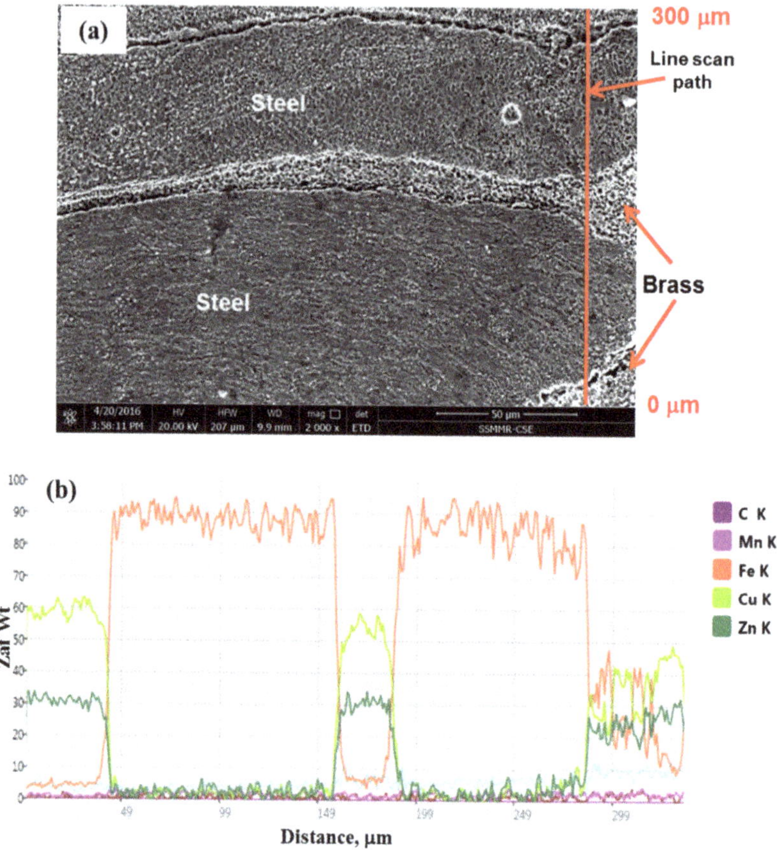

Figure 14. EDS analysis across the joint containing steel extrusion and brass for a region located on the right side of the cross-section through the FSSW joint by stirring at 1000 rpm, dwell time of 20 s (333 R). (**a**) EDS line path on SE SEM image, and (**b**) curves of the EDS elemental analysis.

Figure 15 includes the microstructure of a region from the left side to the keyhole in a cross-section of the brass/steel spot joint produced at 1500 rpm for a dwell time of 10 s, which provides a total of 250 revolutions. Figure 15a shows the steel extrusion in the form of a hock inside brass. Excellent cohesion at the interface between brass and steel can be seen in Figure 15b. In further magnification, microstructural changes of brass can be seen, depending on the degree of deformation and heat input to the joint materials (Figure 15c,d).

Figure 16 shows the EDS line scan analysis of the brass/steel contacting region from the microstructure shown in Figure 15c. It can be noted that at the interface region between the brass and steel parts some traces of elements can be seen in both sides, which indicate the mutual diffusion of elements. Fe peak can be observed in the brass side as well as peaks of Cu and Zn in the steel side. This type of diffusion is expected to occur at this high temperature and high strain experienced during FSSW. Bonding of dissimilar metals during the different joining processes is explained on by many mechanisms [56–58]. In the current work, the attained results implied from Figures 12–16 indicate that the main dominant bonding mechanism could be mechanical bonding and diffusion bonding on short scale at brass/steel interface.

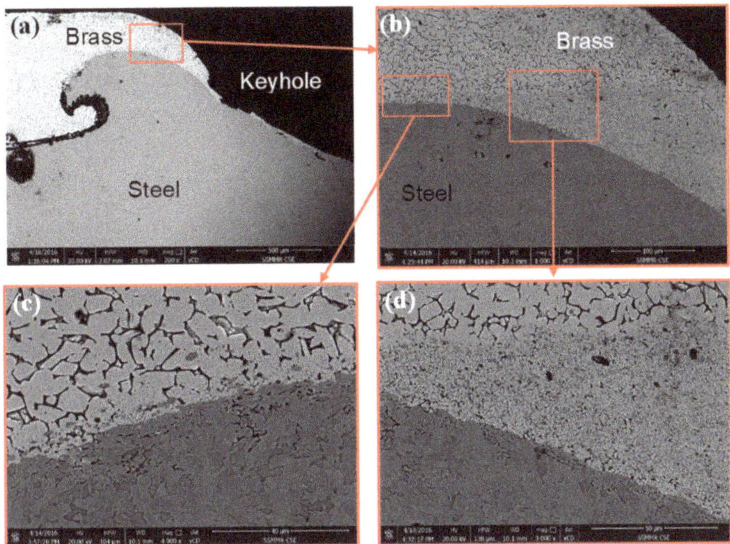

Figure 15. BSE SEM microstructure at different magnifications (**a**) 200×, (**b**) 1000×, (**c**) 4000×, (**d**) 3000× for the FSSW joints showing mechanical mixing and cohesion of brass on steel at the stirred of samples joined at a rotational speed of 1500 rpm, dwell time of 10 s (250R).

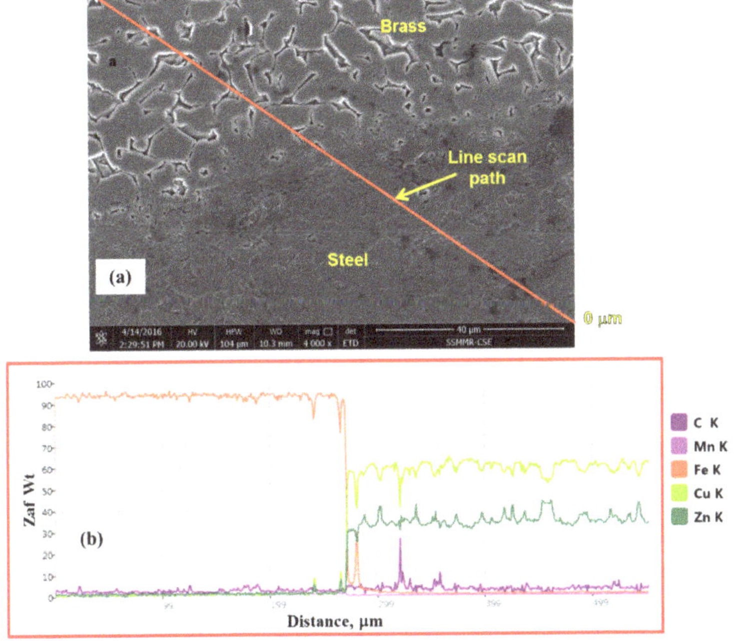

Figure 16. (**a**) EDS line scan from the lower right side to the upper left side of the SE SEM image of FSSW brass/steel joint by stirring at 1500 rpm and dwell time of 10 s (250 R), and (**b**) EDS line scan results.

Table 5 includes the results of EDS analysis of different spots in the recrystallized brass, as shown in Figure 17. The elemental analysis shows that spots 1 and 2 are richer in zinc than spot 3 allocated in the matrix, representing α-brass phase, which has a zinc content of 32 wt.%, while the zinc content of spots 1 and 2 are 42.5 and 40.93 wt.%, respectively. The values of carbon that appeared in this analysis are mainly from contamination from the molding and sticking materials of the sample.

Table 5. EDS elemental analysis of spots 1 and 2 (β-phase) and spot 3 (α-phase).

Element	(wt.%) Spot 1	(wt.%) Spot 2	(wt.%) Spot 3
Fe	0.6	1.6	0.8
Cu	54.1	53.2	63.3
Zn	42.5	40.9	32.1

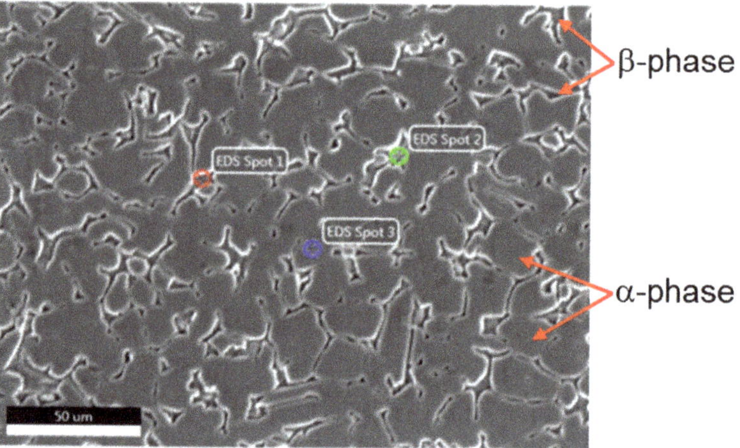

Figure 17. SE SEM microstructure, appearance of β-phase in the recrystallized structure of brass in the HAZ of samples joined at a rotational speed of 1500 rpm, dwell time of 10 s (250 R).

In the next and final section, further investigation of the effect of FSSW conditions on the microstructure of brass are presented. Because brass represents the softer and weaker component of the joint, it determines the join's strength; thus, that understanding its microstructural changes is of great importance. Figure 17 shows the microstructure of brass of the joint welded at 250 R (Figure 15) near the highest temperature region at the keyhole. It can be noticed that there are islands of β-phase distributed on discontinuous grain boundaries, while the matrix is mainly α-phase. This α-phase is dominating in brasses containing zinc content lower than 37 wt.%, according to the Cu–Zn phase diagram [42]. The β-phase appears in the as-received brass (Figure 1a) in the form of mechanical twins. At the regions adjacent to the high temperature, as shown in Figure 15, brass is recrystallized, and no more mechanical twins can be noticed, as shown in Figure 17.

3.6. *General Discussion*

The main contribution of this study is the investigation of the dissimilar spot joints between steel alloy and copper base alloy and the use of new parameters to justify the joint heat input and quality, which is the tool revolutions per spot as one parameter encompasses both the tool rotation rate and dwell time. In this section, some normalized points from the applied process conditions and the results to be applied for other FSSW are presented. In this work, three tool rotation speeds (1000, 1250, and 1500 rpm) and four dwell times (5, 10, 20, and 30 s), which form 12 process conditions, were reduced to seven rotations per spot (83, 104, 125, 167 250, 333, 500 R). The studied materials have melting points very far apart: brass (CuZn40) having a melting point of 900 °C and the relatively harder material represented by low-carbon steel with a much higher melting point (1530 °C). The normalized measured temperature at the applied number of revolutions (Table 4) show that the welding conditions resulted in generating heat, which raised the temperature up to $\approx 0.4\ T_m$ for such steel and $\approx 0.6\ T_m$ for such soft brass, which could be suitable conditions for such material joining. Furthermore, the generated heat at a number of revolutions of 250 or higher (Table 3 and Figure 4) was also a reasonable level to produce a dissimilar joint with good load-carrying capacity of 4 to 7.5 kN, as shown in Figure 10. Whenever such relative temperature is reached, a successful FSSW joint can be achieved for similar combinations of materials with very different melting points. Even the rotational speed and the dwell times were normalized into the number of revolutions per spot, keeping in mind that the rotational speed and the dwell time should be high enough to provide enough higher rate of heat generation. If other FSSW tools are used, the generated heat input by the new tool alone would be estimated using the given equations to see if the heat input value is enough to place the selected new tool geometry or the new process conditions (rotational speed and dwell time) in the heat input range to produce acceptable load-carrying capacity. The new tool geometry (such as pin height) should, of course, be related to thickness of the top sheet materials thickness; here, it was 0.1 mm shorter than the top sheet materials thickness. Heat input calculation for the FSSW of steel–brass was found to be linearly proportional with the number of revolutions per spot joint. Maximum heat input obtained was 11 KJ at the number of revolutions of 500. The temperature measurement during FSSW showed agreement with the heat input dependence on the number of revolutions. However, at the same number of revolutions of 500, it was found that the higher rotation speed of 1500 rpm resulted in higher temperature of 583 °C compared to 535 °C at a rotation speed of 1000 rpm. This implies the significant effect for the rotation speed in the increase of temperature. The macro investigations of the friction stir spot-welded joints transverse sections showed sound joints at the different investigated parameters with significant joint ligament between the steel and brass. The tensile shear load results showed wide scattering at the same value of number of revolution, with maximum tensile shear load of about 7.52 kN attained at a rotation speed of 1500 rpm and 10 s, or the total number of rotations per spot of 250.

Tensile test results are usually scattered in a narrow range. The scattering of TLST results could not be compared with the scattering of the fusion welded or even with the FSW samples. The variation of the TSLT results shown in Figure 10 could be related to variation in behavior of the stirred at the keyhole and the form of extrusion of steel chips and brass around the keyhole. Figures 12–16 show the different forms of cohesion at the steel–brass interface, which results in different load-carrying capacity of the samples. However, sticking to the conditions producing a high heat input range (i.e., at 250 revolutions per spot or higher) produces lap joints with relatively high load-carrying capacity.

The hardness measurement across the transverse section of the spot joint showed an increase in the carbon steel (lower sheet) with a maximum of 248 HV and an increase of brass hardness at the mixed interface between brass and steel and significant reduction in the stir zone. Microstructural investigation of the joint zone showed mechanical mixing between steel and brass, with the steel extruded from the lower sheet into the upper brass sheet. The EDS analysis across the interface showed a concentration gradient for both Fe into brass and Cu into steel, which implies the mutual diffusion at the experienced thermomechanical conditions. The current study showed the importance of parameters optimization using the different optimization software to minimize the number of experiments. Thus, the results of this study can used in an optimization study using Taguchi method, for example. Furthermore, the refill FSSW technology can be used to improve the joint appearance and quality.

4. Conclusions

Based on the obtained results and analysis, the following conclusions can be outlined:

(1) Whenever a generated FSSW heat over 4 kN is secured by any parameters combination, successful joints can be achieved.
(2) Heat input for the FSSW of steel–brass was found to be linearly proportional with the number of revolutions per spot joint, and the maximum heat input obtained is 11 kJ at the number of revolutions of 500.
(3) Application of the acceptable range of FSSW revolution per spot heats the joint zone to a normalized temperature of $\approx$0.4 T_m for steel and $\approx$0.6 T_m for brass and produces good steel/brass adhesion, which is strengthened by the mutual atomic diffusion.
(4) Considering other successful FSSW conditions, an appropriate tool geometry with a height of 0.1 mm less than the top sheet thickness produces good joint load-carrying capacity.
(5) FSSW of steel/brass joints with a number of revolutions ranging between 250 to 500 revolutions per spot at appropriate tool speed range (1000–1500 rpm) produces joints with high load-carrying capacity from 4 kN to 7.5 kN.

Author Contributions: Conceptualization, S.A., M.M.E.-S.S., M.I.A.H. and M.M.Z.A.; Data curation, M.M.Z.A. and A.M.S.; Formal analysis, M.M.Z.A., M.M.E.-S.S., S.A. and F.H.L.; Funding acquisition, K.H. and S.A.; Investigation, M.M.E.-S.S., Y.G.Y.E. and M.I.A.H.; Methodology, M.M.E.-S.S., Y.G.Y.E. and S.A.; Project administration, K.H. and S.A.; Resources, M.M.E.-S.S., K.H. and F.H.L.; Software, K.H., A.M.S. and F.H.L.; Supervision, S.A.; M.M.E.-S.S. and M.I.A.H.; Validation, M.M.Z.A., K.H. and S.A.; Visualization, K.H., S.A. and F.H.L.; Writing—original draft, S.A. and F.H.L., Writing—review and editing, M.M.Z.A., M.I.A.H., A.M.S. and M.M.E.-S.S. All authors have read and agreed to the published version of the manuscript.

Funding: Funded by the Deanship of Scientific Research at Imam Mohammad Ibn Saud Islamic University, KSA (Research Group No. RG-21-12-04).

Institutional Review Board Statement: Not applicable.

Informed Consent Statement: Not applicable.

Data Availability Statement: The data presented in this study are available on request from the corresponding author. The data are not publicly available due to the extremely large size.

Acknowledgments: The authors appreciate the Deanship of Scientific Research at Imam Mohammad Ibn Saud Islamic University for funding this work through Research Group No. RG-21-12-04.

Conflicts of Interest: The authors declare no conflict of interest.

References

1. Venu, B.; BhavyaSwathi, I.; Raju, L.S.; Santhanam, G. A review on Friction Stir Welding of various metals and its variables. *Mater. Today Proc.* **2019**, *18*, 298–302. [CrossRef]
2. Padhy, G.K.; Wu, C.S.; Gao, S. Friction stir based welding and processing technologies—processes, parameters, microstructures and applications: A review. *J. Mater. Sci. Technol.* **2018**, *34*, 1–38. [CrossRef]
3. De Carvalho, W.S.; Vioreanu, M.C.; Lutz, M.R.A.; Cipriano, G.P.; Amancio-Filho, S.T. The influence of tool wear on the mechanical performance of AA6061-T6 refill friction stir spot welds. *Materials* **2021**, *14*, 7252. [CrossRef] [PubMed]
4. Andrade, D.G.; Sabari, S.; Leitão, C.; Rodrigues, D.M. Shoulder related temperature thresholds in fssw of aluminium alloys. *Materials* **2021**, *14*, 4375. [CrossRef]
5. Kumamoto, K.; Kosaka, T.; Kobayashi, T.; Shohji, I.; Kamakoshi, Y. Microstructure and fatigue behaviors of dissimilar a6061/galvannealed steel joints fabricated by friction stir spot welding. *Materials* **2021**, *14*, 3877. [CrossRef]
6. Kubit, A.; Trzepieciński, T.; Gadalińska, E.; Slota, J.; Bochnowski, W. Investigation into the effect of rfssw parameters on tensile shear fracture load of 7075-t6 alclad aluminium alloy joints. *Materials* **2021**, *14*, 3397. [CrossRef]
7. Zlatanovic, D.L.; Balos, S.; Bergmann, J.P.; Rasche, S.; Pecanac, M.; Goel, S. Influence of tool geometry and process parameters on the properties of friction stir spot welded multiple (Aa 5754 h111) aluminium sheets. *Materials* **2021**, *14*, 1157. [CrossRef]
8. Li, M.; Zhang, C.; Wang, D.; Zhou, L.; Wellmann, D.; Tian, Y. Friction stir spot welding of aluminum and copper: A review. *Materials* **2020**, *13*, 156. [CrossRef]
9. Steel, A.; Joints, K.F. Effect of Loading Methods on the Fatigue Properties. *Materials* **2020**, *13*, 4247.

10. Ahmed, M.M.Z.; Abdul-maksoud, M.A.A.; El-sayed Seleman, M.M.; Mohamed, A.M.A. Effect of dwelling time and plunge depth on the joint properties of the dissimilar friction stir spot welded aluminum and steel. *J. Eng. Res. Res.* **2021**, *9*, 1–20. [CrossRef]
11. Tonelli, L.; Morri, A.; Toschi, S.; Shaaban, M.; Ammar, H.R.; Ahmed, M.M.Z.; Ramadan, R.M. Effect of FSP parameters and tool geometry on microstructure, hardness, and wear properties of AA7075 with and without reinforcing B 4 C ceramic particles. *Int. J. Adv. Manuf. Technol.* **2019**, *102*, 3945–3961. [CrossRef]
12. Hamada, A.S.; Järvenpää, A.; Ahmed, M.M.Z.; Jaskari, M.; Wynne, B.P.; Porter, D.A.; Karjalainen, L.P. The microstructural evolution of friction stir welded AA6082-T6 aluminum alloy during cyclic deformation. *Mater. Sci. Eng. A* **2015**, *642*, 366–376. [CrossRef]
13. Ahmed, M.M.Z.; El-Sayed Seleman, M.M.; Zidan, Z.A.; Ramadan, R.M.; Ataya, S.; Alsaleh, N.A. Microstructure and mechanical properties of dissimilar friction stir welded AA2024-T4/AA7075-T6 T-butt joints. *Metals* **2021**, *11*, 128. [CrossRef]
14. Ahmed, M.M.Z.; Ataya, S.; El-Sayed Seleman, M.M.; Mahdy, A.M.A.; Alsaleh, N.A.; Ahmed, E. Heat input and mechanical properties investigation of friction stir welded aa5083/aa5754 and aa5083/aa7020. *Metals* **2021**, *11*, 68. [CrossRef]
15. Shen, Z.; Ding, Y.; Gerlich, A.P. Advances in friction stir spot welding. *Crit. Rev. Solid State Mater. Sci.* **2020**, *45*, 457–534. [CrossRef]
16. Ahmed, M.M.Z.; Abdelazem, K.A.; El-Sayed Seleman, M.M.; Alzahrani, B.; Touileb, K.; Jouini, N.; El-Batanony, I.G.; Abd El-Aziz, H.M. Friction stir welding of 2205 duplex stainless steel: Feasibility of butt joint groove filling in comparison to gas tungsten arc welding. *Materials* **2021**, *14*, 4597. [CrossRef]
17. Bozkurt, Y.; Salman, S.; Çam, G. Effect of welding parameters on lap shear tensile properties of dissimilar friction stir spot welded aa 5754-h22/2024-t3 joints. *Sci. Technol. Weld. Join.* **2013**, *18*, 337–345. [CrossRef]
18. Chowdhury, S.H.; Chen, D.L.; Bhole, S.D.; Cao, X.; Wanjara, P. Lap shear strength and fatigue behavior of friction stir spot welded dissimilar magnesium-to-aluminum joints with adhesive. *Mater. Sci. Eng. A* **2013**, *562*, 53–60. [CrossRef]
19. Zhou, L.; Li, G.H.; Zhang, R.X.; Zhou, W.L.; He, W.X.; Huang, Y.X.; Song, X.G. Microstructure evolution and mechanical properties of friction stir spot welded dissimilar aluminum-copper joint. *J. Alloys Compd.* **2019**, *775*, 372–382. [CrossRef]
20. Sun, Y.F.; Fujii, H.; Takaki, N.; Okitsu, Y. Microstructure and mechanical properties of dissimilar Al alloy/steel joints prepared by a flat spot friction stir welding technique. *Mater. Des.* **2013**, *47*, 350–357. [CrossRef]
21. Ahmed, M.M.Z.; El-Sayed Seleman, M.M.; Shazly, M.; Attallah, M.M.; Ahmed, E. Microstructural Development and Mechanical Properties of Friction Stir Welded Ferritic Stainless Steel AISI 409. *J. Mater. Eng. Perform.* **2019**, *28*, 6391–6406. [CrossRef]
22. Chen, K.; Liu, X.; Ni, J. Keyhole refilled friction stir spot welding of aluminum alloy to advanced high strength steel. *J. Mater. Process. Technol.* **2017**, *249*, 452–462. [CrossRef]
23. Das, H.; Mondal, M.; Hong, S.T.; Lim, Y.; Lee, K.J. Comparison of microstructural and mechanical properties of friction stir spot welded ultra-high strength dual phase and complex phase steels. *Mater. Charact.* **2018**, *139*, 428–436. [CrossRef]
24. Kuang, S.; Kang, Y.-L.; Yu, H.; Liu, R.D. Effect of continuous annealing parameters on the mechanical properties and microstructures of a cold rolled dual phase steel. *Int. J. Miner. Metall. Mater.* **2009**, *16*, 159–164. [CrossRef]
25. Mazar Atabaki, M.; Ma, J.; Liu, W.; Kovacevic, R. Hybrid laser/arc welding of advanced high strength steel to aluminum alloy by using structural transition insert. *Mater. Des.* **2015**, *75*, 120–135. [CrossRef]
26. Sarkar, R.; Sengupta, S.; Pal, T.K.; Shome, M. Microstructure and Mechanical Properties of Friction Stir Spot-Welded IF/DP Dissimilar Steel Joints. *Metall. Mater. Trans. A Phys. Metall. Mater. Sci.* **2015**, *46*, 5182–5200. [CrossRef]
27. Shen, J.J.; Liu, H.J.; Cui, F. Effect of welding speed on microstructure and mechanical properties of friction stir welded copper. *Mater. Des.* **2010**, *31*, 3937–3942. [CrossRef]
28. Emami, S.; Saeid, T. Effects of welding and rotational speeds on the microstructure and hardness of friction stir welded single-phase brass. *Acta Metall. Sin. (Eng. Lett.)* **2015**, *28*, 766–771. [CrossRef]
29. Lee, W.B.; Jung, S.B. The joint properties of copper by friction stir welding. *Mater. Lett.* **2004**, *58*, 1041–1046. [CrossRef]
30. Meran, C. The joint properties of brass plates by friction stir welding. *Mater. Des.* **2006**, *27*, 719–726. [CrossRef]
31. Gao, Y.; Nakata, K.; Nagatsuka, K.; Shibata, Y.; Amano, M. Optimizing tool diameter for friction stir welded brass/steel lap joint. *J. Mater. Process. Technol.* **2016**, *229*, 313–321. [CrossRef]
32. Panagopoulos, C.N.; Georgiou, E.P.; Simeonidis, K. Lubricated wear behavior of leaded αβ brass. *Tribol. Int.* **2012**, *50*, 1–5. [CrossRef]
33. Luo, J.; Xiang, J.; Liu, D.; Li, F.; Xue, K. Radial friction welding interface between brass and high carbon steel. *J. Mater. Process. Technol.* **2012**, *212*, 385–392. [CrossRef]
34. Kimura, M.; Kusaka, M.; Kaizu, K.; Fuji, A. Effect of post-weld heat treatment on joint properties of friction welded joint between brass and low carbon steel. *Sci. Technol. Weld. Join.* **2010**, *15*, 590–596. [CrossRef]
35. Hofmann, U.; El-Magd, E. Behaviour of Cu-Zn alloys in high speed shear tests and in chip formation processes. *Mater. Sci. Eng. A* **2005**, *395*, 129–140. [CrossRef]
36. Bakhtiari Argesi, F.; Shamsipur, A.; Mirsalehi, S.E. Preparation of bimetallic nano-composite by dissimilar friction stir welding of copper to aluminum alloy. *Trans. Nonferrous Met. Soc. China* **2021**, *31*, 1363–1380. [CrossRef]
37. Ilman, M.N. Kusmono Microstructure and mechanical properties of friction stir spot welded AA5052-H112 aluminum alloy. *Heliyon* **2021**, *7*, e06009. [CrossRef]
38. Yapici, G.G.; Ibrahim, I.J. On the fatigue and fracture behavior of keyhole-free friction stir spot welded joints in an aluminum alloy. *J. Mater. Res. Technol.* **2021**, *11*, 40–49. [CrossRef]

39. Feng, X.S.; Li, S.B.; Tang, L.N.; Wang, H.M. Refill Friction Stir Spot Welding of Similar and Dissimilar Alloys: A Review. *Acta Metall. Sin. (Eng. Lett.)* **2020**, *33*, 30–42. [CrossRef]
40. Davis, J.R. (Ed.) *Annual Book of ASTM Standards*; American Society for Testing and Materials: Materials Park, OH, USA, 2001; ISBN 9780803135079.
41. *ASTM E8/E8M-08*; Standard Test Method for Tension Testing of Metallic Materials. ASTM International: West Conshohohocken, PA, USA, 2008; 1–25.
42. Davis, J.R. *ASM Speciality Hand Book: Copper and Copper Alloys*; ASM International: Materials Park, OH, USA, 2001; ISBN 0871707268.
43. Janeczek, A.; Tomków, J.; Fydrych, D. The influence of tool shape and process parameters on the mechanical properties of aw-3004 aluminium alloy friction stir welded joints. *Materials* **2021**, *14*, 3244. [CrossRef]
44. Chupradit, S.; Bokov, D.O.; Suksatan, W.; Landowski, M.; Fydrych, D.; Abdullah, M.E.; Derazkola, H.A. Pin angle thermal effects on friction stir welding of AA5058 aluminum alloy: CFD simulation and experimental validation. *Materials* **2021**, *14*, 7565. [CrossRef]
45. Saini, N.; Pandey, C.; Thapliyal, S.; Dwivedi, D.K. Mechanical Properties and Wear Behavior of Zn and MoS2 Reinforced Surface Composite Al- Si Alloys Using Friction Stir Processing. *Silicon* **2018**, *10*, 1979–1990. [CrossRef]
46. Hwang, J.Y.; Monteiro, S.N.; Bai, C.G.; Carpenter, J.; Cai, M.; Firrao, D.; Kim, B.G. *Characterization of Minerals, Metals, and Materials*; Wiley: Hoboken, NJ, USA, 2012; ISBN 9781118291221.
47. Ahmed, M.M.Z.; Habba, M.I.A.; Jouini, N.; Alzahrani, B.; El-Sayed Seleman, M.M.; El-Nikhaily, A. Bobbin tool friction stir welding of aluminum using different tool pin geometries: Mathematical models for the heat generation. *Metals* **2021**, *11*, 438. [CrossRef]
48. Nandan, R.; Roy, G.G.; Lienert, T.J.; Debroy, T. Three-dimensional heat and material flow during friction stir welding of mild steel. *Acta Mater.* **2007**, *55*, 883–895. [CrossRef]
49. Peel, M.J.; Steuwer, A.; Withers, P.J.; Dickerson, T.; Shi, Q.; Shercliff, H. Dissimilar Friction Stir Welds in AA5083-AA6082. Part I : Process Parameter Effects on Thermal History and Weld Properties. *Metall. Mater. Trans. A* **2006**, *37*, 2183–2193. [CrossRef]
50. Atak, A.; Şık, A.; Özdemir, V. Thermo-Mechanical Modeling of Friction Stir Spot Welding and Numerical Solution With The Finite Element Method. *Int. J. Eng. Appl. Sci.* **2018**, *5*, 70–75.
51. Abdullah, I.T.; Hussein, S.K. Improving the joint strength of the friction stir spot welding of carbon steel and copper using the design of experiments method. *Multidiscip. Model. Mater. Struct.* **2018**, *14*, 908–922. [CrossRef]
52. Figner, G.; Vallant, R.; Weinberger, T.; Schröttner, H.; Pašič, H.; Enzinger, N. Friction Stir Spot Welds between aluminium and steel automotive sheets: Influence of welding parameters on mechanical properties and microstructure. *Weld. World* **2009**, *53*, 13–23. [CrossRef]
53. Fereiduni, E.; Movahedi, M.; Kokabi, A.H. Aluminum/steel joints made by an alternative friction stir spot welding process. *J. Mater. Process. Technol.* **2015**, *224*, 1–10. [CrossRef]
54. Mubiayi, M.P.; Akinlabi, E.T. Evolving properties of friction stir spot welds between AA1060 and commercially pure copper C11000. *Trans. Nonferrous Met. Soc. China* **2016**, *26*, 1852–1862. [CrossRef]
55. Gao, Y.; Nakata, K.; Nagatsuka, K.; Matsuyama, T.; Shibata, Y.; Amano, M. Microstructures and mechanical properties of friction stir welded brass/steel dissimilar lap joints at various welding speeds. *Mater. Des.* **2016**, *90*, 1018–1025. [CrossRef]
56. Bhanu, V.; Fydrych, D.; Gupta, A.; Pandey, C. Study on microstructure and mechanical properties of laser welded dissimilar joint of p91 steel and incoloy 800ht nickel alloy. *Materials* **2021**, *14*, 5876. [CrossRef]
57. Batistão, B.F.; Bergmann, L.A.; Gargarella, P.; de Alcântara, N.G.; dos Santos, J.F.; Klusemann, B. Characterization of dissimilar friction stir welded lap joints of AA5083 and GL D36 steel. *J. Mater. Res. Technol.* **2020**, *9*, 15132–15142. [CrossRef]
58. Campo, K.N.; Campanelli, L.C.; Bergmann, L.; dos Santos, J.F.; Bolfarini, C. Microstructure and interface characterization of dissimilar friction stir welded lap joints between Ti–6Al–4V and AISI 304. *Mater. Des.* **2014**, *56*, 139–145. [CrossRef]

 *materials*

Article

Friction Stir Welding of 2205 Duplex Stainless Steel: Feasibility of Butt Joint Groove Filling in Comparison to Gas Tungsten Arc Welding

Mohamed M. Z. Ahmed [1,2,*], Khaled A. Abdelazem [3,4], Mohamed M. El-Sayed Seleman [2], Bandar Alzahrani [1], Kamel Touileb [1], Nabil Jouini [1,5], Ismail G. El-Batanony [4] and Hussein M. Abd El-Aziz [6]

1 Mechanical Engineering Department, College of Engineering at Al Kharj, Prince Sattam Bin Abdulaziz University, Al Kharj 16273, Saudi Arabia; ba.alzahrani@psau.edu.sa (B.A.); k.touileb@psau.edu.sa (K.T.); n.jouini@psau.edu.sa (N.J.)
2 Department of Metallurgical and Materials Engineering, Faculty of Petroleum and Mining Engineering, Suez University, Suez 43512, Egypt; Mohamed.elnagar@suezuniv.edu.eg
3 PETROJET Company, Cairo 11835, Egypt; kabdelazem@yahoo.com
4 Mechanical Engineering Department, Faculty of Engineering, Al-Azhar University, Cairo 11651, Egypt; ismailghazy2017@azhar.edu.eg
5 Laboratoire de Mécanique, Matériaux et Procédés (LR99ES05), École Nationale Supérieure d'Ingénieurs de Tunis, Université de Tunis, Tunis 1008, Tunisia
6 Mining and Petroleum Engineering Department, Faculty of Engineering, Al-Azhar University, Cairo 11651, Egypt; dr_eng_hussein@yahoo.com
* Correspondence: moh.ahmed@psau.edu.sa; Tel.: +966-115888273

Citation: Ahmed, M.M.Z.; Abdelazem, K.A.; El-Sayed Seleman, M.M.; Alzahrani, B.; Touileb, K.; Jouini, N.; El-Batanony, I.G.; Abd El-Aziz, H.M. Friction Stir Welding of 2205 Duplex Stainless Steel: Feasibility of Butt Joint Groove Filling in Comparison to Gas Tungsten Arc Welding. *Materials* **2021**, *14*, 4597. https://doi.org/10.3390/ma14164597

Academic Editor: Jerzy Winczek

Received: 12 July 2021
Accepted: 10 August 2021
Published: 16 August 2021

Abstract: This work investigates the feasibility of using friction stir welding (FSW) process as a groove filling welding technique to weld duplex stainless steel (DSS) that is extensively used by petroleum service companies and marine industries. For the FSW experiments, three different groove geometries without root gap were designed and machined in a DSS plates 6.5 mm thick. FSW were carried out to produce butt-joints at a constant tool rotation rate of 300 rpm, traverse welding speed of 25 mm/min, and tilt angle of 3° using tungsten carbide (WC) tool. For comparison, the same DSS plates were welded using gas tungsten arc welding (GTAW). The produced joints were evaluated and characterized using radiographic inspection, optical microscopy, and hardness and tensile testing. Electron back scattering diffraction (EBSD) was used to examine the grain structure and phases before and after FSW. The initial results indicate that FSW were used successfully to weld DSS joints with different groove designs with defect-free joints produced using the 60° V-shape groove with a 2 mm root face without root gap. This friction stir welded (FSWed) joint was further investigated and compared with the GTAW joint. The FSWed joint microstructure mainly consists of α and γ with significant grain refining; the GTWA weld contains different austenitic-phase (γ) morphologies such as grain boundary austenite (GBA), intragranular austenite precipitates (IGA), and Widmanstätten austenite (WA) besides the ferrite phase (α) in the weld zone (WZ) due to the used high heat input and 2209 filler rod. The yield strength, ultimate tensile strength, and elongation of the FSWed joint are enhanced over the GTAW weldment by 21%, 41%, and 66% and over the BM by 65%, 33%, and 54%, respectively. EBSD investigation showed a significant grain refining after FSW with grain size average of 1.88 μm for austenite and 2.2 μm for ferrite.

Keywords: groove joint design; friction stir welding; gas tungsten arc welding; 2205 DSS; mechanical properties

1. Introduction

Compared to ferritic and austenitic stainless steels, duplex stainless steel (DSS) has excellent strength and corrosion resistance in heavy-duty engineering applications due to a dual-phase structure (nearly 50% ferritic (α) phase and 50% austenitic (γ) phase) [1–5].

This unique microstructure promotes DSS advantages over each austenitic and ferritic stainless steel. It was reported that [4] stress corrosion cracking and the elastic limit of austenitic stainless steel improves with the presence of the α-phase. Moreover, the presence of the γ-phase leads to toughness enhancement of the duplex stainless steels by hindering grain growth of the α-phase [5]. Thus, DSS is recommended extensively for use in many applications, such as chemical and petrochemical industries, marine industry, water desalination pipes, and liquefied natural gas pipes, besides being used in many mining applications [3,5].

Consequently, DSS is considered a proper choice in many engineering applications instead of ferritic and austenitic stainless steels [6]. Among duplex stainless steels, 2205 DSS is regarded as the most common type of this class of steels. This grade is widely used in oil and gas industries, seawater components, and power generation applications [4,6]. The joint efficiency of the 2205 DSS weldments is mainly structure-sensitive and depends on the type of welding method. The 2205 DSS can be welded using high arc energy fusion welding methods such as gas metal arc welding (GMAW) [7], gas tungsten arc welding (GTAW) [8,9], and shielded metal arc welding (SMAW) [2]. The GTAW process promotes efficient and clean weldment compared with other fusion welding processes [10]. It was reported that the weld zone (WZ) of GTAW welds may show an acceptable γ/α ratio by using a special filler rod. However, the weld microstructure features are not similar to its base metal microstructure [11], and the productivity is low, as the maximum thickness that can be welded in single pass cannot exceed 3 mm [2,12]. The GTAW process using nickel-enriched filler rod ER 2209 provides an acceptable α/γ ratio in the weldments [13]. It was reported that the chemical composition of the filler rod plays a significant role in the γ-phase reformation than the applied cooling rate [14].

FSW, as a solid-state joining technique, has become increasingly prominent in the joining and welding of duplex stainless steels (DSS) [15]. In 2205 DSS, optimum properties are obtained when the material has equal proportions of austenite and ferrite [16]. This desired phase ratio changed during fusion welding processes and promoted more ferrite formation in the weld zone as a result of remelting and solidification of materials [17]. In addition, it is recommended to weld DSSs within the range of 0.2–1.5 kJ/mm to avoid precipitation of brittle intermetallic phases [18]. On the other hand, solid-state welding techniques such as FSW seem to be a suitable technique in this regard [19]. This process minimizes the ferritization problem of DSSs during thermal welding cycles due to their solid-state nature and reduces the common problems associated with fusion welding. During the FSW process, the microstructure of the base material experiences both high temperature and severe deformation caused by the stirring tool. The coexistence of these two factors activates the occurrence of some softening mechanisms throughout the material, which results in a kind of fine microstructure commonly found during the hot deformation processes [20].

Friction stir welding (FSW) of soft materials such as aluminum and magnesium alloys became a common activity of welding due to the ability of the tool steel material to weld such materials at minimum cost and risk of failure [21,22]. However, the risk of failure and cost are still very high in case of tool materials used for FSW of hard metals, such as steel [23], titanium and nickel alloys [24]. This is mainly due to the high temperature and severe stress experienced during FSW [25]. Tool materials for FSW of high softening temperature alloys must exhibit excellent properties at temperatures in excess of 900 °C [25]. There are various types of materials used in FSW of high softening temperature materials, such as WC-based materials [26], W–Re [27], and PCBN materials [28]. However, the FSW tool is subjected to severe stress and high temperatures, particularly for the welding of hard alloys such as steels and titanium alloys and the commercial application of FSW to these alloys is now limited by the high cost and short life of FSW tools [25]. However, due to high temperatures and pressures required in the manufacturing of PCBN, the tool costs are very high [25]. Owing to its low fracture toughness, PCBN also has a tendency to fail during the initial plunge stage [25]. Thus, the WC material represents a cost-effective option for

FSW of hard metals [26] and hard composites [29,30] if the tool life can be extended. An important parameter that can be considered to extend tool life is the geometry of the tool pin and the shoulder [31]. In addition, reducing the forces experienced by the tool, either upon plunging and or during FSW, can improve the WC tool life. In this respect, having grooves along the joint line will significantly reduce these forces. In addition, the idea of groove filling using FSW is almost new and innovative, in respect to filling without using a filler rod. The advantage of FSW over other welding techniques, such as laser beam welding [32] as a welding technique without the need for filler rods, is the lower heat input. Thus, this work aims to introduce FSW as a groove filling joining technique to weld DSS in petroleum services companies instead of the GTAW fusion technique.

2. Materials and Methods

2.1. Material

SAF 2205 DSS plates of 6.5 mm in thickness were used as the base material. The workpieces for welding were cut from the as-received plates for both GTAW and FSW processes, 200 mm in length and 100 mm in width. The chemical composition of the base material DSS is listed in Table 1. The filler rod used in the GTAW process is AWS A5.9 ER2209, which contains 3.68% more nickel than 2205 DSS. The increased nickel content in the filler rod stabilizes the austenite phase in the weldments. The chemical composition of the filler rod is given in Table 2. The chemical compositions of the base material and the filler rod are obtained from the datasheet provided by the supplier.

Table 1. Chemical composition of the 2205 DSS base material (in wt.%).

Cr	Ni	Mo	Mn	Si	N	C	P	S	Fe
22.43	5.74	3.15	1.51	0.38	0.17	0.015	0.025	0.001	Bal.

Table 2. Chemical composition of the E2209 filler rod (in wt.%).

Cr	Ni	Mo	Mn	Si	N	C	P	S	Cu	Fe
22.10	9.42	3.03	0.76	0.86	0.14	0.022	0.015	0.016	0.05	Bal.

2.2. Welding Procedure

For the FSW, three different groove geometries without root gap were designed and machined, as shown in Figure 1: (a) a 60° U-shaped groove with a 2 mm root face, (b) a 60° V-shape groove without a root face, (c) a 60° V-shape groove with a 2 mm root face, and (d) for the GTAW 60° V-shape groove with a root face and root gap. Before welding, the samples were mechanically cleaned using a stainless-steel wire brush to remove surface oxides and contaminations.

The FSW tool was designed with a tapered pin geometry. The shoulder diameters, pin tapered angle, pin tip, and pin length were selected as 20 mm, 30°, 5 mm, and 5.5 mm, respectively. Due to the high frictional stress and heating during FSW of DSS; tool, shoulder, and pin were fabricated from tungsten carbide (WC) [1]. Figure 2a shows the drawings of the WC tool.

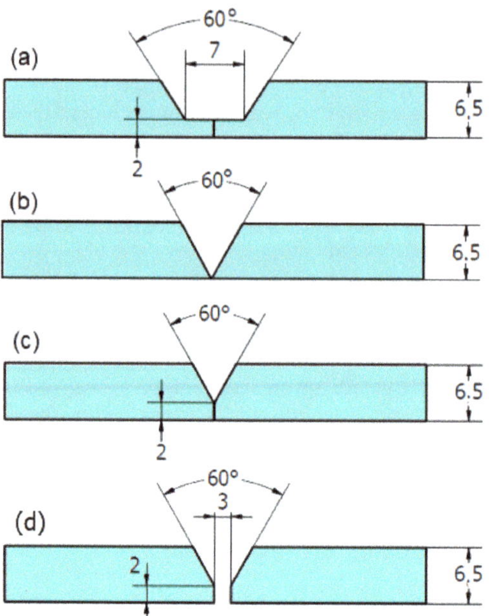

Figure 1. Groove joint designs (**a**–**c**) for the FSW process and (**d**) for the GTAW process [33].

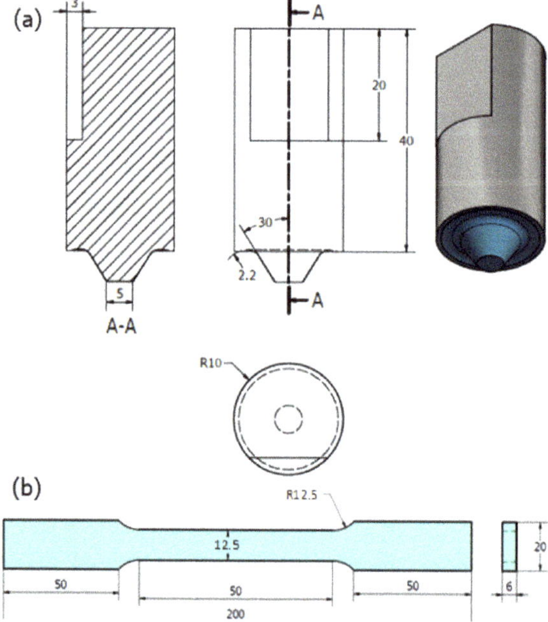

Figure 2. Schematic of (**a**) drawings of the WC FSW tool and (**b**) the tensile test sample (dimensions in mm).

Based on preliminary FSW experimental trials and also the available data in the literature [34], the applied optimum FSW welding parameters were a rotational speed of 300 rpm, a travel speed of 25 mm/min, and a tilt angle of 3°. The FSW experiments were conducted using the FSW machine model (EG-FSW-M1) [35]. For comparison between FSW and the fusion welding technique, gas tungsten arc welding (GTAW) was chosen and applied to weld DSS 2205 in the butt joint. Among fusion welding techniques, GTAW is used extensively to weld duplex materials in the petroleum and petrochemical industries. The used geometry joint design for the GTAW process was a 60° V-shape groove angle with a 2 mm root face and root gap of 2 mm; Figure 1d. The joint was prepared according to a joint design within the approved welding procedure specifications.

The GTAW process was performed using the DCEN-GTAW machine with tungsten electrode 2.4 mm type EWTh-2, arc length 2.4 mm, and argon flow rate 15 l/min. The GTAW process was carried out using the prequalified welding procedure specifications (PQWPS) with welding parameters, as shown in Table 3.

Table 3. GTAW welding parameters.

Pass	Filler rod	Ampere (A)	Volt (V)	Travel Speed (mm/min)	Heat Input (kJ/mm)
1st	ER2209	116	10	90	0.733
2nd		159	13	90	1.378
3rd		155	13.5	81	1.550
Cap1		150	14.5	76	1.717
Cap2		116	10	135	0.516

2.3. *Joints Evaluation and Characteriztion*

Radiographic inspection test (RT) was carried out according to ASME V and ASME VIII on the welded joints using a Gamma-ray camera (Model 880 MAN-027, NSW, Australia) with an Iridium-192 source using D7 radiographic films. Microstructural analysis of the welds was performed using optical microscope (Olympus model: BX41M-LED, Tokyo, Japan). The transverse cross-sections were prepared through basic metallographic steps according to ASTM E-3 and ASTM E-2014 up to 2400 grit and then mechanically polished using alumina suspension of 0.05 μm. This was followed by cleaning with acetone, and drying and electrochemical etching using 20 g KOH in 100 ml H_2O solution. The phase composition in the stir and fusion zones of the FSWed and GTAWed butt joints were examined by X-ray diffraction (XRD, Siemens Incorporating, Munich, Germany) analysis using Siemens-D5000 with Cu Kα radiation (wavelength λ = 0.15406 nm) at 40 kV and 30 mA in the 2θ-range 20–110° and step size of 0.03°. The ferrite number of base material (BM), heat affected zone (HAZ), and weld zone (WZ) was evaluated for the joints welded by FSW and GTAW using a ferrite–scope (FERITSCOPE MP30—Fischer Company, Worcestershire, UK).

The Vickers hardness tester model (Qness model Q10 M, Golling, Austria) was used to obtain the hardness values across the transverse cross-section at a spacing of 1 mm with a load of 98 N and 15 s holding time. In order to evaluate the welding joint efficiency, tensile test specimens were cut perpendicular to the welding direction according to ASTM E8. Figure 2b shows the standard tensile test sample dimensions that were used as a reference in the preparation of the joints tensile samples; the tensile test sample thickness was varied according to the thickness of the weld zone i.e., the thickness at the weld zone was used as the tensile sample thickness in each case, which means that the extra thickness material above that has been machined. The test was carried out at room temperature at a ram head-speed of 0.5 mm/min using a universal test machine (Instron 4208, 300 kN capacity, Norwood, MA, USA).

3. Results and Discussions

3.1. Mechanism of FSW with Groove Filling

It has been known since the invention of the FSW process that this solid state welding process cannot be used for welding with the need for filling of grooves. In this work, a feasibility study for the use of FSW in the welding of butt joints with grooves was carried out. Figure 3 shows the FSW tool plunging into the groove from the front in (a) and from the back in (b). It can be noted that the existence of the groove facilitates the plunging stage and is expected to reduce tool wear and extend tool life. After complete plunging, as shown in Figure 3c, the material starts to flow around the tool and a pole is formed around the tool that becomes stable with only little flow in front of the tool that does not affect the joint behind the tool. It can be observed that this pole becomes stable with the travers of the tool as can be seen from the existing hole pictures shown in Figure 4 for a U-shaped groove with root (a), and a V-shaped groove without root (b). This joint configuration is expected to facilitate tool plunging, reduce tool resistance forces, and reduce the heat input. This will directly enhance the tool life and joint properties. Figure 5a shows the WC FSW tool with the holder arrangement after eight trials of 2205 DSS welding with some trace of material sticking on the WC tool. In addition, the enlarged image of the WC tool in Figure 5b shows the stability of the tool features, which indicate that no wear took place under the condition of having the grooves and only oxidation can be noted on the shoulder outer surface due to the high thermal cycles experienced. The stability of the WC after repeated plunging and welding for eight times can be attributed to the existence of the groove, which reduces the tool resistance forces.

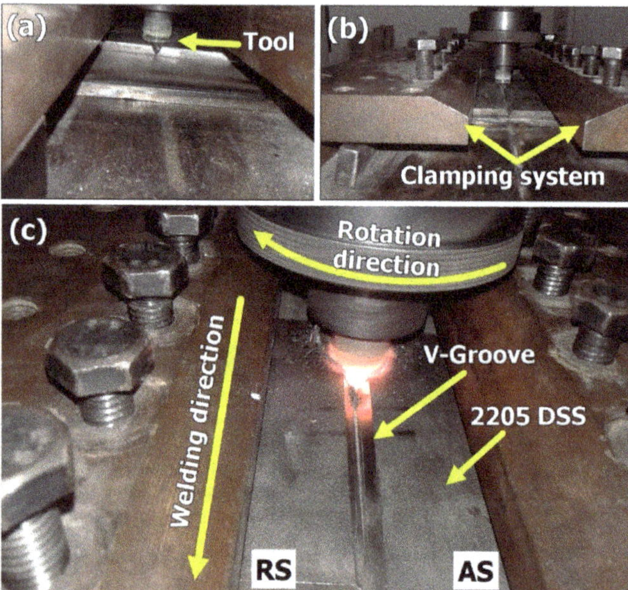

Figure 3. Photographs taken during FSW of 2205 DSS with groove filling. (**a**) Front view during FSW tool plunge, (**b**) back view during FSW tool plunge and (**c**) front view during FSW.

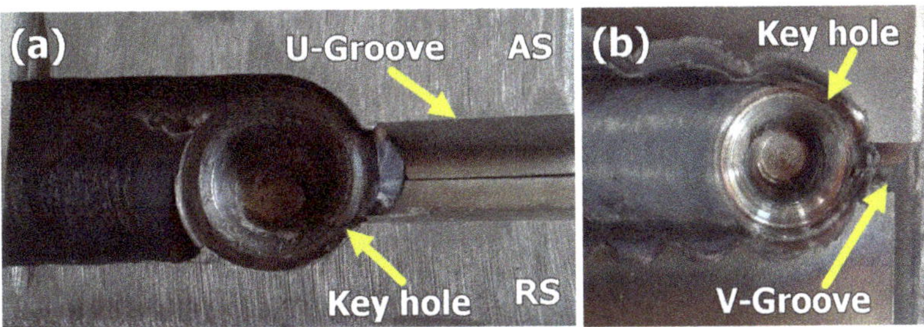

Figure 4. Key hole after FSW of 2205 DSS (**a**) U shape groove with root. (**b**) V shape groove without root.

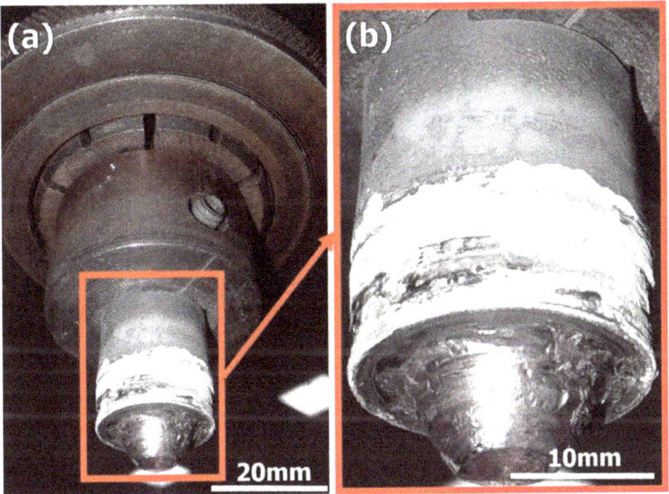

Figure 5. (**a**) WC FSW tool with the holder arrangement after eight trials of duplex stainless steel welding with the WC tool enlarged in (**b**).

3.2. Visual and Radiographic Inspection

Figure 6 shows the top view of the FSWed 2205 DSS butt joints produced with self-filling of different groove designs. It can be observed that in all cases, the grooves filled successfully, with some defects noted in the U- and V-shaped grooves without a root face mainly due to the large volume of the grooves that require large volumes of material to be soundly filled. For example, the U-shaped joint produces a good surface appearance with extensive flash. Moreover, around 2 mm reduction in weld pass thickness was detected, compared to the base material, to achieve filling of the groove; Figure 6a. This loss in thickness is due to the lack of material flow and the large volume of the joint groove design (Figure 1a). The produced weldment of the V-shaped groove design without root reveals a partially good appearance with less flash. Some surface voids along the joint welding length at the retreating side are seen; Figure 6b. Moreover, around 1 mm reduction in weld pass thickness was measured, compared to the DSS BM. These top surface features are also related to the lack of material filling and flow as a result of the large volume of the groove-shaped design (Figure 1b). The obtained joint using the V-shape with root face design showed a good surface appearance, with little flash and minimum reduction in thickness. For comparison purposes, the same lap joint of 2205 DSS was produced using the GTAW method with a V-shaped groove design with root and 3 mm root gap (Figure 1d). Figure 6d shows the top view of the GTAW butt weld of 2205 DSS. It can be considered to be of acceptable appearance without surface defects. The radiographic inspection reveals internal defects for the butt joints produced using U-shape and V-shape without root face designs, as shown in Figure 7a,b, respectively. Furthermore, nearly sound 2205 DSS butt joints were obtained for both FSW butt joints using a V-shaped groove design with root and GTAW butt joint, as given in Figure 7c,d, respectively.

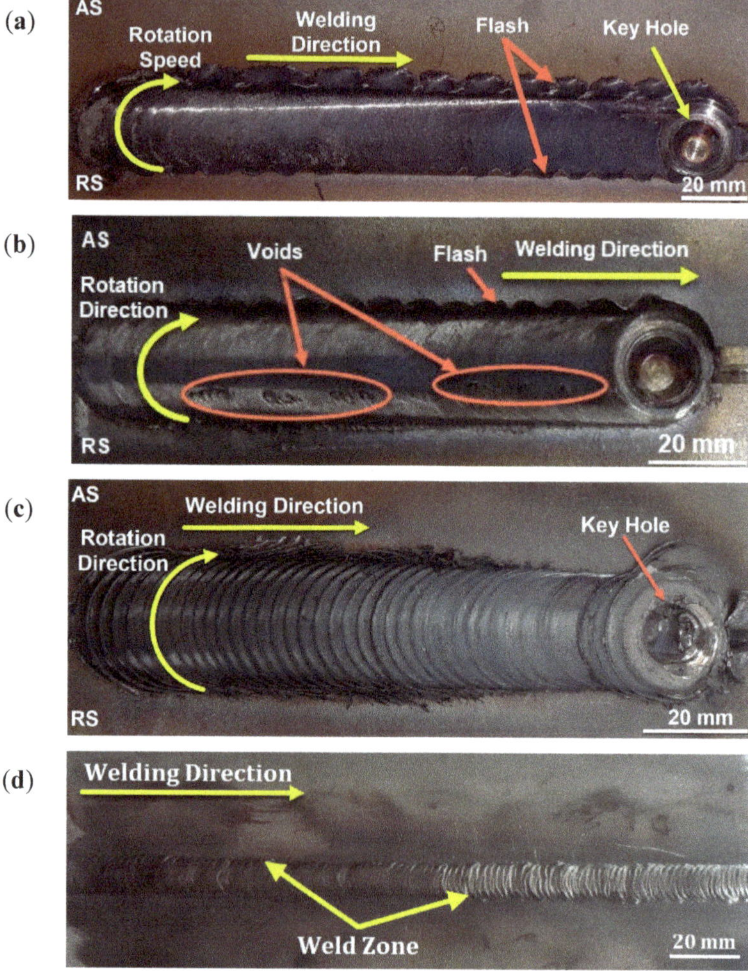

Figure 6. Top view of FSW welds (**a**) FSW, U-shaped groove with root; (**b**) FSW, V-shaped groove without root; (**c**) FSW, V-shaped groove with root face; and (**d**) GTAW, V-shape groove with root face and gap GTAW weld for 2205 DSS butt joints [33].

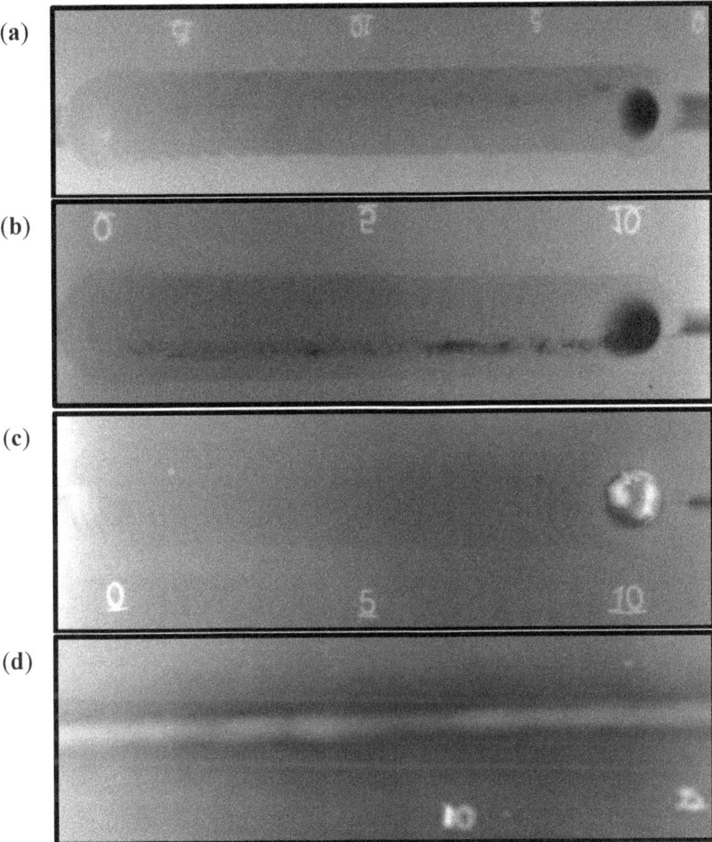

Figure 7. Radiographic films of FSW welds: (**a**) FSW, U-shaped groove with root; (**b**) FSW, V-shaped groove without root; (**c**) FSW, V-shaped groove with root; and (**d**) GTAW, V-shaped groove with root face and root gap GTAW weld for 2205 DSS [33].

3.3. Macrostructure and Microstructure

Figure 8a–c shows the transverse cross-section macrographs of the FSWed butt joints produced using different groove designs. Figure 8d represents the cross-section macrograph of the GTAWed butt joint. For the FSWed joints, three distinguished zones—stir zone (SZ), thermomechanically affected zone (TMAZ), and heat-affected zone (HAZ)—are observed [35]. Furthermore, the interface between the SZ and the TMAZ is very distinguished on the advancing side, while it is more scattered on the retreating side, which is due to the alignment of the rotation direction, transverse direction at this side, and also the more refined grains at the advancing side than that at the retreating side [36]. There are tunnel defects and 2 mm thickness loss of the 2205 DSS butt joint using U-shaped groove design without root, as shown in Figure 8a. These defects are due to the large groove volume of the joint. In addition, there are internal defects observed (Figure 8b) for the butt joint welded using the V-shaped groove without root; these defects are tunnel defects. Moreover, the reduction in thickness for this joint is around half reduction thickness for the joint welded using the U-shaped groove design with root. These defects may also be ascribed to improper groove joint design. For the welded joint using a V-shaped groove with root, it was noted that the cross-section is defect-free and a very slight reduction in thickness was detected. Based on visual inspection, radiographic examination, and macrograph investigation, it can be concluded that a 60° V-shaped groove with 2 mm root face without root gap is a proper joint design for the FSW filling groove process to produce a sound joint. In comparison, a defect-free joint was obtained with a V-shaped groove with 2 mm root face and 3 mm root gap produced using GTAW, as shown Figure 8d. Moreover, the two distinguished zones, WZ and HAZ, can be detected as fusion welding features [37].

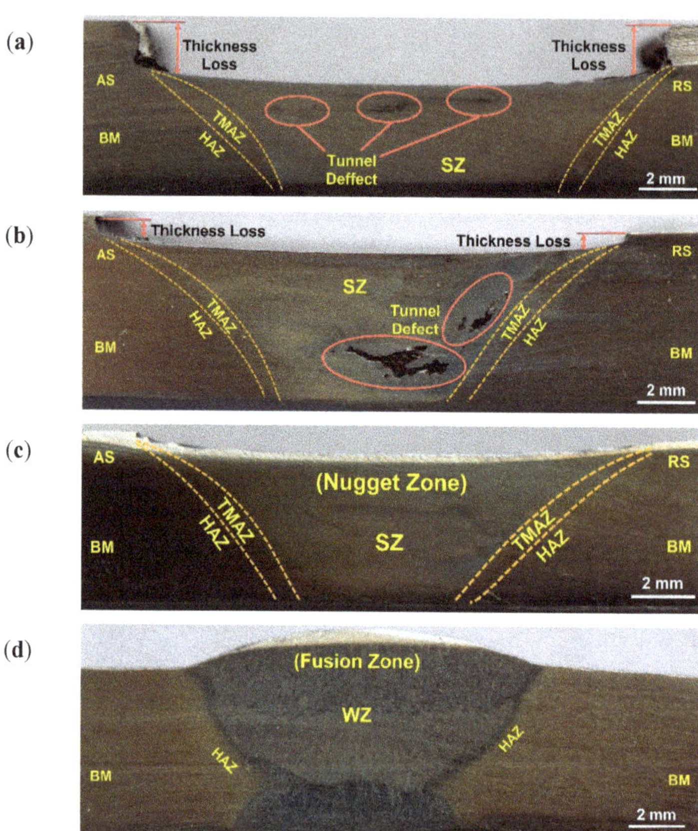

Figure 8. Macrostructure of FSW welds: (**a**) FSW, U-shaped groove with root; (**b**) FSW, V-shaped groove without root; (**c**) FSW, V-shaped groove with root face; and (**d**) GTAW, V-shaped groove with root face and root gap, GTAW weld for 2205 DSS butt joints.

Figure 9 shows the microstructural features at the different zones of the sound FSWed joint produced using the V groove, 60° groove angle, 2 mm root face joint design, (a) base metal (b) TMAZ in the advancing side (AS), (c) SZ, and (d) TMAZ in retreating side (RS). A typical lamellar microstructure of 2205 DSS consists of γ-phase islands in the α-phase matrix. These γ islands have a pronounced orientation in the α matrix, parallel to the rolling direction, as given in Figure 9a. This lamellar structure of the α and γ phases of DSS BM is recrystallized and grain refined to give the equiaxed grains (Figure 9c) due to the severe plastic deformation enforced by the FSW tool and the high frictional heating. It has been reported that SZ is the core of the FSWed area, where the FSW tool causes major material agitation and heating. Consequently, temperature and deformation are at their extreme values in the SZ compared to TMAZ [38]. At a little distance away from the centerline of the 2205 DSS FSW pass, TMAZ is noticeable (Figure 9b,d). It is characterized by lower temperature and deformation, compared to the SZ [39,40]. The refined equiaxed grains in the SZ and the formation of elongated grains in the TMAZ promotes a sharp interface between the two zones (Figure 9b,d). After TMAZ moves toward the BM, HAZ appears, as shown in Figure 9b,d). It is far away from the tool-stirring zone (SZ). Thus, it only suffers from a cycle of heating and cooling without any deformation. Consequently, the microstructure features of this zone are close to BM with a slight grain growth in some FSWed joints [34]. The DSS solidification during the fusion welding is a type of ferrite prior. This means that from the liquid state, the grains of α-phase nucleate and grow preferentially, and different morphologies of γ grains are subsequently formed [41]. Furthermore, the thermal cycle across the DSS welds is highly heterogeneous and complex, especially for the multi-pass welding as in the GTAW process [42]. Figure 10 shows the optical microstructure of the cross-section of the GTAWed joint—the BM (Figure 10a) and the different welding zones: HAZ (Figure 10b,c), and WZ (Figure 10d–f). High and low magnifications of the HAZ are illustrated in Figure 10b,c, respectively. It can be seen that a relatively wide HAZ width was obtained. The HAZ coarse-grained structure zone (Figure 10b) adjacent to the fusion zone line could be stemmed from nearly complete γ-phase dissolution on the heating cycle and subsequent α-phase grain growth. The HAZ microstructure consists of large α grains with continuous networks of γ-phase at the α-grain boundaries (GBA) and tiny ones of Widmanstatten austenite (WA) (Figure 10). However, the heat input at the WZ is higher than that of the HAZ; therefore, different γ grain morphologies are likely to be found, including GBA, the intragranular γ-precipitates (IGA), and WA. Eghlimi et al. [43] reported that the GBA forms at a temperature range of 800–1350 °C. The WA grains precipitated from the grain boundary γ-phase, as shown in the WZ (Figure 10b,d). The WA side laths, nucleating from grain boundary γ-phase, form along grain morphologies orientations in the α matrix. Ramirez et al. [44] reported that the WA forms at a temperature range of 650—800 °C. The IGA nucleates within α grains (Figure 10d,e). It also maintains a crystallographic orientation with the α-phase. The IGA grains are finer compared with the GBA and WA grains. These microstructure features are mainly considered the microstructures of 2205 DSS welded by fusion welding techniques, as reported in many previous works [2]. Figure 9 is simple and consists of fine recrystallized ferrite and austenite. The microstructure after GTAW, shown in Figure 10, is quite complex and consists of different austenite morphologies such as GBA, IGA, and coarse ferrite grains with a continuous network of austenite.

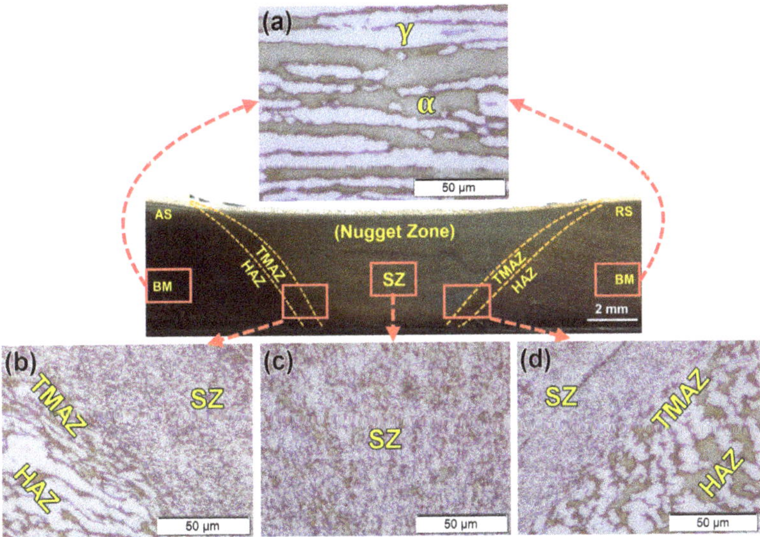

Figure 9. Optical microstructure of the FSWed 2205 DSS butt joint using the V-shaped groove with root face: (**a**) BM, (**b**) SZ/TMAZ/HAZ on the advanced side, (**c**) SZ, and (**d**) SZ/TMAZ/HAZ on the retreating side.

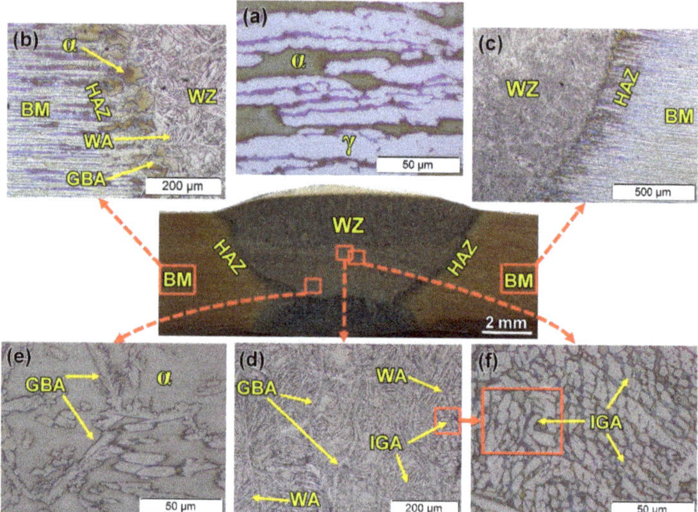

Figure 10. Optical microstructure of the GTAWed 2205 DSS butt joint using the V-shaped groove with root face and root gap: (**a**) BM, (**b**,**c**) WZ/HAZ/BM at high and low magnifications, respectively, (**d**–**f**) WZ.

3.4. XRD Assessment

Figure 11 shows the X-ray diffraction patterns of the FSWed and GTAWed butt joints of the 2205 DSS. Only peaks of α and γ phases can be seen. It seems that sigma phase (σ) was not found in the WZ for both joints or its amount may be lower than 5%, which is well-known to be the detection average limit by XRD. It should be mentioned that σ was also not observed in microstructures' investigation. It was reported that [13,45] the weld thermal cycle of DSS has a crucial influence on the σ-phase precipitation. The σ-phase is normally formed after long holding times at temperatures from 650 to 950 °C.

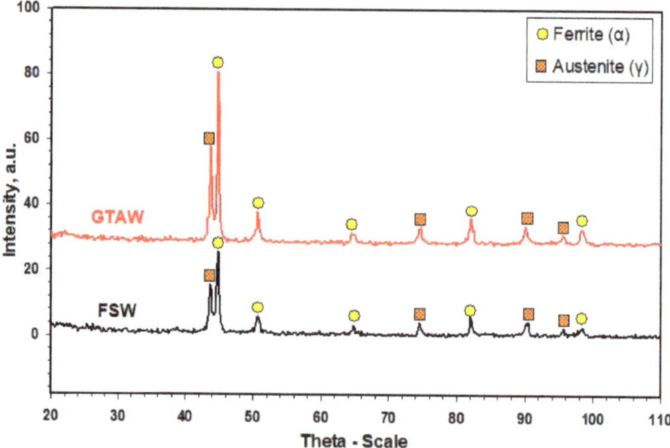

Figure 11. XRD patterns of the 2205 DSS FSWed joint produced using a 60° V groove with root face and GTAWed butt joints.

3.5. Measured Austenite Content

Figure 12 shows the γ content measured using the ferrite–scope type of FERITSCOPE MP30 across the 2205 DSS welded joints produced by GTAW and FSW techniques. The γ-phase contents in BM, HAZ, and WZ of the 2205 DSS GTAW welded joint are 49%, 46%, and 59%, respectively, as given in Figure 12. It can be noted that the γ-phase percentage in the WZ of the 2205 DSS joint is 59%, which is higher than the BM. This increase in γ-phase can be ascribed to the use of filler rod ER2209 during the 2205 DSS GTAW method, containing higher Ni than the 2205 DDS BM [46]. Moreover, the γ content in HAZ is 46%, which is slightly less than that of BM. This decrease in γ-phase percentage can be attributed to the heating/cooling cycle during the GTAW process [46]. The γ-phase content of the SZ in the 2205 DSS welded joints using FSW is shown in Figure 12. The γ-phase contents are 54% across the SZ and 51% across the HAZ/TMAZ zone, which is slightly higher than that of the BM. The increase in the γ-phase content in the SZ is ascribed to the exposure to the high temperature during the FSW process at the applied welding conditions of 300 rpm and 25 mm/min. This high-temperature exposure has resulted in the transformation of a portion of the α-phase into γ-phase upon cooling [47]. It can be said that the α/γ ratios in the weld zone are more suitable for joints welded by FSW than by GTAW.

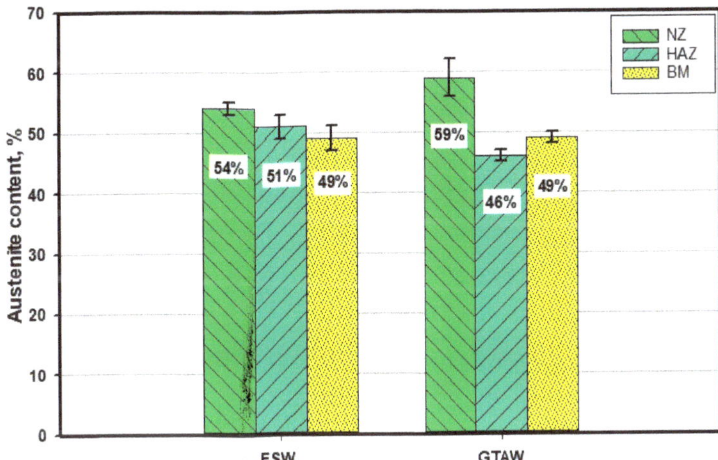

Figure 12. Austenite contents (%) in microstructural zones of the FSWed joint produced using a 60° V groove with root face and GTAWed 2205 DSS obtained using a ferrite-scope type of FERITSCOPE MP30. Note: for the FSWed sample, WZ represents the SZ and HAZ represents both TMAZ/HAZ.

3.6. EBSD Investigation

Figure 13 shows the phase-colored maps (austenite in green and ferrite in red) for the base material (a and b) and the stir zone of the FSWed joint produced using a 60° V groove with root face (c) with high angle boundaries (HABs) >15° in black lines superimposed. It can be observed that the base material microstructure consists of highly elongated grains of both ferrite matrix (red) and austenite islands (green) with volume fractions of 0.46 and 0.54, respectively, as indicated on the map. It should be noted here that the volume fraction is slightly changed after changing the data acquisition step size as after using 0.5 μm step in (Figure 13b), it becomes 0.49 and 0.51, respectively, which almost reaches the ideal phase fraction of 1:1. The elongated ferrite grains are more continuous while the elongated austenite grains are divided into more fine grains with HABs. Figure 14 shows the grain size distribution of the base material austenite (a) and ferrite (b). It can be observed that the austenite has almost random distribution and grain size ranging from 3 μm up to 22 μm with an average of 6.5 μm. While the ferrite has non-random distribution and wider grain size range from 3 μm up to 60 μm with an average of 7.2 μm.

Figure 13c shows the phase-colored map for the NG zone after FSW. Clearly the grain morphology in the NG zone is completely different than that observed in the base material. Significant grain refining can be observed with equiaxed grain morphology for both ferrite and austenite. The stir zone in the FSW experiences very high degree of plastic strain at high temperatures both generated by the FSW tool stirring of the material surrounding the tool [19,48]. This condition has resulted in a dynamic recrystallization process to take place in the stir zone and the formation of the equiaxed fine grain structure. The grain size distribution after FSW is shown in Figure 14c,d for austenite and ferrite phases, respectively. Both phases show grain size random distribution with range between 1.2 µm to 4.7 µm for austenite with an average of 1.88 µm and 1.2 to 6 µm for the ferrite phase with an average of 2.2 µm. The volume fraction of the two phases are 0.62 and 0.38 for austenite and ferrite, respectively. It can be noted that there is an increase in the austenite percentage in the stir zone, which is consistent with the obtained result using the ferrite–scope device with slight higher percentage obtained using EBSD. This might need a smaller step size to reduce the percentage austenite, as observed when two different step sizes are used in the case of BM. The results obtained in terms of grain refining and the percentage of phases are in agreement with the results reported in the literature [49].

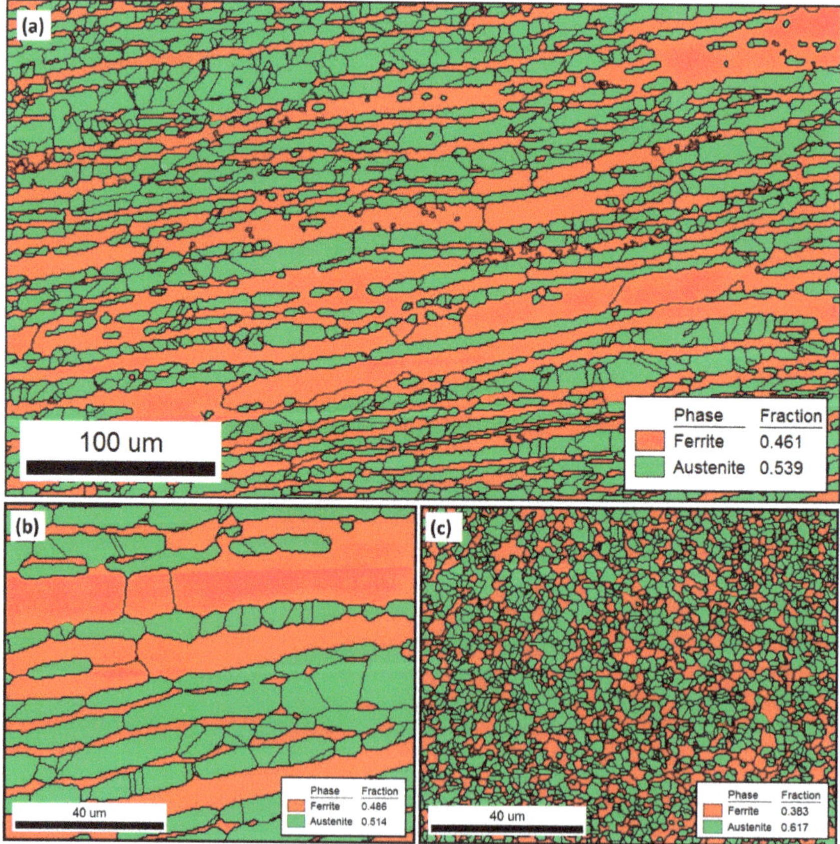

Figure 13. Phase-colored maps (austenite in green and ferrite in red) for the base material (**a**,**b**) and the FSWed joint stir zone (**c**) with the high angle boundaries (HABs) >15° in black lines superimposed. The data in (**a**) is obtained using a 1 µm step size while the data in (**b**,**c**) are obtained using a 0.5 µm step size.

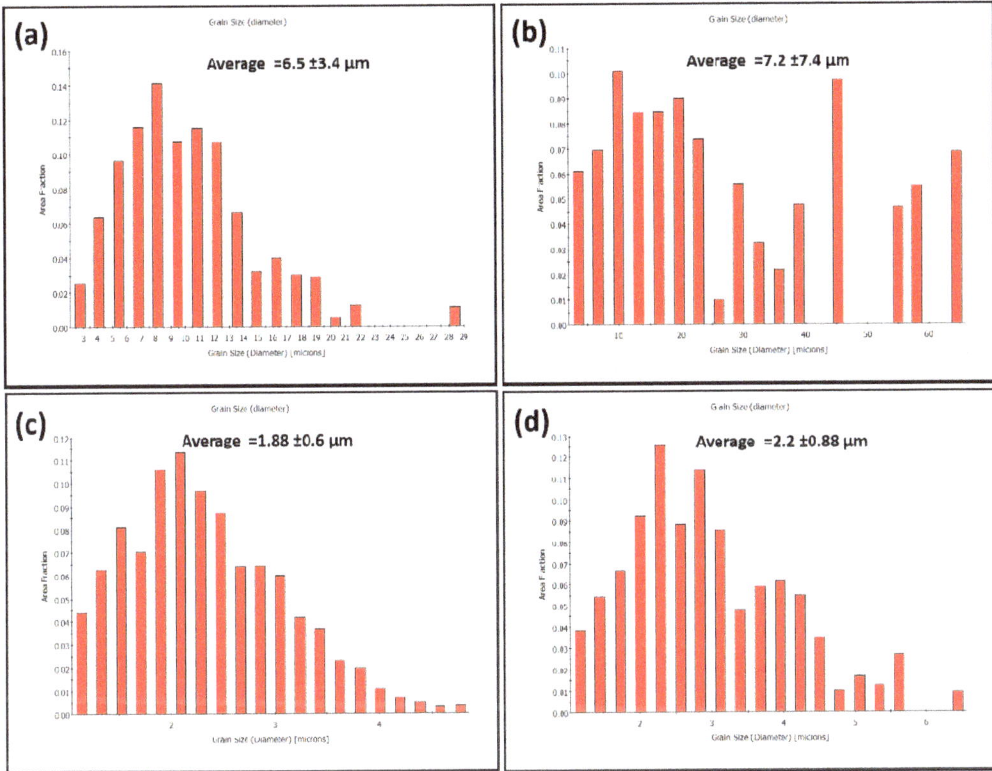

Figure 14. Grain size distribution of BM (**a**) austenite, (**b**) ferrite and NG of FSWed joint (**c**) austenite, (**d**) ferrite.

From the misorientation angle distribution shown in Figure 15, it can be noted that the BM is dominated by HABs with non-random distribution for both phases, austenite (a) and ferrite (b). A high density of twin boundaries at 60° can be observed in the austenite, as it can also be observed in the GB maps superimposed in Figure 13. After FSW, both phases (c) austenite and (d) ferrite misorientation angle distributions show random distribution, which is the typical distribution of the recrystallized material. The density of twin boundaries in the austenite can still be observed at the same misorientation angle of around 60°.

Crystallographic texture before and after FSW is presented in Figure 16 as 001, 101, and 111 pole figures (a,b for BM austenite and ferrite, respectively), (c,d for FSWed NG zone austenite and ferrite, respectively). Both austenite and ferrite of BM show the typical fcc and bcc rolling texture with more strong texture (~ 11 times random) in case of the ferrite than that of the austenite (~ 4 times random). After FSW, the texture for both phases is weak, with only about two times random, resembling the shear texture of both phases.

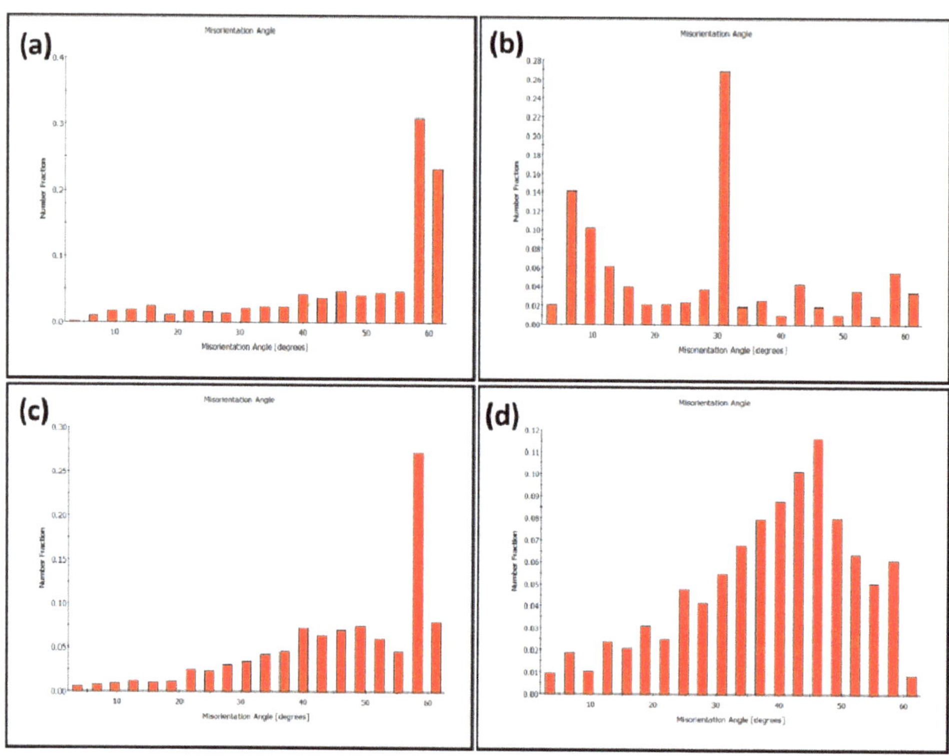

Figure 15. Misorientation angle distribution of BM (**a**) austenite, (**b**) ferrite and NG of FSWed joint (**c**) austenite, (**d**) ferrite.

3.7. Hardness Distribution

Figure 17 exhibits the variation of hardness across the 2205 DSS butt joints welded by (a) FSW and (b) GTAW. In general, it can be seen that the hardness values in the WZ of FSWed joint is higher than that of the joint welded by the GTAW process. For the FSWed joint, it can be seen from Figure 17a that the hardness reaches a maximum value of 280 HV at the SZ and gradually decreases by passing from TMAZ and HAZ to be 260 HV and 250 HV, respectively. The severe plastic deformation accompanied by the maximum temperature in the SZ during the FSW process leads to the most effective microstructure modification in terms of grain refining (Figure 9c) due to dynamic recrystallization [35,50], which justifies the highest hardness value in this region. In the TMAZ, lower plastic deformation and temperature are introduced during the FSW, leading to less grain refinement and lower hardness values. In HAZ, deformation is absent, and the DSS material only undergoes a thermal cycle [51]. This validates the lowest hardness values close to that of the BM in this region. For the GTAWed joint, it can be seen from Figure 17b that the hardness reaches a maximum value of 265 HV at the WZ and gradually decreases by passing from HAZ and BM to be 250 HV and 240 HV, respectively. The fluctuation in hardness values can be ascribed to the microstructural features observed [13]. The existence of more intergranular γ (Figure 10d) enhances the hardness of WZ [52], while the presence of much coarser α grains decreases the hardness of HAZ. The same trends were obtained in other works [46].

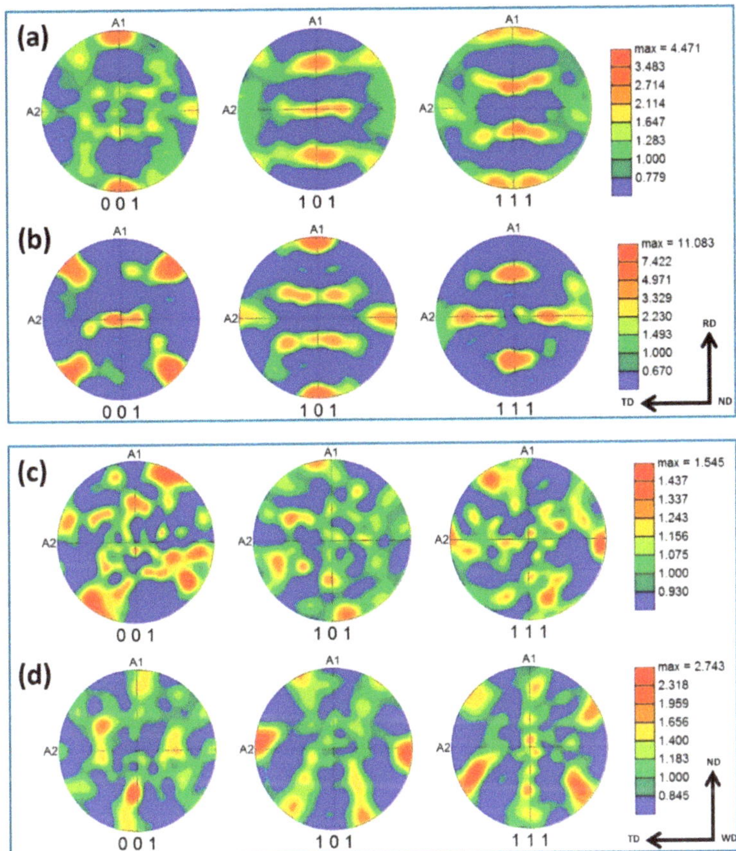

Figure 16. 001, 101, and 111 pole figures of BM (**a**) austenite, (**b**) ferrite and NG of FSWed joint (**c**) austenite, (**d**) ferrite.

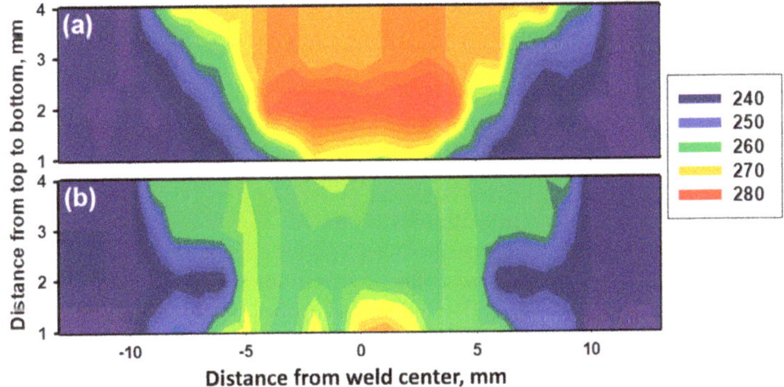

Figure 17. Vickers hardness maps of the DSS 2205 welds (**a**) FSWed joint produced using a 60o V groove with root face and (**b**) GTAWed joint.

3.8. Tensile Properties

Figure 18 shows the tensile properties in terms of yield stress (Ys), ultimate tensile stress (UTS), and elongation (E%) of the welded joints by FSW and GTAW processes compared to the BM. It is obvious that the FSWed joints have much higher tensile properties (Ys, UTS, and E%) than that given by both GTAWed joints and the BM. This is in agreement with the measurements of hardness shown in Figure 16. It can be seen in Figure 17 that the Ys, UTS, and E% of FSWed joints are enhanced over the BM by 65%, 33%, and 54%, and over the GTAW weldments by 21%, 41%, and 66%, respectively. This enhancement of the tensile properties is mainly ascribed to the main features of grain modification during the FSW process [53]. Ghadar et al. [38] reported significant improvement of ultimate tensile strength and fracture strain for friction stir processed 3.5 mm 2205 DSS and ascribed this enhancement in tensile properties to the grain refining. It can be said that the grain size in the WZ is a dominant factor governing mechanical properties. Furthermore, only around 16% improvement in the Ys is detected for the GTAWed joints over the Ys of the BM with no remarkable increase in the UTS. This enhancement in yield stress may also be ascribed to the grain refining in the WZ [42].

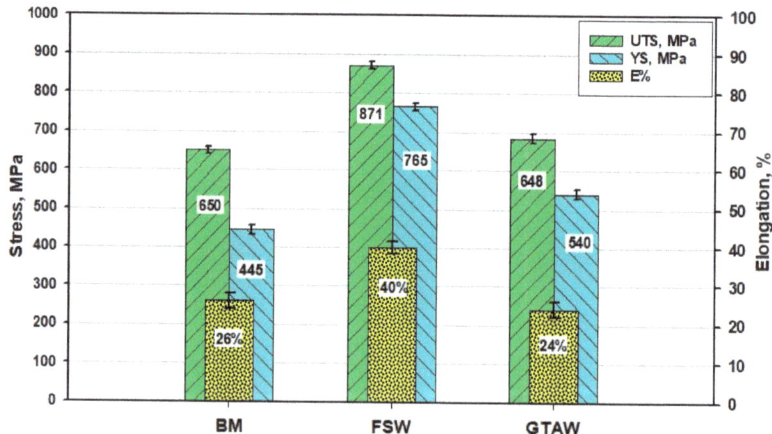

Figure 18. Tensile properties of DSS 2205 base material and FSWed joint produced using a 60° V groove with root face and a GTAWed joint.

4. Conclusions

The effect of groove joint design on the microstructure and mechanical properties of 6.5 mm plates butt joints 2205 DSS welded by FSW and GTAW techniques were examined and evaluated. The following conclusions can be drawn:

1. The FSW used to weld 2205 DSS of 6.5 mm thick with groove filling is shown to be feasible.
2. The optimum groove design to produce a sound joint with groove filling is the V-shaped with 60° groove angle and 2 mm root face.
3. The use of the groove in the butt joint of DSS 2205 extended WC tool life due to the reduction in tool resistance forces during plunging and welding.
4. The heat input and 2209 filler rod used in the GTAW process produce different γ-phase morphologies, such as GBA, IGA, and WA, besides the α phase in the WZ. In comparison, the friction stir weld has only two-phase microstructures: α and γ with significant grain refining.
5. The α/γ ratio in the weld zone is influenced by welding techniques (solid-state welding and fusion welding) and their parameters in terms of welding temperature, joint groove design, and the presence or absence of filler rod. Compared with the

BM that attains an α/γ ratio of 51/49, GTAW promotes a ratio of 41/59 while FSW produces a ratio of 44/56.
6. Grain refining in the NG zone of the FSWed joint has been quantified to be 1.88 μm for austenite and 2.2 μm for ferrite from about 6.5 μm and 7.2 μm in the BM phases, respectively.
7. Significant hardness improvements have been detected for the weld joint produced by FSW than that produced by GTAW.
8. The FSWed 2205 DSS butt joint shows higher tensile properties than both BM and GTAWed joints. The Ys, UTS, and E% of the FSWed joint are enhanced over BM by 65%, 33%, and 54%, and over the GTAW weldment by 21%, 41%, and 66%, respectively.

Author Contributions: Conceptualization, M.M.Z.A. and K.A.A.; methodology, M.M.Z.A.; M.M.E.-S.S., and K.A.A. software, M.M.Z.A. and K.A.A.; validation, M.M.Z.A.; M.M.E.-S.S., and K.A.A.; formal analysis, K.A.A.; H.M.A.E.-A., and M.M.E.-S.S.; investigation, K.A.A. and H.M.A.E.-A. resources, B.A., N.J., and K.T. data curation, M.M.Z.A.; K.A.A., and H.M.A.E.-A.; writing—original draft preparation, K.A.A. and M.M.E.-S.S.; writing—review and editing, M.M.Z.A. and M.M.E.-S.S.; visualization, H.M.A.E.-A. and K.T; supervision, M.M.Z.A., I.G.E.-B., and H.M.A.E.-A.; project administration, M.M.Z.A., I.G.E.-B., and H.M.A.E.-A.; funding acquisition, B.A., N.J., and K.T. All authors have read and agreed to the published version of the manuscript.

Funding: This research received no external funding.

Institutional Review Board Statement: Not applicable.

Informed Consent Statement: Not applicable.

Acknowledgments: The technical support provided by Mohamed Habba and Yousef Gamal (Suez University) in the operation of the FSW machine and preparation of some figures is acknowledged and highly appreciated.

Conflicts of Interest: The authors declare no conflict of interest.

References

1. Chaudhari, A.N.; Dixit, K.; Bhatia, G.S.; Singh, B.; Singhal, P.; Saxena, K.K. Welding Behaviour of Duplex Stainless Steel AISI 2205: A Review. *Mater. Today Proc.* **2019**, *18*, 2731–2737. [CrossRef]
2. Vinoth Jebaraj, A.; Ajaykumar, L.; Deepak, C.R.; Aditya, K.V.V. Weldability, Machinability and Surfacing of Commercial Duplex Stainless Steel AISI2205 for Marine Applications—A Recent Review. *J. Adv. Res.* **2017**, *8*, 183–199. [CrossRef]
3. Verma, J.; Taiwade, R.V. Effect of Welding Processes and Conditions on the Microstructure, Mechanical Properties and Corrosion Resistance of Duplex Stainless Steel Weldments—A Review. *J. Manuf. Process.* **2017**, *25*, 134–152. [CrossRef]
4. Rahimi, S.; Konkova, T.N.; Violatos, I.; Baker, T.N. Evolution of Microstructure and Crystallographic Texture During Dissimilar Friction Stir Welding of Duplex Stainless Steel to Low Carbon-Manganese Structural Steel. *Metall. Mater. Trans. A Phys. Metall. Mater. Sci.* **2019**, *50*, 664–687. [CrossRef]
5. Shokri, V.; Sadeghi, A.; Sadeghi, M.H. Effect of Friction Stir Welding Parameters on Microstructure and Mechanical Properties of DSS–Cu Joints. *Mater. Sci. Eng. A* **2017**, *693*, 111–120. [CrossRef]
6. Antony, P.J.; Singh Raman, R.K.; Kumar, P.; Raman, R. Corrosion of 2205 Duplex Stainless Steel Weldment in Chloride Medium Containing Sulfate-Reducing Bacteria. *Metall. Mater. Trans. A Phys. Metall. Mater. Sci.* **2008**, *39*, 2689–2697. [CrossRef]
7. Múnez, C.J.; Utrilla, M.V.; Ureña, A.; Otero, E. Influence of the Filler Material on Pitting Corrosion in Welded Duplex Stainless Steel 2205. *Weld. Int.* **2010**, *24*, 105–110. [CrossRef]
8. Using, S.; Metallography, C. Duplex Stainless Steels Represent an Important and Expanding Class of Superior to Those of the 300-Series Austenitic Stainless Steels Such Property Combinations Have Led to an Increased Application of These Alloys Cations in Which Duplex Stainless. *Current* **1998**, *225*, 215–225.
9. Muthupandi, V.; Bala Srinivasan, P.; Seshadri, S.K.; Sundaresan, S. Effect of Weld Metal Chemistry and Heat Input on the Structure and Properties of Duplex Stainless Steel Welds. *Mater. Sci. Eng. A* **2003**, *358*, 9–16. [CrossRef]
10. Geng, S.; Sun, J.; Guo, L.; Wang, H. Evolution of Microstructure and Corrosion Behavior in 2205 Duplex Stainless Steel GTA-Welding Joint. *J. Manuf. Process.* **2015**, *19*, 32–37. [CrossRef]
11. Devakumar, D.; Jabaraj, D.B.; Bupesh Raja, V.K.; Periyasamy, P. Characterization of Duplex Stainless Steel/Cold Reduced Low Carbon Steel Dissimilar Weld Joints by GTAW. *Appl. Mech. Mater.* **2015**, *766*, 780–788. [CrossRef]
12. Magudeeswaran, G.; Nair, S.R.; Sundar, L.; Harikannan, N. Optimization of Process Parameters of the Activated Tungsten Inert Gas Welding for Aspect Ratio of UNS S32205 Duplex Stainless Steel Welds. *Def. Technol.* **2014**, *10*, 251–260. [CrossRef]

13. Mourad, A.H.I.; Khourshid, A.; Sharef, T. Gas Tungsten Arc and Laser Beam Welding Processes Effects on Duplex Stainless Steel 2205 Properties. *Mater. Sci. Eng. A* **2012**, *549*, 105–113. [CrossRef]
14. Bhattacharya, A.; Singh, P.M. Stress Corrosion Cracking of Welded 2205 Duplex Stainless Steel in Sulfide-Containing Caustic Solution. *J. Fail. Anal. Prev.* **2007**, *7*, 371–377. [CrossRef]
15. Emami, S.; Saeid, T.; Abdollah-zadeh, A. Effect of Friction Stir Welding Parameters on the Microstructure and Microtexture Evolution of SAF 2205 Stainless Steel. *J. Alloys Compd.* **2019**, *810*, 151797. [CrossRef]
16. Emami, S.; Saeid, T.; Khosroshahi, R.A. Microstructural Evolution of Friction Stir Welded SAF 2205 Duplex Stainless Steel. *J. Alloys Compd.* **2018**, *739*, 678–689. [CrossRef]
17. Saeid, T.; Abdollah-zadeh, A.; Assadi, H.; Malek Ghaini, F. Effect of Friction Stir Welding Speed on the Microstructure and Mechanical Properties of a Duplex Stainless Steel. *Mater. Sci. Eng. A* **2008**, *496*, 262–268. [CrossRef]
18. Gideon, B.; Ward, L.; Biddle, G. Duplex Stainless Steel Welds and Their Susceptibility to Intergranular Corrosion. *J. Miner. Mater. Charact. Eng.* **2008**, *7*, 247–263. [CrossRef]
19. Ahmed, M.M.Z.; Ataya, S.; Seleman, M.M.E.; Mahdy, A.M.A.; Alsaleh, N.A.; Ahmed, E. Heat Input and Mechanical Properties Investigation of Friction Stir Welded AA5083/AA5754 and AA5083/AA7020. *Metals* **2021**, *11*, 68. [CrossRef]
20. Cui, L.; Fujii, H.; Nogi, K. Friction Stir Welding of a High Carbon Steel. *Scr. Mater.* **2007**, *56*, 637–640. [CrossRef]
21. Ahmed, M.M.Z.; Wynne, B.P.; Rainforth, W.M.; Addison, A.; Martin, J.P.; Threadgill, P.L. Effect of Tool Geometry and Heat Input on the Hardness, Grain Structure, and Crystallographic Texture of Thick-Section Friction Stir-Welded Aluminium. *Metall. Mater. Trans. A* **2019**, *50*, 271–284. [CrossRef]
22. Shazly, M.; Ahmed, M.M.Z.; El-Raey, M. Friction stir welding of polycarbonate sheets. In *Characterization of Minerals, Metals, and Materials 2014*; John Wiley& Sons: Hoboken, NJ, USA, 2014; ISBN 9781118887868.
23. Ahmed, M.M.Z.; Seleman, M.M.E.S.; Shazly, M.; Attallah, M.M.; Ahmed, E. Microstructural Development and Mechanical Propertiesof Friction Stir Welded Ferritic Stainless Steel AISI 40. *J. Mater. Eng. Perform.* **2019**, *28*, 6391–6406. [CrossRef]
24. Ahmed, M.M.Z.; Wynne, B.P.; Martin, J.P. Effect of Friction Stir Welding Speed on Mechanical Properties and Microstructure of Nickel Based Super Alloy Inconel 718. *Sci. Technol. Weld. Join.* **2013**, *18*, 680–687. [CrossRef]
25. Rai, R.; De, A.; Bhadeshia, H.K.D.H.; DebRoy, T. Review: Friction Stir Welding Tools. *Sci. Technol. Weld. Join.* **2011**, *16*, 325–342. [CrossRef]
26. Ahmed, M.M.Z.; Barakat, W.S.; Mohamed, A.Y.A.; Alsaleh, N.A. The Development of WC-Based Composite Tools for Friction Stir Welding of High-Softening-Temperature Materials. *Metals* **2021**, *11*, 285. [CrossRef]
27. Ragab, M.; Liu, H.; Yang, G.; Ahmed, M.M.Z. Applied Sciences Friction Stir Welding of 1Cr11Ni2W2MoV Martensitic Stainless Steel: Numerical Simulation Based on Coupled Eulerian Lagrangian Approach Supported with Experimental Work. *Appl. Sci.* **2021**, *11*, 3049. [CrossRef]
28. Zhang, Y.; Sato, Y.S.; Kokawa, H.; Hwan, S.; Park, C.; Hirano, S. Stir Zone Microstructure of Commercial Purity Titanium Friction Stir Welded Using PcBN Tool. *Mater. Sci. Eng. A* **2008**, *488*, 25–30. [CrossRef]
29. Tonelli, L.; Morri, A.; Toschi, S.; Shaaban, M.; Ammar, H.R.; Ahmed, M.M.Z.; Ramadan, R.M. Effect of FSP Parameters and Tool Geometry on Microstructure, Hardness, and Wear Properties of AA7075 with and without Reinforcing B 4 C Ceramic Particles. *Int. J. Adv. Manuf. Technol.* **2019**, *102*, 3945–3961. [CrossRef]
30. Bakkar, A.; Ahmed, M.M.Z.; Alsaleh, N.A.; Seleman, M.M.E.; Ataya, S. Microstructure, Wear, and Corrosion Characterization of High TiC Content Inconel 625 Matrix Composites. *J. Mater. Res. Technol.* **2019**, *8*, 1102–1110. [CrossRef]
31. Janeczek, A. The Influence of Tool Shape and Process Parameters on the Mechanical Properties of AW-3004 Aluminium Alloy Friction Stir Welded Joints. *Materials* **2021**, *14*, 3244. [CrossRef]
32. Zhang, R.; Buchanan, C.; Matilainen, V.; Daskalaki-mountanou, D.; Britton, T.B.; Piili, H.; Salminen, A.; Gardner, L. Mechanical Properties and Microstructure of Additively Manufactured Stainless Steel with Laser Welded Joints. *Mater. Des.* **2021**, *208*, 109921. [CrossRef]
33. Abdelazem, K.A.; El-aziz, H.M.A.; Ahmed, M.M.Z. Characterization of Mechanical Properties and Corrosion Resistance of SAF 2205 Duplex Stainless Steel Groove Joints Welded Using Friction Stir Welding Process. *Int. J. Recent Technol. Eng.* **2020**, *8*, 3428–3435. [CrossRef]
34. Trinh, D.; Frappart, S.; Rückert, G.; Cortial, F.; Touzain, S. Effect of Friction Stir Welding Process on Microstructural Characteristics and Corrosion Properties of Steels for Naval Applications. *Corros. Eng. Sci. Technol.* **2019**, *54*, 353–361. [CrossRef]
35. Ahmed, M.M.Z.; Seleman, M.M.E.; Zidan, Z.A.; Ramadan, R.M.; Ataya, S.; Alsaleh, N.A. Microstructure and Mechanical Properties of Dissimilar Friction Stir Welded AA2024-T4/AA7075-T6 T-*Butt Joints*. *Metals* **2021**, *11*, 128. [CrossRef]
36. Moreira, P.M.G.P.; Santos, T.; Tavares, S.M.O.; Richter-Trummer, V.; Vilaça, P.; de Castro, P.M.S.T. Mechanical and Metallurgical Characterization of Friction Stir Welding Joints of AA6061-T6 with AA6082-T6. *Mater. Des.* **2009**, *30*, 180–187. [CrossRef]
37. Balaram Naik, A.; Chennakesava Reddy, A. Macro and Microstructure Evaluation of TIG Weld on DSS (2304) with Bead Geometry at Three Positions. *Mater. Today Proc.* **2020**, in press. [CrossRef]
38. Ghadar, S.; Momeni, A.; Khademi, E.; Kazemi, S. Effect of Rotation and Traverse Speeds on the Microstructure and Mechanical Properties of Friction Stir Processed 2205 Duplex Stainless Steel. *Mater. Sci. Eng. B Solid-State Mater. Adv. Technol.* **2021**, *263*, 114813. [CrossRef]

39. Hamada, A.S.; Järvenpää, A.; Ahmed, M.M.Z.; Jaskari, M.; Wynne, B.P.; Porter, D.A.; Karjalainen, L.P. The Microstructural Evolution of Friction Stir Welded AA6082-T6 Aluminum Alloy during Cyclic Deformation. *Mater. Sci. Eng. A* **2015**, *642*, 366–376. [CrossRef]
40. El-Mahallawi, I.; Ahmed, M.M.Z.; Mahdy, A.A.; Abdelmotagaly, A.M.M.; Hoziefa, W.; Refat, M. Effect of Heat Treatment on Friction-Stir-Processed Nanodispersed AA7075 and 2024 Al Alloys. In *Friction Stir Welding and Processing IX*; The Minerals, Metals & Materials Series; Hovanski, Y., Mishra, R., Sato, Y., Upadhyay, P., Yan, D., Eds.; Springer: Cham, Switzerland, 2017; pp. 297–309. [CrossRef]
41. Makhdoom, M.A.; Ahmad, A.; Kamran, M.; Abid, K.; Haider, W. Microstructural and Electrochemical Behavior of 2205 Duplex Stainless Steel Weldments. *Surf. Interfaces* **2017**, *9*, 189–195. [CrossRef]
42. Xie, X.F.; Li, J.; Jiang, W.; Dong, Z.; Tu, S.T.; Zhai, X.; Zhao, X. Nonhomogeneous Microstructure Formation and Its Role on Tensile and Fatigue Performance of Duplex Stainless Steel 2205 Multi-Pass Weld Joints. *Mater. Sci. Eng. A* **2020**, *786*, 139426. [CrossRef]
43. Eghlimi, A.; Shamanian, M.; Raeissi, K. Effect of Current Type on Microstructure and Corrosion Resistance of Super Duplex Stainless Steel Claddings Produced by the Gas Tungsten Arc Welding Process. *Surf. Coat. Technol.* **2014**, *244*, 45–51. [CrossRef]
44. Ramirez, A.J.; Lippold, J.C.; Brandi, S.D. The Relationship between Chromium Nitride and Secondary Austenite Precipitation in Duplex Stainless Steels. *Metall. Mater. Trans. A Phys. Metall. Mater. Sci.* **2003**, *34*, 1575–1597. [CrossRef]
45. Luo, J.; Yuan, Y.; Wang, X.; Yao, Z. Double-Sided Single-Pass Submerged Arc Welding for 2205 Duplex Stainless Steel. *J. Mater. Eng. Perform.* **2013**, *22*, 2477–2486. [CrossRef]
46. Wang, L.; Zhao, P.; Pan, J.; Tan, L.; Zhu, K. Investigation on Microstructure and Mechanical Properties of Double-Sided Synchronous TIP TIG Arc Butt Welded Duplex Stainless Steel. *Int. J. Adv. Manuf. Technol.* **2021**, *112*, 303–312. [CrossRef]
47. Calliari, I.; Breda, M.; Gennari, C.; Pezzato, L.; Pellizzari, M.; Zambon, A. Investigation on Solid-State Phase Transformations in a 2510 Duplex Stainless Steel Grade. *Metals* **2020**, *10*, 967. [CrossRef]
48. Zayed, E.M.; El-Tayeb, N.S.M.; Ahmed, M.M.Z.; Rashad, R.M. *Development and Characterization of AA5083 Reinforced with SiC and Al_2O_3 Particles by Friction Stir Processing*; Springer International Publishing: New York, NY, USA, 2019; Volume 92.
49. Wang, W.; Hu, Y.; Zhang, M.; Zhao, H. Microstructure and Mechanical Properties of Dissimilar Friction Stir Welds in Austenitic-Duplex Stainless Steels. *Mater. Sci. Eng. A* **2020**, *787*, 139477. [CrossRef]
50. Ahmed, M.M.Z.; Wynne, B.P.; El-Sayed Seleman, M.M.; Rainforth, W.M. A Comparison of Crystallographic Texture and Grain Structure Development in Aluminum Generated by Friction Stir Welding and High Strain Torsion. *Mater. Des.* **2016**, *103*, 259–267. [CrossRef]
51. Santos, T.F.A.; Idagawa, H.S.; Ramirez, A.J. Thermal History in UNS S32205 Duplex Stainless Steel Friction Stir Welds. *Sci. Technol. Weld. Join.* **2014**, *19*, 150–156. [CrossRef]
52. Nowacki, J.; Łukojć, A. Structure and Properties of the Heat-Affected Zone of Duplex Steels Welded Joints. *J. Mater. Process. Technol.* **2005**, *164*, 1074–1081. [CrossRef]
53. Sorger, G.; Sarikka, T.; Vilaça, P.; Santos, T.G. Effect of Processing Temperatures on the Properties of a High-Strength Steel Welded by FSW. *Weld. World* **2018**, *62*, 1173–1185. [CrossRef]

Article

Effect of Process Factors on Tensile Shear Load Using the Definitive Screening Design in Friction Stir Lap Welding of Aluminum–Steel with a Pipe Shape

Leejon Choy [1], Seungkyung Kim [1], Jeonghun Park [1], Myungchang Kang [1,*] and Dongwon Jung [2,*]

1 Graduate School of Convergence Science, Pusan National University, Busan 46241, Korea; leejonee@hanmail.net (L.C.); kskyeong9970@gmail.com (S.K.); jeonghun.park@ctr.co.kr (J.P.)
2 Faculty of Mechanical, Jeju National University, Jeju-si 63243, Korea
* Correspondence: kangmc@pusan.ac.kr (M.K.); jdwcheju@jejunu.ac.kr (D.J.); Tel.: +82-10-3852-2337 (M.K.); +82-10-3459-2467 (D.J.)

Abstract: Recently, friction stir welding of dissimilar materials has emerged as one of the most significant issues in lightweight, eco-friendly bonding technology. In this study, we welded the torsion beam shaft—an automobile chassis component—with cast aluminum to lighten it. The study rapidly and economically investigated the effects of friction stir welding and process parameters for A357 cast aluminum and FB590 high-strength steel; 14 decomposition experiments were conducted using a definitive screening design that could simultaneously determine the effects of multiple factors. Friction stir welding experiments were conducted using an optical microscope to investigate the tensile shear load behavior in the welding zone. In addition to understanding the interactions between tool penetration depth and plunge speed and tool penetration depth and dwell time, we investigated and found that tool penetration depth positively affected the size of the hooking area and contributed to the stabilization and size reduction of the cavity. The experimental results showed that the plunge depth and tool penetration depth effects were most important; in this case, the plunge depth negatively affected the magnitude of tensile shear load, whereas the tool penetration depth had a positive effect. Therefore, when selecting a tool, it is important to consider the plunge depth and tool penetration depth in lap welding.

Keywords: friction stir lap welding; definitive screening design (DSD); tensile shear load; tool penetration depth; plunge depth

Citation: Choy, L.; Kim, S.; Park, J.; Kang, M.; Jung, D. Effect of Process Factors on Tensile Shear Load Using the Definitive Screening Design in Friction Stir Lap Welding of Aluminum–Steel with a Pipe Shape. *Materials* **2021**, *14*, 5787. https://doi.org/10.3390/ma14195787

Academic Editors: Józef Iwaszko and Jerzy Winczek

Received: 2 September 2021
Accepted: 30 September 2021
Published: 3 October 2021

1. Introduction

Recently, environmental protection and energy-saving strategies—including weight reduction and miniaturization—have emerged as critical issues in the automobile and other industries [1–3]. Reducing a vehicle's weight improves its overall performance and fuel economy; therefore, replacing steel with Al alloys is a possible alternative. However, the complete replacement of steel with Al alloys has technical limitations, such as the fact that Al alloys do not have finite fatigue limits. The high strength, excellent creep resistance, and formability of steel and the low density, high thermal conductivity, and excellent corrosion resistance of aluminum alloys can be combined into one hybrid structure. There are four main aspects of the bonding difficulty [4]. The current welding methods—such as low energy input fusion welding and brazing—can induce the formation of an intermetallic compound (IMC) layer with a certain thickness. The evolution of brittle IMCs, which are generated during the interfacial reaction between solid steel and liquid aluminum, can significantly influence the mechanical properties of Al/steel joints. Therefore, new methods must be developed to realize the rapid development of dissimilar aluminum alloy and steel welding [5]. Invented and patented in 1991 by The Welding Institute in Cambridge, UK, friction stir welding (FSW) is an energy-efficient, environmentally friendly, and versatile bonding technology [6]. FSW is a solid-state bonding method that has low thermal deformation and high bonding efficiency in joining dissimilar materials such as aluminum and steel. Because of its eco-friendliness and economic benefits, it is actively being introduced to the manufacture of automobile parts. A coupled torsion beam axle (CTBA) is installed at the rear end of a vehicle, connecting the tire to the body. This component absorbs vibrations or shocks on the road surface to improve ride comfort, supports braking and lateral forces received from tires, and adjusts the roll angle of the vehicle when cornering with torsional stiffness of the body [7]. In automobiles and other areas, the two materials must be combined efficiently to replace steel with Al alloys. Park et al. [8] investigated and found that proper mixing of materials occurred in the nugget region when the stronger base material was placed on the advancing side in the butt joint of Al and Steel. As an applied study of dissimilar Al and steel FSW, Park et al. [9] performed FSW between a 3 mm thick A357 cast Al flat plate and FB590 high-strength steel flat plate and reported achieving 72.8% strength compared to the Al base material. A characteristic feature of lap FSW (FSLW) is the geometric defect called a "hook" that occurs at the interface of two weld sheets. FSW is a complex thermomechanical process; the final performance—including mechanical strength—of FSW joints is closely related to the tool geometry and welding parameters.

To reduce defects and increase the mechanical strength of these FSW joints, many researchers have conducted studies on tool shape and process parameters [10–19]. Research has also been conducted on the importance of the tool shoulder [20–26]. Trimble et al. [20] considered the effect of tool shape and rotational speed on the AA2024-T3 flat plate material. The results indicated that it is possible to achieve good weld quality at speeds up to 355 mm/min by welding with a scroll shoulder and triflute pin at a rotational speed of 450 rpm. The recently developed "scroll"-shaped tool shoulder is particularly desirable for curved joints. Trueba Jr. et al. [21] reported that the FSW tool shoulder with a raised helical design produces the best welds in terms of surface quality and mechanical properties in butt-welded aluminum 6061-T6 plates using six differently designed tools of Ti-6Al-4V material. Research has been conducted on process-parameter-related tools to reduce defects and increase the mechanical strength of FSW joints [27,28]. The normal force must be varied in the following four stages: tool plunge stage, dwell stage, welding stage, and tool retreating stage. Process parameters affect the normal force and temperature in FSW processes. In addition to tools, research has been conducted on process parameters to reduce defects and increase the mechanical strength of FSW joints [29–34]. Baskoro et al. [29] conducted experiments and analyses for plunge speeds of 2, 3, and 4 mm/min and dwell times of 0, 2, and 4 s in the high-speed micro-friction stir spot welding (μFSSW) of Al A1100 with a thickness of 0.4 mm and revealed that a dwell time of 2 s was the most important welding parameter in the μFSSW process. Li et al. [30] investigated the correlation between microstructure and mechanical properties by performing FSSW at a plunge depth of 0.1 mm with a dwell time between 1 and 9 s on dissimilar lap joints of 2 mm thick 1060 aluminum and T2 copper plates. At shorter dwell times, a discontinuous layer of $CuAl_2$ forms and mixes with the CuAl formed at the interface due to insufficient heat input. Longer dwell times can lead to higher peak temperatures but lower plunging forces and torques. They are characterized by the formation of a continuous $CuAl_2$-CuAl-Al_4Cu_9 stacked layer at the interface, resulting in microcracks. They also reported the need for an appropriate dwell time. Zheng et al. [31] reported that the pin failed at low loads that did not reach the nickel alloy surface; in contrast, the maximum tensile shear strength of the lap joint was obtained at a plunge depth of 0.3 mm and reached 7.9 kN. The plunge depth of the joint was determined to have a significant effect on the strength of FSW lap joints of Al alloy (2A70) and nickel-based alloys through experiments at various plunge depths from 0 to 0.5 mm. Devanathan et al. [32] investigated the effect of plunge depths of 0 and 0.2 mm and reported that increasing the plunge depth reduced mechanical performance, leading to defective welds in 6063Al butt welding; their study used a single tool with a shoulder diameter of 24 mm, pin diameter of 4 mm, and pin length of 5 mm. Wei et al. [33] performed FSLW of a 3 mm thick Al (1060Al) sheet and a 1 mm thick stainless steel (SUS321, austenite) sheet using a 6 mm diameter pin, with an insertion depth of 2.8 and 3.2 mm, and found that the greater the tool penetration depth, the greater the tensile strength. Regensburg et al. [34] investigated hooking and IMC formation with the pin lengths at the junctions being 1.8, 2.0, 2.2, 2.4, and 2.6 mm at the lap joint of EN AW1050/CW024A material with a thickness of 2 mm. It was found that plunging into the lower copper sheet of about 0.2 mm by a pin length of 2.2 mm yielded the highest breaking load. The further increase in pin length led to the formation of hooking defects, which resulted in void formation at the interface and failure within the thin aluminum sheet area. However, the effect of plunge depth and tool penetration depth could not be considered simultaneously. FSLW is more difficult than butt FSW. Many studies have been conducted on FSLW between Al alloys [11,13,28,34]. Significant research has been conducted on FSLW of Al–Mg [12], Al–Cu [30,35], Al–Ni [31], etc.; however, in practical cases, many studies have been conducted on FSLW of Al alloy and steel [5,9,33,36–39]. Choy et al. [37] performed FSW of 3 mm thick A357 cast Al pipes and FB590 high-strength steel to investigate the significance of the influence of process parameters. As a result, the plunge depth effect was most dominant. However, it was not possible to isolate the effects of plunge depth and tool penetration depth using a single tool. Because FSLW is a different structure from butt FSW, geometrical defects occur [11,34,39,40]. FSW of pipes is tedious owing to its complex geometry, and therefore, research papers are rare. The butt FSW between Al alloy pipe [41–46] and butt FSW between steel pipes are mainly performed [47,48]. Choy et al. [37] performed FSLW of 3 mm thick A357 cast Al and FB590 high-strength steel pipes. Two-factor analysis has been used to identify the factors that affect characteristic values using the design of experiment (DOE) method [49,50].

In previous research, although it was possible to conduct experiments on some of the individual factors, it was difficult to conduct experiments on the relative importance of multiple process factors, owing to time and cost. However, DSD is an innovative experimental method that can reduce time and cost, enabling researchers to select and test the relative importance of multiple process factors. Choy et al. [37] reported the effect of each process factor on tensile shear load (TSL) by performing FSW based on the DSD experimental design for five process factors with three levels for the lap joint of each 3 mm thick pipe-shaped A357 cast Al and FB590 high-strength steel, using a tool of a single dimension. However, the change in the plunge depth affects the tool penetration depth owing to the experiment with a single dimension tool pin; hence, the exact effect of the plunge depth cannot be obtained. Most studies to date have attempted to establish correlations between post-welding properties and the main FSW process variables, welding speed, and tool rotation speed, and one or two individual variables. Additionally, studies on the correlation of multiple process variables are limited.

Therefore, this study considers the influences on the TSL at the time of dissimilar FSLW bonding of pipe-shaped A357 cast Al and FB590 high-strength steel. To isolate the mutual effect of plunge depth and tool penetration depth, which were not considered at all in previous papers, four types of tools with different pin lengths according to the number of levels were selected and tested accordingly. To investigate the relative importance of a number of process factors, which were rarely addressed in previous papers, the DOE method of DSD was adopted to examine the selection of the multiple process factors for TSL, their relative importance, and the effects of linear and curved relationships. Minitab (Ver. 19, Minitab Ltd., State College, PA, USA) was used for DSD design. In addition, the characteristics of the microstructure and TSL were evaluated according to factor 6 and level 3 (depth of plunge was level 2 [51]).

2. Experimental Preparation and Design Methods

2.1. Materials and Tools

The pipes used in this experiment were A357 cast Al and FB590 high-strength steel. The chemical composition of each material is shown in Table 1 [9]. The test specimen, A357 cast Al pipe, was manufactured to have an outer diameter of 111 mm, length of

155 mm, joint thickness of 3 mm, and non-joint thickness of 6 mm. A357 cast Al pipe is subjected to T6 heat treatment after casting and surface treatment through shot peening. FB590 is a high-strength steel pipe with an outer diameter of 105 mm, length of 110 mm, and thickness of 3 mm. Before welding, the aluminum oxide layer of the aluminum alloy was removed with a brush and sandpaper. The two materials were joined by FSLW, as illustrated in Figure 1. Figure 1a shows the four stages of the FSW process, and Figure 1b shows the plunge depth and tool penetration depth of FSLW. The FSW process is primarily classified into four stages: the plunge stage, where the tool descends to the depth of the workpiece; the dwell stage, where the tool stays to provide a constant temperature; the welding stage to join the workpiece; and the retreating stage, where the tool exits after welding is finished.

Table 1. Chemical composition of FB590 (high-strength steel) and A357 (cast Al) pipes [9].

Material	C	Si	Mn	P	S	Cr	Ni
FB590	0.076	0.094	1.472	0.013	0.001	0.019	0.008
Material	Si	Mg	Cu	Zn	Fe	Mn	Ti
A357	6.937	0.507	0.034	0.017	0.181	0.007	0.116

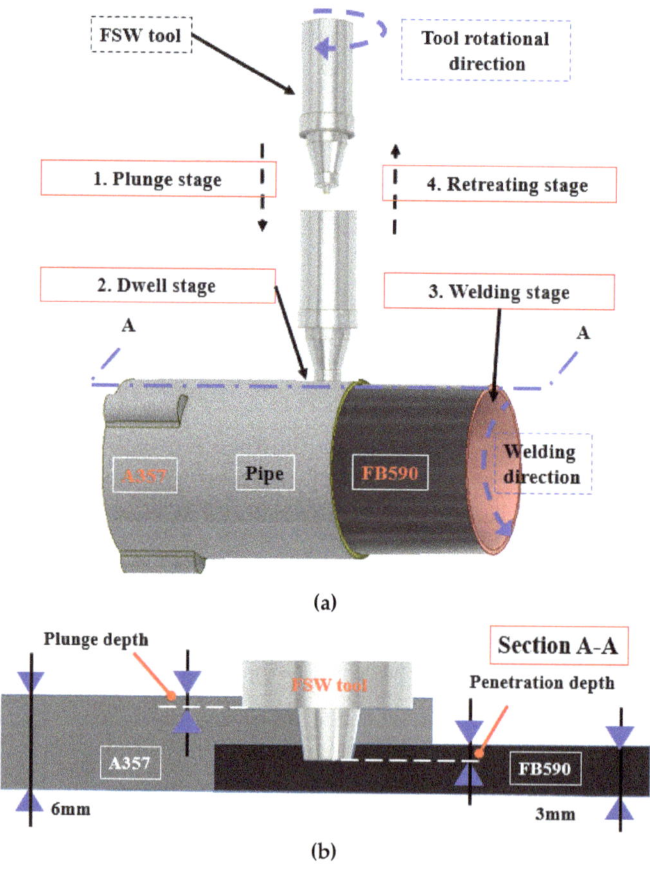

Figure 1. Schematic of FSW experimental configuration. (**a**) Four stages of FSW process; (**b**) plunge depth and penetration depth of FSLW.

During the FSW process, heat is generated by friction between the tool and the workpiece, which causes plastic deformation of the workpiece [52]. While the heat softens the material in the shear layer around the tool, the plastic material flow in the shear layer produces localized viscous dissipation heat energy. The combination of the tool rotation and translation leads the softened material to flow from the front of the tool (advancing side (AS)) to the back of the tool (retreating Side (RS)), where it is forged into a joint. In the FSW process, both the heat generation and material flow have crucial effects on the metallurgical characteristics and mechanical properties of the weld joints [53,54]. Furthermore, the preheating effects of the plunge and dwell stages significantly affect the welding force and tool wear [16]. Moreover, the plunge depth and tool penetration depth due to the pin length of the initially selected tool have important effects on welding force and tool wear. Therefore, a complete understanding of both the heat generation and material flow at different stages of the FSW process is imperative in optimizing the process and controlling the microstructures and joint properties.

The experimental device, shown in Figure 2a, comprises a Winxen milling device that supplies the rotational force of the spindle up to 2000 rpm, a chuck that fixes both sides of FB590 high-strength steel and A 357 cast Al pipe for FSW processing, and a fixing jig consisting of supporting bearings to secure both sides of the FB590 high-strength steel and A357 cast Al pipes for welding. Figure 2b shows the tool used in the FSW processing and the enlarged picture of the scroll shape of the shoulder used to investigate the effect of plunge depth and tool penetration depth. The material and shape dimensions of the tool were selected through a literature review and experiments. The material of the FSW tool was manufactured using W–Ni–Fe alloy, which is a type of heavy alloy. The tool's pin was processed into a threaded shape with cylindrical tape, and the tool's shoulder was processed into a parallel scroll shape to increase the z-axis vertical force, improving frictional heat and stirring during the FSW joining process.

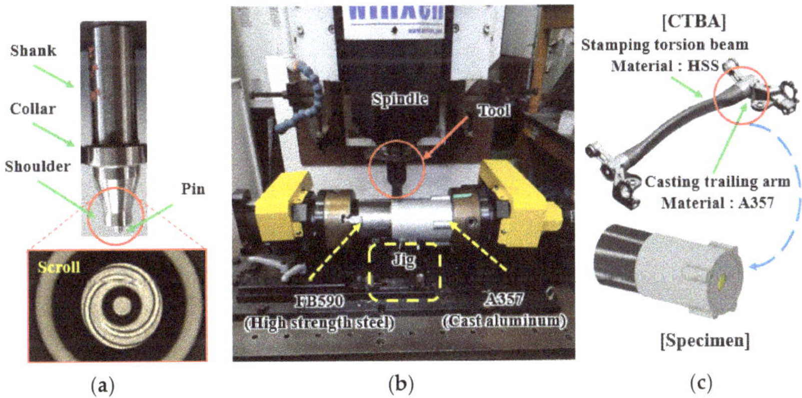

Figure 2. Overall configuration of FSW experiment. (**a**) Photograph of experimental equipment; (**b**) tool and close view of shoulder; (**c**) CTBA and pipe specimen.

The shoulder, pin root, and pin diameter of the tool are 10, 5, and 4 mm, respectively. To exclude the correlation between plunge depth and tool penetration depth, four types of tools with a pin length of 2.5, 3, 3.5, and 4 mm were used.

The experiment was conducted by selecting a tool with a pin length according to the plunge depth and tool penetration depth according to the order and levels of the experiment. Figure 2c shows a CTBA, which is the rear wheel suspension of a vehicle, composed of a trailing arm made of casting (cast Al) and a stamping torsion beam made of high-strength steel; also shown is the pipe specimen for the experiment—a part of the trailing arm and a part of the stamping torsion beam are separated.

2.2. Definitive Screening Design and Analysis

The DOE table of DSD has $2m + 1$ runs for m factors plus m pairs of fold-overs and total centroids. Each run (except all centroids) has a centroid at exactly one factor level, and all other factor levels are designed at the vertices [51,55]. Using the design structure of DSD, a DOE was performed with 14 runs with six factors. Table 2 shows the factors and the number of levels for the DSD.

Table 2. Process parameters (factors) and their levels.

Level \ Factors	Tool Speed (rpm)	Welding Speed (rpm)	Plunge Speed (mm/min)	Dwell Time (s)	Plunge Depth (mm)	Penetration Depth (mm)
−	1700	0.10	5	3	0	0
0	1800	0.15	7	5	-	0.5
+	1900	0.20	9	7	0.5	1

The following FSW process factors were selected: a tool rotation speed of 1700–1900 rpm, pipe welding speed of 0.1–0.2 rpm, plunge speed of 5–9 mm/min, dwell time of 3–7 s, plunge depth of 0–0.5 mm, and tool penetration depth of 0–1.0 mm. Among the number of levels for the six important factors, only the plunge depth was set to two levels, and all others were set to three levels; the microstructure and TSL characteristics of the joint were evaluated by performing DOE with a standard number of runs of 14 using the DSD method. For structural observation and evaluation of mechanical properties of the joint, tensile test specimens and microstructure observation specimens were prepared through wire processing. The test specimen of TSL was manufactured according to the ASTM E8 standard, and TSL was measured using a tensile tester (AGS-X Shimadzu, Japan). The microstructure observation specimen was polished and then observed through an optical microscope (KH-8700, HIROX, Japan).

3. Results and Discussion

3.1. Tensile Shear Load Characteristics of Dissimilar Friction Stir Joints

The tools that have an important influence on the FSW characteristics of A357 cast Al and FB 590 high-strength steel pipes were selected through preliminary experiments with reference to the cited papers. Among the process variables that have an important influence—other than tools—six factors were selected as the factors affecting the TSL of dissimilar materials for FSLW, including those not selected by the previous researchers in the experimental plan. Table 3 shows the results of TSL after FSLW of dissimilar materials; in the case of A357 cast Al raw material, the maximum TSL was 7912 N. The highest TSL was 2672.21 N under the conditions of a tool rotational speed of 1800 rpm, welding speed of 0.2 rpm, plunge speed of 5 mm/min, dwell time of 3 s, plunge depth of 0 mm, and tool penetration depth of 0 mm. The TSL value was 897.35 N under the conditions of a tool rotational speed of 1900 rpm, welding speed of 0.2 rpm, plunge speed of 7 mm/min, dwell time of 7 s, plunge depth of 0 mm, and tool penetration depth of 0 mm. In the study by Choy et al. [37], a TSL value of 3500–4500 N could be obtained by using a single tool with a pin length of 3.3 mm and adding the normal force due to the plunge depth.

Table 3. Experiment level with run order.

No.	Tool Speed (rpm)	Welding Speed (rpm)	Plunge Speed (mm/min)	Dwell Time (s)	Plunge Depth (mm)	Penetration Depth (mm)	Tensile Shear Load (N)
Symbol	A	B	C	D	E	F	TSL
1	1900	0.2	5	3	0.5	0.5	1440.28
2	1800	0.15	7	5	0.5	0.5	1414.87
3	1800	0.2	9	7	0.5	1	1112.12
4	1900	0.2	7	7	0	0	807.35
5	1700	0.1	9	7	0	0.5	860.50
6	1800	0.15	5	5	0	0.5	1779.25
7	1900	0.15	5	3	0	1	2000.99
8	1700	0.15	7	7	0.5	0	0
9	1700	0.2	7	3	0.5	0	1238.16
10	1900	0.1	5	5	0.5	0	0
11	1900	0.1	5	7	0	1	2034.85
12	1900	0.1	7	3	0.5	1	1182.08
13	1800	0.1	5	3	0	0	2677.21
14	1700	0.2	5	5	0	1	1984.72

3.2. *Definitive Screening Design Model and Main Effect Plot*

The model summary in Table 4 shows how well the model explains the observed response variation through process variables with R-Square and R-Square (modified) representing 96.90% and 93.27%. The error (%) of the pre-correction model is 100 (%) − [R-sq] (%) = 3.10 (%) and the error (%) of the modified model is 100 (%) − [R-sq (adj)] (%) = 6.73 (%). The ANOVA table in Table 5 shows the influence of the model terms on the response variable, and statistical significance or influence is judged by the F-statistic and P-value. The appropriate TSL model is a hierarchical model with a small error under the significance level of 10%, and important factors were selected by the DOE method of the DSD step. The influence of the factors on TSL was determined using the coded coefficients and F-statistics and P-values of ANOVA. The regression equation of the TSL value in uncoded units is expressed as Equation (1):

$$TSL = 1193 + 916 \times C - 483.1 \times D - 406.1 \times E - 3370 \times F - 80.0 \times C \times C + 194.7 \times C \times F + 512.6 \times D \times F \quad (1)$$

where C is the plunge speed, D is the dwell time, E is the plunge depth, and F is the tool penetration depth.

Table 4. Model summary.

S	R-sq	R-sq (adj)	R-sq (pred)
197.261	96.90%	93.27%	81.18%

Figure 3 shows the main effect plots for the tensile shear load. This figure shows the average value of tensile shear load according to the number of levels of each of the six factors—tool rotational speed, welding speed, plunge speed, dwell time, plunge depth, and tool penetration depth.

Table 5. Analysis of variance.

Source	DF	Adj SS	Adj MS	F-Value	*p*-Value
Model	7	7,287,025	1,041,004	26.75	0.000
Linear	4	5,644,173	1,411,043	36.26	0.000
Plunge speed	1	432,946	432,946	11.13	0.016
Dwell time	1	1,995,350	1,995,350	51.28	0.000
Plunge depth	1	2,111,423	2,111,423	54.26	0.000
Penetration depth	1	749,284	749,284	19.26	0.005
Square	1	237,734	237,734	6.11	0.048
Plunge speed x Plunge speed	1	237,734	237,734	6.11	0.048
Two-way interactions	2	1,638,507	819,253	21.05	0.002
Plunge speed x Penetration depth	1	280,384	280,384	7.21	0.036
Dwell time x Penetration depth	1	1,603,050	1,603,050	41.20	0.001
Error	6	233,471	38,912		
Total	13	7,520,496			

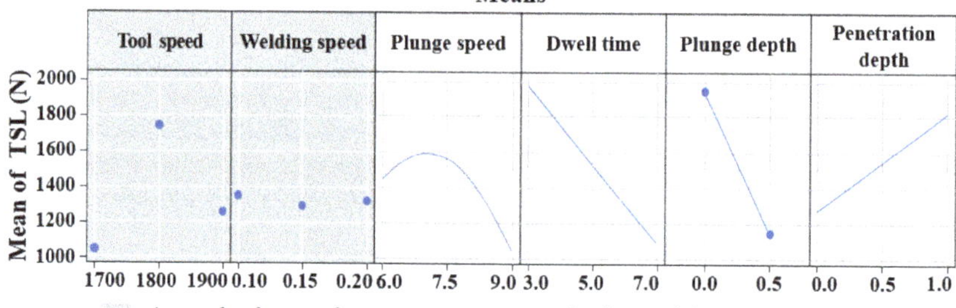

Figure 3. Main effect plots for tensile shear load.

At a given level, it is possible to screen for factors that do not affect TSL with DSD.

Tool rotational speed (A) and welding speed (B) are grayed out in the main effects plot in Figure 3 as they have no effect at any given level. Therefore, the tool rotational speed (A) and welding speed (B) are excluded from the process of finding the maximum value of TSL, and the maximum value of the TSL is found as the remaining four factors that affect it. At a given level, the plunge speed is indicative of the curvature, which is optimal. It was found that dwell time and plunge depth had a negative effect, and only the tool penetration depth had a positive effect. The desired result is high tensile shear load, so the maximum TSL is achieved at a plunge speed of 6.5 mm/min, a dwell time of 3.0 s, a plunge depth of 0.0 mm, and a tool penetration depth of 1.0 mm. This result is inconsistent with the study by Zheng et al. [31], where the plunge depth of dissimilar materials played an important role in the joint tensile strength and increased the maximum breaking load. This does not separate the plunge depth from the tool penetration depth, and the change in the tool penetration depth occurs simultaneously when the plunge depth is changed, resulting in an overall increase in the load when the plunge depth is increased. In addition, Li et al. [30] demonstrated that if the dwell time is short, microcracks occur owing to insufficient heat input, and the hardness may decrease slightly if the dwell time is long; however, the results do not significantly affect the tensile strength. However, this is a result of welding in a steady state, and in the initial stage of welding, the longer the dwell time and the lower the tensile strength. Baskoro et al. [29] presented a result different from this, showing the optimum value at 6.5 mm/min within a given section; their investigation of the effects between process variables showed that the TSL increases as the plunge speed decreases during Al thin plate welding. This is thought to be due to the difference between single materials and dissimilar materials and the difference between butt FSW and FSLW.

3.3. Characteristics between the Factors of a Definitive Screening Design

Figure 4 shows the interaction for TSL. This is according to the interaction between factors in the regression equation of the TSL of Equation (1). Figure 4a is a diagram with the interaction between the plunge speed and the tool penetration depth (C × F) for TSL. In the equation, it is expressed as a curved surface as an effect of the square term of the plunge speed, and it can be seen that an optimal value exists within a given range. It shows the positive effect of increasing TSL with increasing tool penetration depth over the entire range at a given level. Figure 4b is a diagram of the interaction between dwell time and tool penetration depth (D × F) for TSL, where TSL increases with increasing tool penetration depth for dwell times greater than 4 s. On the other hand, when the dwell time is less than 4 s, it shows that the TSL decreases as the tool penetration depth increases.

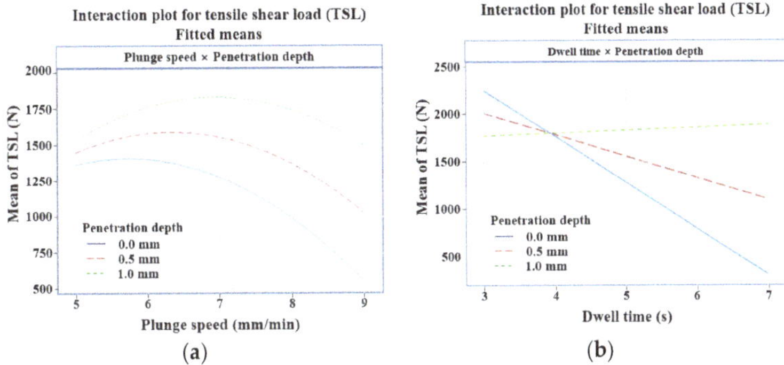

Figure 4. Interaction plot for tensile shear load. (**a**) Penetration depth and plunge speed; (**b**) penetration depth and dwell time.

The plot also shows that the interaction between dwell time and tool penetration depth (D × F) is relatively low for plunge speed and tool penetration depth (C × F) without intersection. Figure 5 shows a contour plot for TSL, and in order to examine the effect of a given factor, fixed values of the remaining factors were selected by considering the main effects plot of Figure 3. Figure 5a shows the plane contour plot of the plunge speed and the tool penetration depth for TSL.

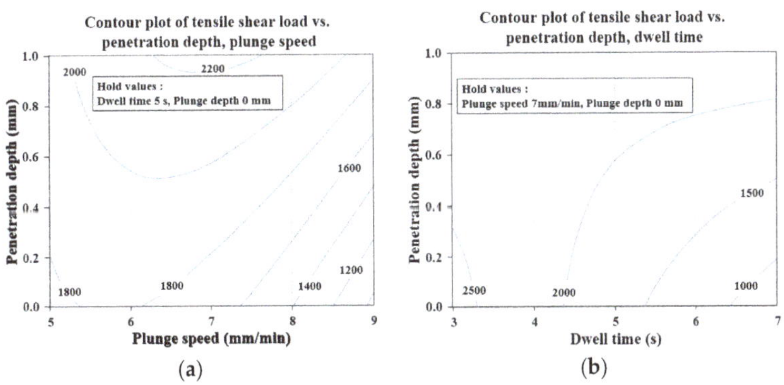

Figure 5. *Cont.*

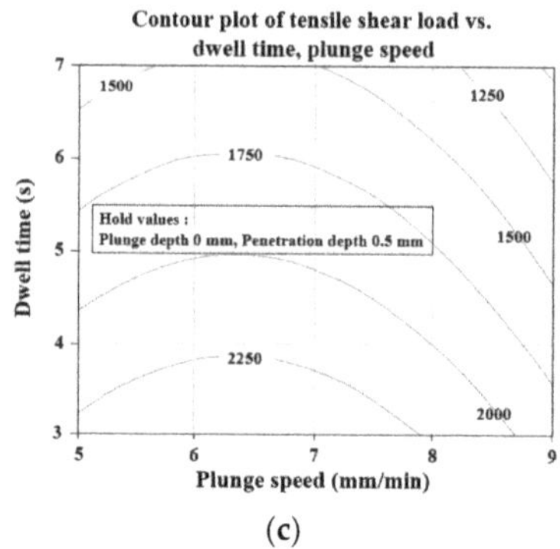

(c)

Figure 5. Contour plots of tensile shear load. (**a**) Penetration depth and plunge speed; (**b**) penetration depth and dwell time; (**c**) dwell time and plunge speed.

A dwell time of 5 s and a plunge depth of 0 mm were chosen for the fixed values of the remaining factors. On the contour plot, you can find the dwell time and tool penetration depth that maximize the TSL. Over the entire range of a given plunge speed, TSL increases with increasing tool penetration depth. Over the entire range of a given tool penetration depth, TSL increases with increasing plunge speed and decreases after reaching a maximum value. The high TSL value at low plunge speed and high tool penetration can also be seen in the main effect plot in Figure 3. Figure 5b shows the plane contour plot of the dwell time and the tool penetration depth for TSL. A plunge speed of 7 mm/min and a plunge depth of 0 mm were chosen from the main effects plots in Figure 3 as fixed values for the remaining factors. On the contour plot, you can find the dwell time and tool penetration depth that maximizes TSL. Over the entire range of a given tool penetration depth, increasing the dwell time decreases the TSL; for dwell times of less than 4 s in a given range, the TSL decreases with increasing tool penetration depth. For dwell times greater than 4 s in a given range, TSL increases with increasing tool penetration depth. It can be seen that in a given range, a maximum TSL value of about 2600 N appears at a dwell time of 3 s and a tool penetration depth of 0 mm. Figure 5c shows a plane contour plot of plunge speed and dwell time for TSL. A tool penetration depth of 0.5 mm and a plunge depth of 0 mm were chosen from the main effects plots in Figure 3 as fixed values for the remaining factors. On the surface plot, you can find the dwell time and plunge speed that maximize the TSL. Over the entire range of a given plunge speed, increasing the dwell time decreases the TSL. Over the entire range of a given dwell time, as the plunge speed increases, the TSL increases, reaching a maximum value and decreasing again after reaching the maximum value. The high TSL value at low dwell time and maximum plunge speed can also be seen from the main effect plots in Figure 3. It can be seen that a maximum TSL of about 2450 N in a given range appears at a dwell time of 3 s and a plunge speed of 6.3 mm/min.

Figure 6 shows a surface plot for TSL. In order to examine the effect of a given factor, the fixed values of the remaining factors were selected in consideration of the main effect plot in Figure 3. Figure 6a shows a three-dimensional surface plot of plunge speed and tool penetration depth for TSL. A dwell time of 5 s and a plunge depth of 0 mm were chosen for the fixed values of the remaining factors. On the surface plot, you can find the plunge speed and tool penetration depth that maximize TSL. Over the entire range of a given plunge speed, TSL increases with increasing tool penetration depth. It can be seen that at a tool penetration depth of 0 mm, TSL increases with increasing plunge speed and decreases after reaching a maximum of about 1800 N. It can be seen that at a tool penetration depth of 0.5 mm, the TSL increases with increasing plunge speed and decreases after reaching a maximum value of about 2000 N. It can be seen that at a tool penetration depth of 1 mm, TSL increases with increasing plunge speed and decreases after reaching a maximum value of about 2200 N. A maximum TSL value of about 2200 N is achieved at a plunge speed of 7 mm/min and a tool penetration depth of 1.0 mm. Figure 6b shows a three-dimensional surface plot of dwell time and tool penetration depth for TSL. A plunge speed of 7 mm/min and a plunge depth of 0 mm were chosen from the main effects plots in Figure 3 as fixed values for the remaining factors. On the surface plot, you can find the dwell time and tool penetration depth that maximize the TSL. At a tool penetration depth of 0 mm, the TSL decreases with increasing dwell time. It can be seen that at a tool penetration depth of 0.5 mm, when the dwell time increases, the TSL decreases to the folding point and increases again after the folding point. At a tool penetration depth of 1 mm, the TSL increases with increasing dwell time. At a dwell time of 3 s, the TSL is decreasing with increasing tool penetration depth. At a dwell time of 5 s, as the tool penetration depth increases, the TSL decreases to the folding point and increases again after the folding point. It can be seen that at a dwell time of 7 s, increasing the tool penetration depth increases the TSL. It can be seen that in a given range, a maximum TSL value of about 2500 N appears at a dwell time of 3 s and a tool penetration depth of 0 mm. Figure 6c shows a three-dimensional surface plot of plunge speed and dwell time for TSL. A tool penetration depth of 0.5 mm and a plunge depth of 0 mm were chosen from the main effects plots in Figure 3 as fixed values for the remaining factors. On the surface plot, you can find the dwell time and plunge speed that maximize the TSL. Over the entire range of a given plunge speed, TSL decreases with increasing dwell time. Over the entire range of a given dwell time, as the plunge speed increases, the TSL increases, reaches a maximum value, and decreases again after reaching the maximum value. It can be seen that a maximum TSL of about 2250 N in a given range appears at a dwell time of 5 s and a plunge speed of 3 mm/min. In order to examine the effect of a given factor, the fixed values of the remaining factors were selected in consideration of the main effect plot in Figure 3.

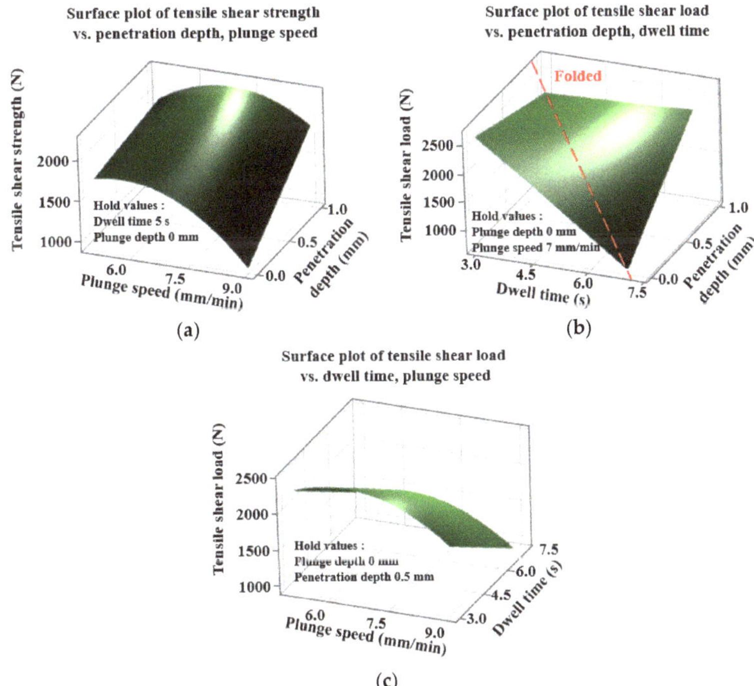

Figure 6. Surface plots of tensile shear load. (**a**) Penetration depth and plunge speed; (**b**) penetration depth and dwell time; (**c**) dwell time and plunge speed.

The individual satisfaction function that maximizes the response d_i is as follows.

$$d_i = 0, \quad \hat{y}_i < L_i \tag{2}$$

$$d_i = \left(\frac{(\hat{y}_i - L_i)}{(T_i - L_i)}\right)^{r_i}, \quad L_i \leq \hat{y}_i \leq T_i \tag{3}$$

$$d_i = 1, \quad \hat{y}_i > T_i \tag{4}$$

where d_i is the individual desirability of the i th response; $\hat{y}_i$ is the expected response value of the i th response; T_i is the target value of the i th response; L_i is the minimum value of the i th response; r_i is the weight of the desirability function of the i th response. The composite desirability function D is expressed as Equation (5).

$$D = \left(\prod (d_i)^{w_i}\right)^{\frac{1}{W}} \ \sum W_i = W \tag{5}$$

where w_i is the importance of the i th response and W is $\sum W_i$.

If there is one response variable and the importance is set to 1 as in Equation (5), individual desirability and composite desirability are the same. Figure 7 shows the reaction optimization for TSL. Among the factors in Figure 3, the tool rotation speed (A) and welding speed (B), which are factors that do not affect TSL, are not considered in the reaction optimization of Figure 7. Therefore, the response optimization value of TSL is found as the remaining four factors affecting TSL. The maximum and minimum levels in the experimental range for each factor are shown, and the maximum value of TSL is found within the combination of these four factors. Using the response optimization tool and the overall satisfaction function = 1 in Equation (5), the optimal conditions for factors maximizing TSL while satisfying the lower limit [2300 N] are shown. In the response optimization analysis, the optimal values of the derived response variables of four factors that satisfy the optimal conditions (mean TSL 2775.49 N, maximizing overall satisfaction (1.0)) are shown. The optimum value of Cur for each factor maximizing TSL is at a plunge speed of 5.7273 mm/min, a dwell time of 3.0 s, a plunge depth of 0 mm, and a tool penetration depth of 0 mm. This value lies between the maximum and minimum values of each factor level. It can be seen that the closer to the maximum value of the plunge speed, the shorter the dwell time, the lower the plunge depth, and the deeper the tool penetration depth, the closer the composite satisfaction is to 1 and the higher the TSL.

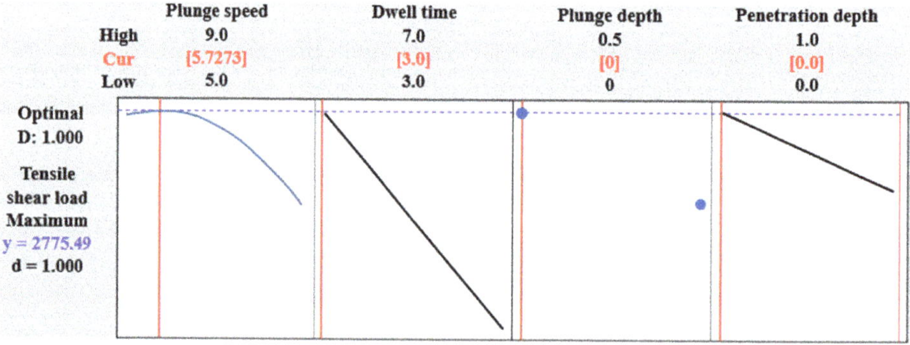

Figure 7. Response optimization of tensile shear load.

3.4. Microstructure Characteristics of Friction Stir Welding Joint

Figure 8 illustrates a magnified view of the microstructure photograph of the FSLW of the A357 cast Al and FB590 high-strength steel pipes and the region of interest around the interface of the dissimilar material to observe the effects of the plunge depth and tool penetration depth.

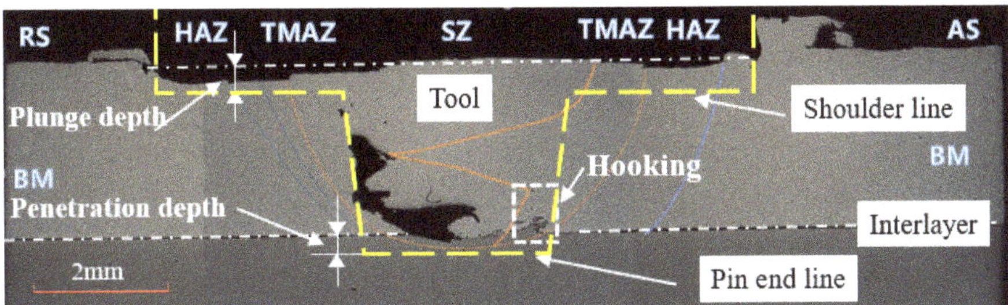

Figure 8. Optical microscope image of the cross-section of lap joints and the dimensions of plunge depth and tool penetration depth.

The shape of the tool pin is indicated by a thick dotted line, and the interlayer is indicated by a dashed-dotted line where A390 Al and FB590 high-strength steel are in contact.

The plunge depth represents the arrow gap between the outer diameter of the A390 Al pipe and the tool shoulder, and the tool penetration depth represents the arrow gap between the boundary layer and the end line of the tool pin. After FSW, there is a stir zone (SZ) around the center of the joint, a thermo-mechanically affected zone (TMAZ) in which grains are increased by the plastic flow on the outside of the SZ, and a heat-affected zone (HAZ) that is heat-affected but has no plastic deformation on the outside of the TMAZ. These zones were observed to have a wider area width than the RS in the tool's AS. In the hooking area, it can be observed that some particles of the steel are raised toward aluminum in a hook shape, indicating that the two materials are physically bonded, and steel fractures are visible around the hooking area. Most of the steel fragments adhered near to AS of the tool pin. In addition, it was observed that steel particles were deposited away from the interface along AS, whereas they were near the interface along RS [40]. Surface defects such as weld flashes and surface grooves were observed at different plunge lengths and tool penetration depths under experimental conditions. The volume of the welding flash increased further as the welding progressed. Similar results were observed by Das et al. [56]; that is, the penetration depth of the tool pin into the workpiece (also known as the target depth) is important for producing a sound weld. They reported that if the plunge depth of the tool is too shallow, the shoulder of the tool will not touch the original workpiece surface, creating a weld with internal channels or surface grooves, and if the plunge depth of the tool is too deep, the shoulder of the tool will enter the workpiece, and cause excessive flash. In contrast, excessive flash and internal cavities were found with and without plunge depth in this study.

Figure 9 shows the microstructure of the specimen after the friction stir test according to the order determined by the corresponding test No. 1–14 for plunge depth and tool penetration depth (a) 0 mm, 0 mm; (b) 0 mm, 0.5 mm; (c) 0 mm, 1 mm; (d) 0.5 mm, 0 mm; (e) 0.5 mm, 0.5 mm; and (f) 0.5 mm, 1 mm, respectively. The picture was taken in the direction of 90 degrees perpendicular to the progress direction of the welding part. Internal welding defects occurred in most of the experimental areas. Major internal defects occur in the form of wormholes, cavities, tunnel defects, and voids and are caused by the lack of material to fill the cavity formed by the flash in the weld area. The cause of the internal cavity is that the region where excessive heat is generated was selected in the process of finding a region where external defects do not occur in different tool pin lengths. The reason for choosing this experimental area is that it is reasonable to select it based on the experimental area of the tool speed and welding speed of Choy et al. [37], where there were no external defects using a tool with fixed single pin-length. In the process of selecting the experimental area with few external defects around this area, the area with similar tool rotation speed and low welding speed was selected.

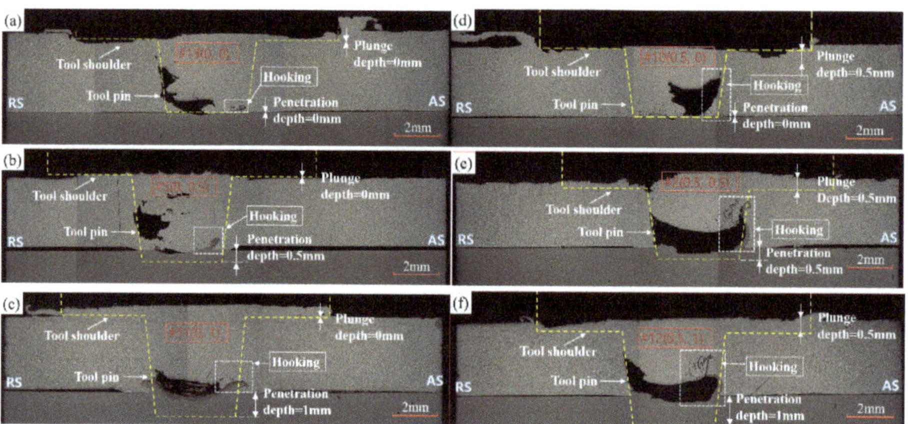

Figure 9. Comparison of the plunge depth and penetration depth on the interface structure. (**a**) (0 mm, 0 mm) (test No. 13); (**b**) (0 mm, 0.5 mm) (test No. 5); (**c**) (0 mm, 1 mm) (test No. 11); (**d**) (0.5 mm, 0 mm) (test No. 10); (**e**) (0.5 mm, 0.5 mm) (test No. 2) (**f**) (0.5 mm, 1 mm) (test No. 12).

As a result, the FSW experiment was conducted in an area where excessive heat was generated. The high heat generated in the weld excessively softens the material inside the weld, and the plasticized material (caused by the continuous stirring action of the tool) is expelled out of the weld in the form of a flash. This leads to the loss of material for completely filling the cavity formed by the rotating tool and results in voids inside the weld. As the process is continuous, the voids extend along the length of the weld to form a tunnel defect. The flow of plasticized material in the FSW process occurs from the advancing side to the RS toward the trailing edge of the tool. The heat and constant rotation of the tool forge the plasticized material toward RS and deposits it in the RS, filling the cavity formed around the tool. Consistent with Arbegast's [57] results, internal weld defects due to insufficient material for the trailing edge of the tool because of over-softening under high-temperature conditions are observed. TSL is affected by plunge depth and tool penetration depth. It was found that as the plunge depth decreased and the tool penetration depth increased, the TSL value increased. On the other hand, according to the geometrical characteristics of FSLW, in experiments No. 13, 5, and 11, a hooking part could be found on the boundary layer, and it was found that the size of the hooking part was affected by the tool penetration depth. Experiment No. 2 and 12 show that, owing to the size and location of the cavity, the hooking part was formed following the shape of the cavity from the boundary layer. However, in experiment No. 10, the hooking part did not form. This seems to be the effect of the difference in the plunge depth when compared with No. 13, which had a penetration depth of 0 mm. In No. 13, where the plunge depth is 0 mm, there is no flash effect by the plunge depth; hence, even if there is no penetration depth, hooking occurs in the process of forming the joint by transmitting the rotational force of the tool. In the case of No. 10, with a plunge depth of 0.5 mm, the material for the hooking part disappears into the cavity, owing to the lack of material to fill the cavity due to the flash effect by the plunge depth; hence, the effect of hooking does not appear. It can be seen that the formation position of the cavity tends to form stably from a one-way bias away from the center of the tool as the tool penetration depth increases and the size of the cavity decreases.

4. Conclusions

FSLW with pipe-type A357 cast Al and FB590 high-strength steel was investigated. Optimized tools were used, and the tool rotational speed, tool welding speed, plunge speed, dwell time, plunge depth, and tool penetration depth were selected among the process variables; and the response was studied through the TSL values. After conducting a total of 14 experiments with 6 factors and 3 levels (excepted for plunge depth, which had 2 levels), the following major results were obtained through tensile test measurements:

(1) Using DSD techniques, which allow for simultaneous identification of the effects of multiple factors with low cost and time, we managed to identify the effects of various process factors in addition to rotational and weld speeds—which are known as the most important factors in FSW—and identified the factors affecting TSL.

(2) Among the process factors selected for the friction stir welding of pipe-type Al and steel, the impact of flange depth and tool penetration depth was most significant to TSL, and an independent relationship with no interaction between plunge depth and tool penetration depth was identified. The plunge depth has a negative effect, and the tool penetration depth has a positive effect on the magnitude of the TSL.

(3) The weak interaction effect between the plunge speed and tool penetration depth and the strong interaction effect between the dwell time and tool penetration depth were confirmed. The folding phenomenon of the interaction between tool penetration depth and dwell time was found to have an opposite effect on TSL depending on the direction of increase or decrease of the factor.

(4) The depth of tool penetration and plunge depth affected the cavity size; the tool penetration depth contributed to the stabilization and size reduction of the cavity.

(5) It was found that the tool penetration depth had the greatest influence on the size of the hooking part in the lap welding of the pipe; moreover, the hooking part, which was not created on the boundary layer by the increase in the plunge depth, was distributed in a certain size along the shape of the cavity.

By confirming the influence of process factors through DSD experiments, we reached the important conclusion that when selecting a tool, the plunge depth and tool penetration depth should be considered, especially in FSLW. This will play an important role in future dissimilar FSLW experiments and can be utilized for optimization through the additional selection of levels.

Author Contributions: Methodology, L.C.; validation, L.C.; formal analysis, L.C.; investigation, L.C.; resources, L.C.; data curation, L.C.; writing—original draft preparation, L.C.; writing—review and editing, L.C., S.K. and J.P.; supervision, M.K. and D.J. All authors have read and agreed to the published version of the manuscript.

Funding: This research was funded by a National Research Foundation of Korea (NRF) grant funded by the Korean Government (MOE) (No. NRF- 2018R1D1A1B07051302).

Institutional Review Board Statement: Not applicable.

Informed Consent Statement: Not applicable.

Data Availability Statement: Not Applicable.

Conflicts of Interest: The authors declare no conflict of interest.

References

1. Joo, Y.H.; Park, Y.C.; Lee, Y.M.; Kim, K.H.; Kang, M.C. The weldability of a thin friction stir welded plate of Al5052-H32 using high frequency spindle. *J. Korean Soc. Manuf. Process Eng.* **2017**, *16*, 90–95. [CrossRef]
2. Laska, A.; Szkodo, M. Manufacturing parameters, materials, and welds properties of butt friction stir welded joints–overview. *Materials* **2020**, *13*, 4940. [CrossRef] [PubMed]
3. Park, J.H.; Jeon, S.T.; Lee, T.J.; Kang, J.D.; Kang, M.C. Effect on drive point dynamic stiffness and lightweight chassis component by using topology and topography optimization. *J. Korean Soc. Manuf. Process Eng.* **2018**, *17*, 141–147. [CrossRef]
4. Ryabov, V.R. *Aluminizing of Steel*; Oxonian Press: New Delhi, India, 1985.

5. Wan, L.; Huang, Y. Friction stir welding of dissimilar aluminum alloys and steels: A review. *Int. J. Adv. Manuf. Technol.* **2018**, *99*, 1781–1811. [CrossRef]
6. Thomas, W.M.; Nicholas, E.D.; Needham, J.C.; Murch, M.G.; Temple-Smith, P.; Dawes, C.J. Friction Welding. U.S. Patent 5,460,317, 1991. Available online: https://patents.google.com/patent/US5460317 (accessed on 2 September 2021).
7. Sampath, K. An understanding of HSLA-65 plate steels. *J. Mater. Eng. Perform.* **2006**, *15*, 32–40. [CrossRef]
8. Park, S.K.; Hong, S.T.; Park, J.H.; Park, K.Y.; Kwon, Y.J.; Son, H.J. Effect of material locations on properties of friction stir welding joints of dissimilar aluminium alloys. *Sci. Technol. Weld. Joi.* **2010**, *15*, 331–336. [CrossRef]
9. Park, J.H.; Park, S.H.; Park, S.H.; Joo, Y.H.; Kang, M.C. Evaluation of mechanical properties with tool rotational speed in dissimilar cast aluminum and high-strength steel of lap jointed friction stir welding. *J. Korean Soc. Manuf. Process Eng.* **2019**, *18*, 90–96. [CrossRef]
10. Dubourg, L.; Merati, A.; Jahazi, M. Process optimization and mechanical properties of friction stir lap welds of 7075-T6 stringers on 2024-T3 skin. *Mater. Des.* **2010**, *31*, 3324–3330. [CrossRef]
11. Wang, M.; Zhang, H.; Zhang, J.; Zhang, X.; Yang, L. Effect of pin length on hook size and joint properties in friction stir lap welding of 7B04 aluminum alloy. *J. Mater. Eng. Perform.* **2014**, *23*, 1881–1886. [CrossRef]
12. Cao, X.; Jahazi, M. Effect of tool rotational speed and probe length on lap joint quality of a friction stir welded magnesium alloy. *Mater. Des.* **2011**, *32*, 1–11. [CrossRef]
13. Salari, E.; Jahazi, M.; Khodabandeh, A.; Ghasemi-Nanesa, H. Influence of tool geometry and rotational speed on mechanical properties and defect formation in friction stir lap welded 5456 aluminum alloy sheets. *Mater. Des.* **2014**, *58*, 381–389. [CrossRef]
14. Mishra, R.S.; Ma, Z.Y. Friction stir welding and processing. *Mater. Sci. Eng.* **2005**, *50*, 1–78. [CrossRef]
15. Kumar, K.; Kailas, S.V. The role of friction stir welding tool on material flow and weld formation. *Mater. Sci. Eng.* **2008**, *485*, 367–374. [CrossRef]
16. Rai, R.; De, A.; Bhadeshia, H.; DebRoy, T. Review: Friction stir welding tools. *Sci. Technol. Weld. Joi.* **2011**, *16*, 325–342. [CrossRef]
17. Khalilabad, M.M.; Zedan, Y.; Texier, D.; Jahazi, M.; Bocher, P. Effect of tool geometry and welding speed on mechanical properties of dissimilar AA2198–AA2024 FSWed joint. *J. Manuf. Process Soc. Manuf. Eng. A* **2018**, *34*, 86–95. [CrossRef]
18. Reza-E-Rabby, M.; Tang, W. Effects of thread interruptions on tool pins in friction stir welding of AA6061. *Sci. Technol. Weld. Joi.* **2017**, *23*, 114–124. [CrossRef]
19. Dialami, N.; Cervera, M.; Chiumenti, M.; Segatori, A.; Osikowicz, W. Experimental validation of an FSW model with an enhanced friction law: Application to a threaded cylindrical pin tool. *Metals* **2017**, *7*, 491. [CrossRef]
20. Trimble, D.; O'Donnell, G.E.; Monaghan, J. Characterisation of tool shape and rotational speed for increased speed during friction stir welding of AA2024-T3. *J. Manuf. Process.* **2015**, *17*, 141–150. [CrossRef]
21. Trueba, L., Jr.; Heredia, G.; Rybicki, D.; Johannes, L.B. Effect of tool shoulder features on defects and tensile properties of friction stir welded aluminum 6061-T6. *J. Mater. Process. Technol.* **2014**, *219*, 271–277. [CrossRef]
22. Zhang, H.J.; Wang, M.; Zhu, Z.; Zhang, X.; Yu, T.; Wu, Z.Q. Impact of shoulder concavity on non-tool-tilt friction stir welding of 5052 aluminum alloy. *Int. J. Adv. Manuf. Technol.* **2018**, *96*, 1497–1506. [CrossRef]
23. Uygur, I. Influence of shoulder diameter on mechanical response and microstructure of FSW welded 1050 Al-alloy. *Arch. Metall. Mater.* **2012**, *57*, 53–60. [CrossRef]
24. Khan, N.Z.; Khan, Z.A.; Siddiquee, A.N. Effect of shoulder diameter to pin diameter (D/d) ratio on tensile strength of friction stir welded 6063 aluminium alloy. *Mater. Today Proc.* **2015**, *2*, 1450–1457. [CrossRef]
25. Imani, Y.; Guillot, M.; Tremblay, A. Optimization of FSW tool design and operating parameters for butt welding of AA6061-T6 at right angle. In Proceedings of the Canadian Society for Mechanical Engineering International Congress, Toronto, ON, Canda, 1–4 June 2014; pp. 1–6.
26. Zhang, Y.N.; Cao, X.; Larose, S.; Wanjara, P. Review of tools for friction stir welding and processing. *Can. Metall. Q.* **2012**, *51*, 250–261. [CrossRef]
27. Shah, L.H.; Midawi, A.R.; Walbridge, S.; Gerlich, A. Influence of tool offsetting and base metal positioning on the material flow of AA5052-AA6061 dissimilar friction stir welding. *J. Mech. Eng. Sci.* **2020**, *14*, 6393–6402. [CrossRef]
28. Barlas, Z. The influence of tool tilt angle on 1050 aluminum lap joint in friction stir welding process. *Acta Phys. Pol. A* **2017**, *132*, 679–681. [CrossRef]
29. Baskoro, A.S.; Suwarsono; Habibullah, M.D.; Arvay, Z.; Kiswanto, G.; Winarto; Chen, Z.W. Effects of dwell-time and plunge speed during micro friction stir spot welding on mechanical properties of thin aluminum A1100 welds. *Appl. Mech. Mater.* **2015**, *758*, 29–34. [CrossRef]
30. Li, G.; Zhou, L.; Zhou, W.; Song, X.; Huang, Y. Influence of dwell time on microstructure evolution and mechanical properties of dissimilar friction stir spot welded aluminum–copper metals. *J. Mater. Res. Technol.* **2019**, *8*, 2613–2624. [CrossRef]
31. Zheng, Q.; Feng, X.; Shen, Y.; Huang, G.; Zhao, P. Effect of plunge depth on microstructure and mechanical properties of FSW lap joint between aluminum alloy and nickel-base alloy. *J. Alloys Compd.* **2017**, *695*, 952–961. [CrossRef]
32. Devanathan, C.; Babu, A.S. Effect of plunge depth on friction stir welding of Al 6063. In Proceedings of the 2nd International Conference on Advanced Manufacturing and Automation, Srivilliputhur, India, 28–30 March 2013; pp. 482–485.
33. Wei, Y.; Li, J.; Xiong, J.; Zhang, F. Effect of tool pin insertion depth on friction stir lap welding of aluminum to stainless steel. *J. Mater. Eng. Perform.* **2013**, 22, 3005–3013. [CrossRef]

34. Aldanondo, E.; Vivas, J.; Álvarez, P.; Hurtado, I. Effect of tool geometry and welding parameters on friction stir welded lap joint formation with AA2099-T83 and AA2060-T8E30 aluminium alloys. *Metals* **2020**, *10*, 872. [CrossRef]
35. Regensburg, A.; Schürer, R.; Weigl, M.; Bergmann, J.P. Influence of pin length and electrochemical platings on the mechanical strength and macroscopic defect formation in stationary shoulder friction stir welding of aluminium to copper. *Metals* **2018**, *8*, 85. [CrossRef]
36. Safeen, M.W.; Spena, P.R. Main issues in quality of friction stir welding joints of aluminum alloy and steel sheets. *Metals* **2019**, *9*, 610. [CrossRef]
37. Choy, L.J.; Park, S.H.; Lee, M.W.; Park, J.H.; Choi, B.J.; Kang, M.C. Tensile strength application using a definitive screening design method in friction stir welding of dissimilar cast aluminum and high-strength steel with pipe shape. *J. Korean Soc. Manuf. Process Eng.* **2020**, *19*, 98–104. [CrossRef]
38. Shen, Z.; Chen, Y.; Haghshenas, M.; Gerlich, A.P. Role of welding parameters on interfacial bonding in dissimilar steel/aluminum friction stir welds. *Eng. Sci. Technol.* **2015**, *18*, 270–277. [CrossRef]
39. Mahto, R.P.; Kumar, R.; Pal, S.K. Characterizations of weld defects, intermetallic compounds and mechanical properties of friction stir lap welded dissimilar alloys. *Mater. Charact.* **2020**, *160*, 110115. [CrossRef]
40. Zhao, Y.; Zhou, L.; Wang, Q.; Yan, K.; Zou, J. Defects and tensile properties of 6013 aluminum alloy T-joints by friction stir welding. *Mater. Des.* **2014**, *57*, 146–155. [CrossRef]
41. Sabry, I.; El-Kassas, A.M. New quality monitoring system for friction stir welded joints of aluminium pipes. *Int. J. Eng. Technol.* **2019**, *11*, 78–87. [CrossRef]
42. Hattingh, D.G.; von Welligh, L.G.; Bernard, D.; Susmel, L.; Tovo, R.; James, M.N. Semiautomatic friction stir welding of 38 mm OD 6082-T6 aluminium tubes. *J. Mater. Process. Technol.* **2016**, *238*, 255–266. [CrossRef]
43. El-Kassas, A.M.; Sabry, I. Using multi criteria decision making in optimizing the friction stir welding process of pipes: A tool pin diameter perspective. *Int. J. Appl. Eng. Res.* **2019**, *14*, 3668–3677.
44. Iqbal, M.P.; Vishwakarma, R.K.; Pal, S.K.; Mandal, P. Influence of plunge depth during friction stir welding of aluminum pipes. *Proc. Inst. Mech. Eng. B J. Eng. Manuf.* **2020**, *23*, 0954405420949754. [CrossRef]
45. Prakash, M.; Das, A.D. Investigation on effect of FSW parameters of aluminium alloy using full factorial design. *Mater. Today Proc.* **2021**, *37*, 608–613. [CrossRef]
46. Akbari, M.; Asadi, P. Optimization of microstructural and mechanical properties of friction stir welded A356 pipes using Taguchi method. *Mater. Res. Express* **2019**, *6*, 066545. [CrossRef]
47. Giorjão, R.A.; Pereira, V.F.; Terada, M.; Fonseca, E.B.; Marinho, R.R.; Garcia, D.M.; Tschiptschin, A.P. Microstructure and mechanical properties of friction stir welded 8mm pipe SAF 2507 super duplex stainless steel. *J. Mater. Res. Technol.* **2019**, *8*, 243–249. [CrossRef]
48. Fairchild, D.; Kumar, A.; Ford, S.; Nissley, N.; Ayer, R.; Jin, H.; Ozekcin, A. Research concerning the friction stir welding of linepipe steels. In Proceedings of the 8th International Conference on Trends in Welding Research, Pine Mountain, GA, USA, 1–6 June 2008; ASM International: Almere, The Netherlands, 2009; pp. 371–380. [CrossRef]
49. Bitondo, C.; Prisco, U.; Squilace, A.; Buonadonna, P.; Dionoro, G. Friction-stir welding of AA 2198 butt joints: Mechanical characterization of the process and of the welds through DOE analysis. *Int. J. Adv. Manuf. Technol.* **2011**, *53*, 505–516. [CrossRef]
50. Filippis, L.A.; Serio, L.M.; Palumbo, D.; Finis, R.; Galiett, U. Optimization and characterization of the friction stir welded sheets of AA 5754-H111: Monitoring of the quality of joints with thermographic techniques. *Materials* **2017**, *10*, 1165. [CrossRef] [PubMed]
51. Nachtsheim, A.C.; Shen, W.; Lin, D.K. Two-level augmented definitive screening designs. *J. Qual. Technol.* **2017**, *49*, 93–107. [CrossRef]
52. Wu, C.S.; Zhang, W.B.; Shi, L.; Chen, M.A. Visualization and simulation of the plastic material flow in friction stir welding of aluminium alloy 2024 plates. *Trans. Nonferr. Met. Soc. China* **2012**, *22*, 1445–1451. [CrossRef]
53. Soundararajan, V.; Zekovic, S.; Kovacevic, R. Thermo-mechanical model with adaptive boundary conditions for friction stir welding of Al 6061. *Int J. Mach. Tools Manuf.* **2005**, *45*, 1577–1587. [CrossRef]
54. Cui, S.; Chen, Z.W.; Robson, J.D. A model relating tool torque and its associated power and specific energy to rotation and forward speeds during friction stir welding/processing. *Int. J. Mach. Tools Manuf.* **2010**, *50*, 1023–1030. [CrossRef]
55. Jones, B.; Nachtsheim, C.J. A class of three-level designs for definitive screening in the presence of second-order effects. *J. Qual. Technol.* **2011**, *43*, 1–15. [CrossRef]
56. Das, B.; Pal, S.; Bag, S. Weld defect identification in friction stir welding using power spectral density. *Mater. Sci. Eng.* **2018**, *346*, 012049. [CrossRef]
57. Arbegast, W.J.; Mishra, R.S.; Mahaney, M.W. *Friction Stir Welding and Processing*; Chapter 13; ASM Int.: Materials Park, OH, USA, 2007; ISBN 978-0-87170-840-3.

Article

Bobbin Tool Friction Stir Welding of Aluminum Thick Lap Joints: Effect of Process Parameters on Temperature Distribution and Joints' Properties

Mohamed M. Z. Ahmed [1,2], Mohamed I. A. Habba [3], Mohamed M. El-Sayed Seleman [2], Khalil Hajlaoui [4], Sabbah Ataya [4,*], Fahamsyah H. Latief [4] and Ahmed E. EL-Nikhaily [3]

1 Mechanical Engineering Department, College of Engineering at Al Kharj, Prince Sattam Bin Abdulaziz University, Al Kharj 16273, Saudi Arabia; moh.ahmed@psau.edu.sa
2 Department of Metallurgical and Materials Engineering, Faculty of Petroleum and Mining Engineering, Suez University, Suez 43512, Egypt; mohamed.elnagar@suezuniv.edu.eg
3 Mechanical Department, Faculty of Technology and Education, Suez University, Suez 43518, Egypt; Mohamed.Atia@suezuniv.edu.eg (M.I.A.H.); ahmed.eassa@ind.suezuni.edu.eg (A.E.E.-N.)
4 Department of Mechanical Engineering, College of Engineering, Imam Mohammad Ibn Saud Islamic University, Riyadh 11432, Saudi Arabia; kmhajlaoui@imamu.edu.sa (K.H.); fhlatief@imamu.edu.sa (F.H.L.)
* Correspondence: smataya@imamu.edu.sa

Abstract: Bobbin tool friction stir welding (BT-FSW) is characterized by a fully penetrated pin and double-sided shoulder that promote symmetrical solid-state joints. However, control of the processing parameters to obtain defect-free thick lap joints is still difficult and needs more effort. In this study, the BT-FSW process was used to produce 10 mm AA1050-H14 similar lap joints. A newly designed bobbin tool (BT) with three different pin geometries (cylindrical, square, and triangular) and concave shoulders profile was designed, manufactured, and applied to produce the Al alloy lap joints. The experiments were carried out at a constant tool rotation speed of 600 rpm and a wide range of various welding travel speeds of 200, 400, 600, 800, and 1000 mm/min. The generated temperature during the BT-FSW process was recorded and analyzed at the joints' center line, and at both advancing and retreating sides. Visual inspection, macrostructures, hardness, and tensile properties were investigated. The fracture surfaces after tensile testing were also examined. The results showed that the pin geometry and travel speed are considered the most important controlling parameters in BT-FSW thick lap joints. The square (Sq) pin geometry gives the highest BT-FSW stir zone temperature compared to the other two pins, cylindrical (Cy) and triangular (Tr), whereas the Tr pin gives the lowest stir zone temperature at all applied travel speeds from 200 to 1000 mm/min. Furthermore, the temperature along the lap joints decreased with increasing the welding speed, and the maximum temperature of 380 °C was obtained at the lowest travel speed of 200 mm/min with applying Sq pin geometry. The temperature at the advancing side (AS) was higher than that at the retreating side (RS) by around 20 °C. Defect-free welds were produced using a bobbin tool with Cy and Sq pin geometries at all the travel welding speeds investigated. BT-FSW at a travel speed of 200 mm/min leads to the highest tensile shear properties, in the case of using the Sq pin. The hardness profiles showed a significant effect for both the tool pin geometry and the welding speed, whereas the width of the softened region is reduced dramatically with increasing the welding speed and using the triangular pin.

Keywords: bobbin tool; AA1050-H14; pin geometry; travel speed; welding temperature; mechanical properties

Citation: Ahmed, M.M.Z.; Habba, M.I.A.; El-Sayed Seleman, M.M.; Hajlaoui, K.; Ataya, S.; Latief, F.H.; EL-Nikhaily, A.E. Bobbin Tool Friction Stir Welding of Aluminum Thick Lap Joints: Effect of Process Parameters on Temperature Distribution and Joints' Properties. *Materials* **2021**, *14*, 4585. https://doi.org/10.3390/ma14164585

Academic Editor: Józef Iwaszko

Received: 31 July 2021
Accepted: 12 August 2021
Published: 15 August 2021

1. Introduction

In the last few years, in engineering applications, the focus is directed more on Bobbin Tool Friction Stir Welding (BT-FSW) due to various advantages over Conventional Tool

Friction Stir Welding (CT-FSW) [1]. The name "bobbin tool; BT" refers to the shape of a tool that consists of two shoulders (upper and lower) connected with a pin [2]. In fact, the bottom shoulder in BT design replaced the backing plate used in the CT-FSW [1]. The BT-FSW process starts with a slow speed until plastic deformation starts, followed by increasing travel speed to the required value. The two shoulders generate frictional and deformation heat from both sides to the workpiece during welding [3]. In friction stir welding (FSW) processes, using BT has many advantages over the use of a conventional tool (CT) such as the welded structure is symmetric in thickness, low distortion of weld joint can be obtained, the elimination of root for welds, a backing plate is not required, and high force is not required for fixing the weld plates and possibility welding a hollow section (U and H shapes) [3]. Various designs of the shoulder were suggested and applied in CT-FSW and BT-FSW processes [4]. Flat and concave are the primary design features among all shapes of shoulders [5]. Concentric circles, ridges, grooves, and knurling are secondary features that may be added for welding thick parts [6,7]. For the concave shoulder, the concavity angle is another important parameter to the weld joint properties. Compared with the flat shoulder, the concave shoulder has been demonstrated to be able to minimize weld flash and improve the joint properties [4]. Whilst the tool shoulder stimulates bulk material flow, the stirring pin during the welding process fosters a layer-by-layer material to flow [8]. Furthermore, the heat generation during the FSW has a strong effect on the weld quality, and it comes from specific tool surfaces, designed tool shoulder, and pin geometry [9,10].The geometry of the tool pin affects the flow of plasticized material and the joint efficiency [10,11]. In fact, the pin geometry is an important parameter in the friction stir welding (FSW) process for temperature history, material flow, and grain size, as well as the quality of the FSW joints [12,13]. Fujii et al. [14] investigated the effect of CT-FSW shape (column without threads, column with threads, and triangular prism shape) on mechanical properties of welding 5 mm AA1050-H14-H24 Al alloy in similar butt joints at different rotation and traverse welding speeds and found that column pin without threads offers butt weld joint with best mechanical properties and this tool shape prompted defects less than the other two tools. Chandrashekar et al. [15] carried out CT-FSW on 6 mm Al plates using three pin profiles: tapered, cylindrical, straight square, and straight cylindrical. The results showed that the square pin profile produced better properties for the joints in comparison with the other two pin profiles. Zhao et al. [16] examined the influence of four conventional tools (CTs) having different pin geometries on the AA2024 Al alloy joints quality and reported that the screw-pitched taper pin produced friction stir welds with minimum defects. Elangovan et al. [17] studied the role of five pin profiles: tapered cylindrical, straight cylindrical, threaded cylindrical, triangular, and square to friction stir weld of AA 6061 Al alloy and found that the CT with the square pin geometry produced defect-free welds for all the applied downward forces used. Mahmoud et al. [18] studied the dispersion of SiC in AA1050-H14-H24 Al surface to fabricate surface composites based on friction stir process using four CTs with different pins: triangular, circular with thread, circular without thread, and square. The square pin geometry produced more homogeneous dispersion of the SiC particles than the other pins, whereas the circular pin profile showed more wear resistance than the flat-faced pin profiles. Wen et al. [9] studied the mechanical properties of the BT-FSW 4 mm thick AA2219-T78 Al alloy welds at various welding speeds from 130 to 350 mm/min and reported a maximum tensile strength obtained at 210 mm/min travel speed with joint efficiency 70%. Li et al. [19] investigated the effect of BT-FSW rotation speed on 4 mm thick 6082-T6 butt joints using a smooth cylindrical pin and concluded that the temperature at the retreating side (RS) was higher than the advanced side (AS) by 20 °C. Moreover, the maximum ultimate tensile strength of 262.7 MPa was obtained at a rotation speed of 800 rpm. Wang et al. [20] studied the influence of rotation speed on the microstructure and mechanical properties of 3.2 mm thick AA2198-T851 BT-FSW butt welds using a cylindrical featureless pin and reported symmetrical hardness distribution along the cross-sections for all welds, and the maximum tensile strength of 370 MPa was reached at 800 rpm. In spite of these achievements in

CT-FSW parameters, the BT-FSW parameters needs more research to clarify the effect of pin geometry and travel welding speed, especially for a thick weld for a material possesses low deformation resistance to avoid the role of precipices or solution hardening. Thus, the aims of this work are placed on to determine the influence of the pin geometry and the welding travel speed on the heat generation and the joint quality of 10 mm thick AA1050-H14 similar lab joints welded via BT-FSW. The mechanical properties in terms of hardness and tensile properties are also examined and analyzed.

2. Materials and Methods

AA1050-H14 Al-sheets of 5 mm thickness, 1000 mm length, and 1000 mm width were cut into plate specimens having the dimension of 120 mm length and 110 mm width to BT-FSW purposes to produce similar lap joints. Tables 1 and 2 show the nominal chemical composition and mechanical properties of the as-received AA1050-H14 Al alloy, respectively.

Table 1. Chemical composition of AA1050-H14 Al alloy.

Composition, in wt.%							
Fe	Si	Cr	Zn	Mg	Cu	Mn	Al
0.50	0.25	0.1	0.07	0.05	0.05	0.05	Balance

Table 2. Mechanical properties of AA1050-H14 Al alloy.

Mechanical Properties			
Tensile strength, MPa	Yield Strength, MPa	Elongation, %	Hardness, HV
110	103	10	28–30

Figure 1 shows example of lap joint arrangement; two sheets 5 mm thick, 100 mm wide, and 120 mm long were used with 40 mm overlap to develop lab joints.

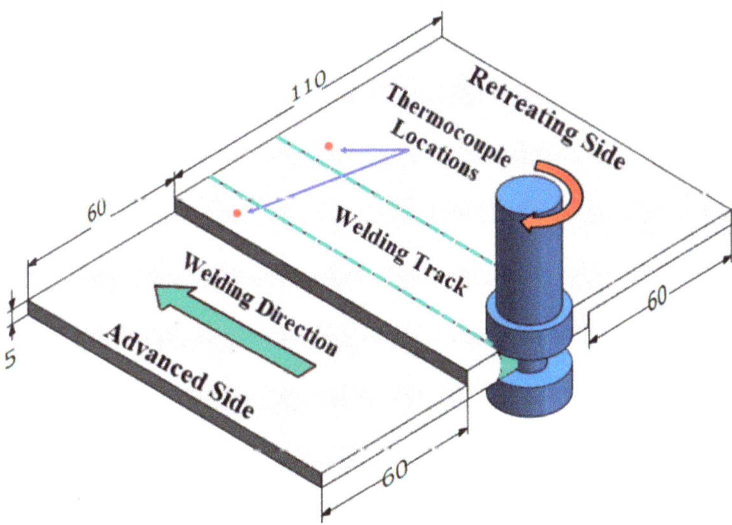

Figure 1. Dimensions of BT-FSW lap joint (all dimensions in mm).

A newly designed and fabricated bobbin tool with different pin geometries was tried in the current study to explore the influence of BT-FSW pin geometries on the joint properties. Figure 2a,b shows the dimensions of the bobbin tool assembly and its various parts, upper shoulder, holder, and lower shoulder. The three different pins, cylindrical (Cy), square (Sq), and triangular (Tr) (Figure 2c,d), were used, with the same upper and lower shoulder dimensions. The new bobbin tools with their pin facilities were machined from cold worked H13 tool steel and were heat-treated to obtain hardness of 52 HRC. The shoulders gap is kept constant to be 25 mm. Both shoulders have concave (6°) and cavities features as given in Figure 2c.

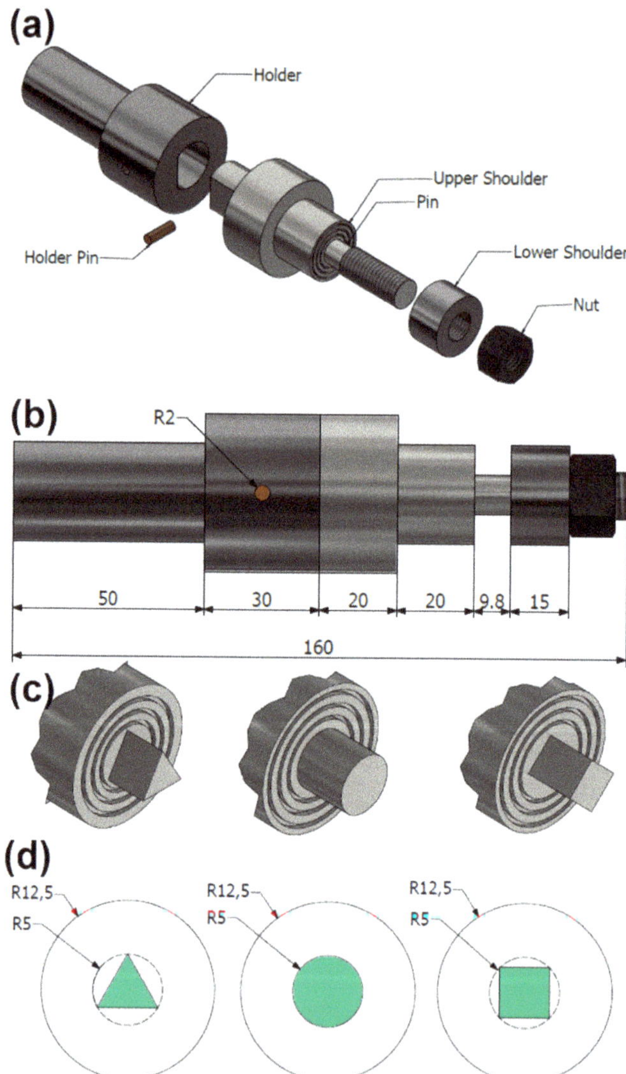

Figure 2. Dimension of BT assembly (all dimensions in mm). (**a**) Exploded assembly of BT parts, (**b**) Dimensions of BT, (**c**) 3D view of used BT pin profiles, and (**d**) dimension of BT pins and shoulders.

The BT-FSW was carried out at various travel speeds "Ts" of 200, 400, 600, 800, and 1000 mm/min using the three different pin geometries (Cy, Sq, and Tr), and constant rotation speed "Rs" of 600 rpm at 0 tilt angle. Table 3 summarizes the BT-FSW process parameters. BT-FSW similar lap joints were carried out using the FSW machine (EG-FSW-

M1) [21–26]. Adaptive fixture was designed and fabricated from steel for BT-FSW purposes. Figure 3a shows overview the BT-FSW using this fixture and Figure 3b shows the 3D drawing with identification of all components.

Table 3. BT-FSW process parameters of AA1050-H14 lap joints.

BT-FSW Process Parameters	
BT pin profile	Cylindrical pin, square pin, triangle pin
Rotation speed, rpm	600
Travel speed, mm/min	200, 400, 600, 800, 1000

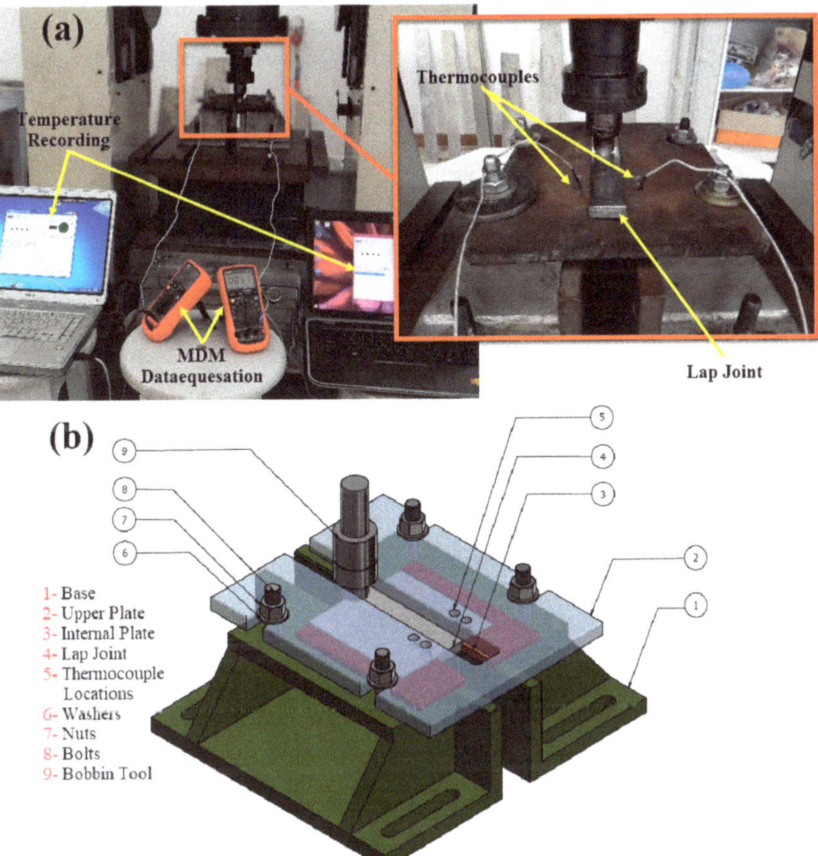

Figure 3. BT-FSW fixture setup configuration to weld AA1050-H14 in similar lap joints at the different welding parameters. (**a**) overview the BT-FSW using this fixture and (**b**) the 3D drawing with identification of all components.

The generated temperatures in the stirring zone, advancing side, and retreating side during the BT-FSW process were measured using two techniques. The temperature distribution along the lap joint during BT-FSW has been measured at several points. The temperature at the weld surface behind the tool was measured using an infrared thermometer (Quicktemp 860-T3, Testo Company—Berlin, Germany). The temperature was measured at different stirring zones during the traveling of BT at the centerline during welding specimens for all processing parameters in terms of various traverse welding speeds and different pin geometries. The mean values were recorded for each welding condition for comparison purposes. A special setup was created using Modern Digital Multimeter (MDM) model (UT61B—Zhejiang, China) with thermocouple type "K" in order to measure the temperature at the AS and RS as given in Figure 3a. With MDM device, two thermocouples were used to collect the temperatures by placing them in two holes close to the weld pass in the heat-affected zone (HAZ) with 3 mm diameters and 3 mm depth drilled on both: retreating side and advanced side.

The produced BT-FSW joints were evaluated at the beginning via visual inspection. Then, the transverse cross-section of welds was macroscopically examined after applying the standard metallographic grinding and polishing. The polished surface was then etched for about 180 s in a solution of 5 mL nitric acid, 3 mL hydrochloric acid, and 2 mL hydrofluoric acid in 190 mL distilled water. Furthermore, the tensile-shear test specimens were cut along the joint's transverse directions with the dimensions as shown in Figure 4. The test was carried out using Instron tensile test machine (Instron-4208-300 kN capacity, Norwood, MA, USA). The velocity of the moving head of the machine was 0.05 mm/min at room temperature. Additionally, the fractures parts were investigated using SEM-Quanta FEG 250 (FEI company, Hillsboro, OR, USA). Finally, the hardness profiles through three lines on transverse cross-sections of the BT welded joints were measured to evaluate the hardness values on the upper, middle, and lower parts of the weld and surrounding regions (Figure 5) using Vickers hardness tester (HWDV-75, TTS Unlimited, Osaka, Japan) under 0.5 kg load for 15 sec dwell time.

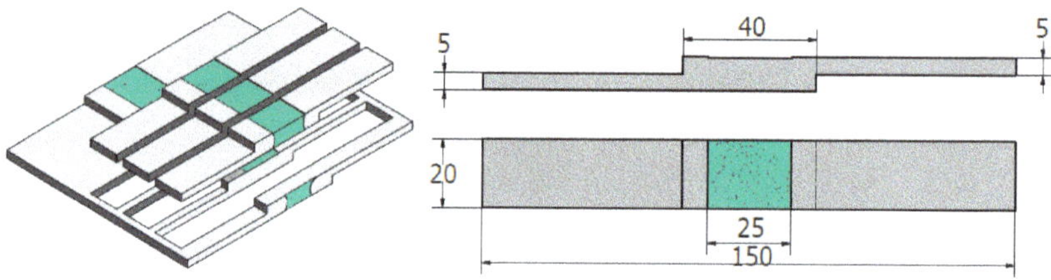

Figure 4. Schematic of the tensile-shear test sample (all dimensions in mm).

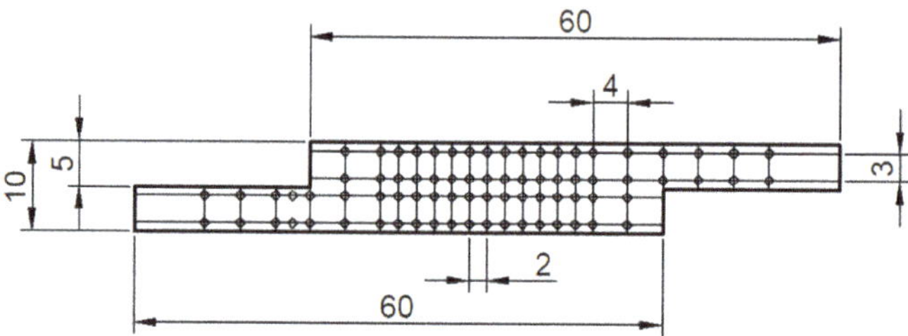

Figure 5. Schematic of measurement points of Vickers hardness profiles on the cross-section perpendicular the welding direction (all dimensions in mm).

3. Results and Discussions

3.1. *Temperature Measured at Weld Center*

The mean values of the measured temperature during the welding process were recorded and plotted versus different BT-FSW travel speeds of 200, 400, 600, 800, and 1000 mm/min using the three different pins Cy, Sq, and Tr at a constant rotation speed of 600 rpm and 0° tilt angle as shown in Figure 6. Among all the produced 10 mm similar AA1050-H14 lap joints, the Sq pin geometry shows the highest stir zone temperature compared to the other two pins, Cy and Tr, whereas the Tr pin gives the lowest stir zone temperature at the various applied travel speeds from 200 to 1000 mm/min using BT-FSW,

Figure 6. BT with square pin geometry produces higher weld temperature around 380 °C at 200 mm/min due to its surface shape that creates more friction heat during BT-FSW. The highest temperatures are 380 °C, 367 °C, and 351 °C, at the travel speed of 200 mm/min using BT with pin geometry of Sq, Cy, and Tr, respectively, whereas at the fast weld travel speed (1000 mm/min), the lowest temperatures are 276 °C, 265 °C, and 240 °C for the Sq, Cy, and Tr pins, respectively. Additionally, it can be seen that as the welding travel speed increases, the recoded sir zone temperature decreases for all the used pin geometries. This happens due to the reduced heat input per unit length and dissipation of heat over a wider region of the workpiece at a higher welding speed. In fact, the introduced temperature values in the stir zone during the FSW expresses the amount of heat input. The decrease of FSW heat input in the stir zoon as a function of increasing travel welding speeds was reported in many research with different weld joints for different materials [24]. The effects of pin geometry on the generated heat during the BT-FSW are clear (Figure 6). First, for the triangular pin, the frictional area between the pin surface and the workpiece material is limited to near the three edges, which is very smaller than that of the Cy and Sq pin geometries. Since a larger frictional area must generate a larger amount of friction heat, the friction heat generated by the triangular profile should be smaller than that by the Cy and Sq geometries. Second, for Tr and Sq pins, the weld temperature is directly proportional to the number of edges of the pin profiles because the increasing of edges leads to increasing of the frictional area in such a way that the weld temperature in the geometries increases from the Tr to Sq pin geometry during BT-FSW.

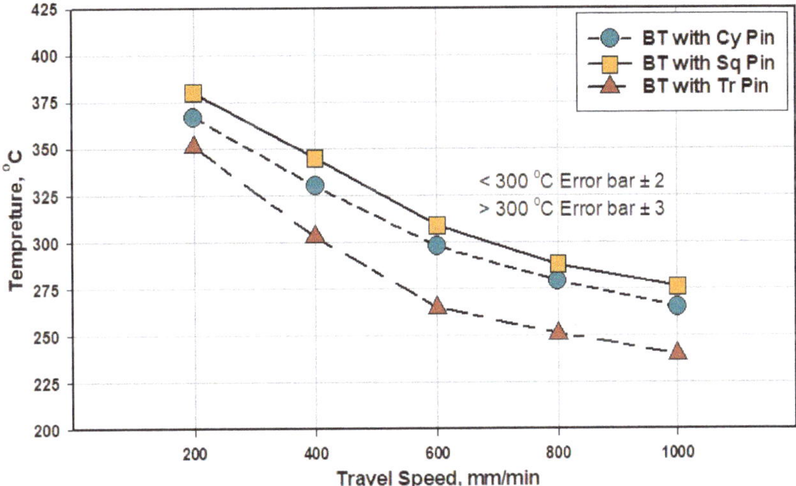

Figure 6. Effects of BT pin geometries and BT travel speed on the welding temperature in the stir zone to produce 10 mm AA1050-H14 similar lap joints.

3.2. *Recorded Thermal Cycles and Temperatures at the AS and RS*

The thermal cycles at both the AS and the RS in terms of the measured working temperatures as a function of time at the different welding speeds during the BT-FSW using the Sq pin are recorded with the embedded thermocouples and plotted in Figure 7a,c, respectively, with their associated enlarged areas for the overlapped peaks in Figure 7b,d, respectively. It should be mentioned here that the same trend of thermal cycle curves was obtained during the BT-FSW of AA1050-H14 using the Cy and Tr pins at the same applied welding conditions with a little difference in the recorded peak temperatures at each travel speed. In the current work, the BT-welding process can be divided into three steps. First, by inserting the tool to the center line of the lap joint and applying a rotation speed of 600 rpm and slow travel speed of 20 mm/min, the working material can be heated to a certain temperature that allows the initial softening of the material. Second, by applying a holding time of around 15 secs to heat the plates under the same rotation speed of 600 rpm, the temperature of the first and second steps is lower than 100 °C. Third, applying the actual travel speed to start achieving the welding path, the temperature of these steps is increasing gradually to the peak temperature. Finally, the process ended by air-cooling the workpiece with loss in the heat gradually. These steps appear in Figure 7. Using BT-FSW with Sq pin geometry, the maximum temperature experienced about 360 °C at the 200 mm/min welding speed at the AS that slightly higher at the RS. The minimum temperature experienced about 206 °C at the RS that slightly smaller at the AS using Tr pin geometry.

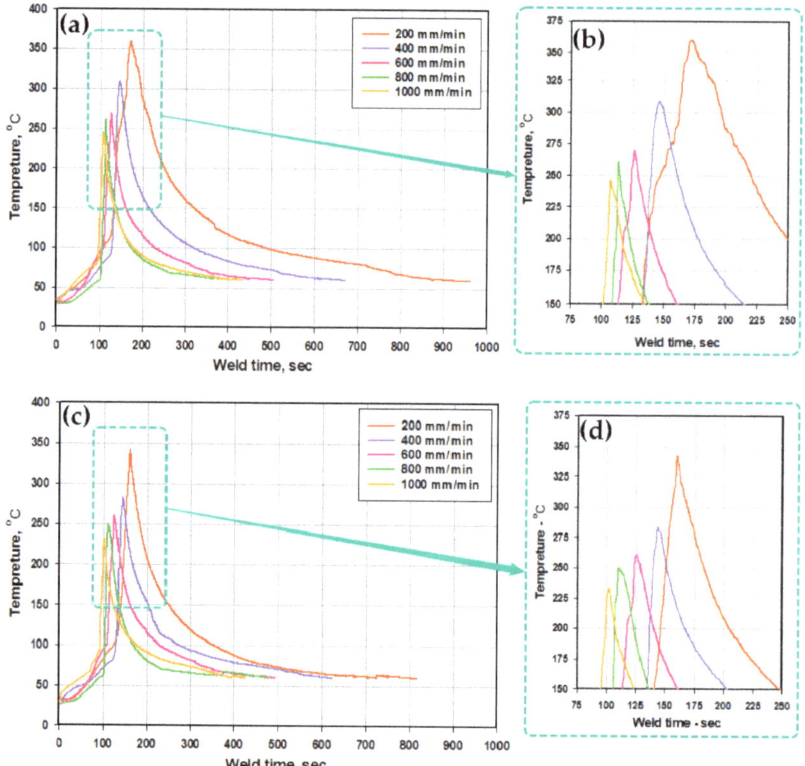

Figure 7. An example of the recorded temperature profiles obtained with the embedded thermocouples during BT-FSW of aluminum using the Sq pin using a constant rotation speed of 600 rpm at (**a**,**b**) advancing side and (**c**,**d**) retreating side.

Figure 8 shows a comparative histogram for the temperatures measured during BT-FSW using the different pin geometries and at the different welding speeds for both the AS (Figure 8a) and RS (Figure 8b). The top surface temperature at both the AS and RS during BT-FSW decreases with increasing travel speed. This can be attributed to the reduction in the number of rotations per unite length by increasing the welding speed at a constant rotation rate. For example, 3 rev/mm at 200 mm/min is reduced to 0.6 rev/mm at 1000 mm/min. This significantly reduces the amount of heat generated by increasing the welding speed and also allows fast cooling after welding. On both sides, the BT with Sq pin generated a higher temperature, and the triangular pin generated the lowest temperature; this trend is the same trend for the temperature produced in the weld center at the top surface. The temperature increases along the direction of the weld and towards the weld center, and the weld temperatures recorded on the retreating side are slightly lower than that at the advancing side. This happens because the advancing side in the BT-FSW process is the location from which the softened solid material starts to flow around the tool pin plunged into the material, which moved from the front of the tool to the back of the tool to achieve the joint behind the tool in the solid state. The AS represents the start point for material stirring that requires the highest torque to achieve the required plastic deformation and heat to soften and move the material around the tool. Thus, it is expected to have a higher temperature at this side than the RS at which the material is left behind the tool and starts cooling.

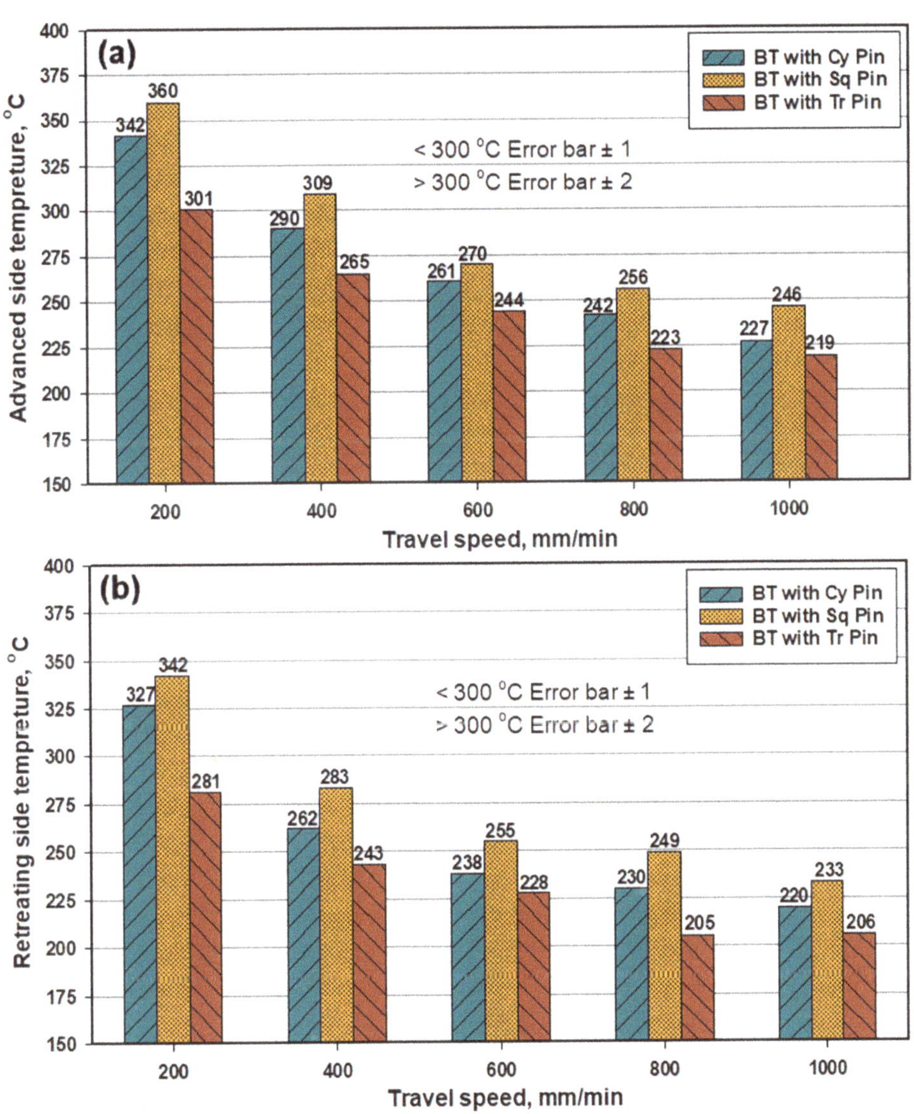

Figure 8. Maximum temperature measured at (**a**) the advancing side and (**b**) the retreating side with different pin profiles.

3.3. BT-FSW Torque

There are many parameters controlling the stirring process in BT-FSW [2]. One of these is the torque. However, few works are developed to study and predict the torque during FSW, which can be used to optimize the tool design as well as the process [27]. Hence, predicting, monitoring, and controlling the BT torques are highly important to predict the tool life. In fact, the torque value given in the monitor of the full-automatic FSW machine in the present work can be used as an indicator for the resistance of materials to move around the pin during stirring. Figure 9 shows the measured torque values during BT-FSW at a travel speed of 200 mm/min and rotation speed of 600 rpm using the different pin geometries. The different regions of torque curves obtained during the BT-FSW process are tool entry, dwell time, welding achievement, and tool exit. At tool entry, the BT moves forward to touch the cold workpiece, and then the torque sharply increased to reach maximum values for all the used pins. In the dwell time region, the values of torque decrease to the minimum value of 10 Nm as a result of softening the stirring material around the tool. During the welding process, the torque increases sharply again as the material highly resists the rotating pin in the stir zone. The resistance of materials to move around the tool increases with increasing sharp pin edges. The highest value of measured torque (66 Nm) was obtained using BT with Sq pin geometry whilst, the lowest value (52 Nm) is given using the Cy pin. Finally, the torque decreases sharply in the tool exit region as a result of BT leaving the welded workpiece.

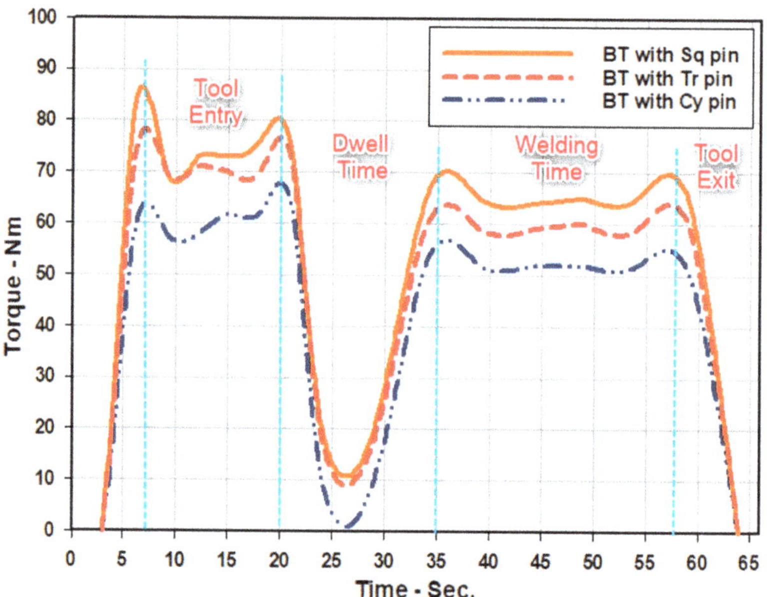

Figure 9. Torque during BT-FSW with different pins at 200 mm/min and 600 rpm.

The torque value was recorded for the used three pin geometries as a function of travel speed and plotted in Figure 10. It can be seen that as the travel speed increases, the measured torque values increase for all the pin geometries. By increasing the welding travel speed, the heat input decreases. Thus, the torque value increases due to the difficulty of material flow [6]. Moreover, material flow with Cy pin is easier during the BT-FSW process compared to the other two used pins. On the other hand, the tool faces more resistance with the Sq, and Tr pins, especially at higher welding speeds due to the colder material.

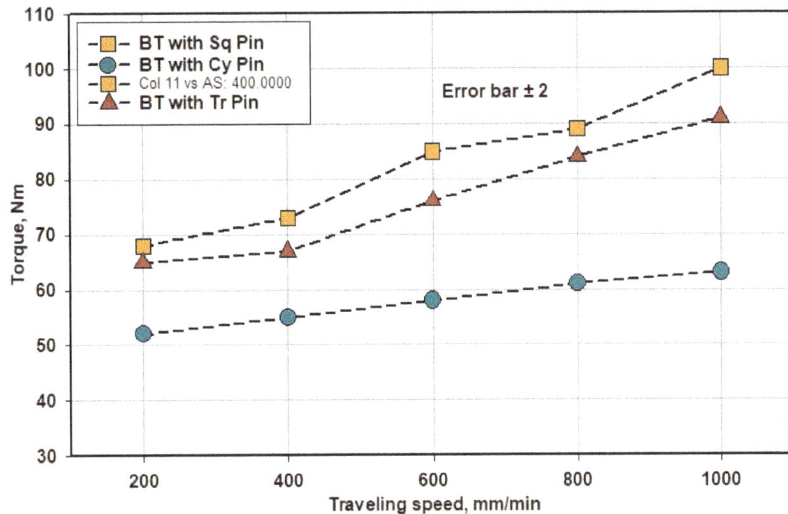

Figure 10. Torque during BT-FSW of AA1050-H14 with various pin geometries.

3.4. Visual Inspection

Three groups of AA1050-H14 Al alloy similar lap joints were friction stir-welded using BT. Two welding parameters, pin geometry and travel speed, were considered with keeping the other parameters constant. The pin geometries were Cy, Tr, and Sq, and the travel welding speeds were varied from 200 to 1000 mm/min. Both the upper and the lower surfaces of the welded laps were visually inspected. No defects were observed for the first and the second welded lap joints groups using the Cy and Sq pins at all the applied welding speeds, as can be seen from the top and bottom surface views illustrated in Figures 11 and 12, respectively. The third group of BT-FSW lap joints welded using Tr pin showed defect-free joints at 200, 400, and 600 mm/min travel speeds as given in Figure 13, whereas at 800 and 1000 travel speeds, uncompleted joints were observed on the joint top surfaces due to appearance of a tunnel defect. A tunnel may be formed during FSW due to insufficient heat input. The decrease in the stir zone temperature leads to a decrease in materials plasticity and hence hinders the flowability of the stirring material around the tool. Furthermore, the length of the uncompleted joining defect increases with increasing the travel speed from 800 to 1000 mm/min, as shown in Figure 14a,b. Some moving layers tend to stick on the rotating tool using Tr pin geometry, as given in Figure 15c,d.

Figure 15a,b shows the keyhole in the exit regions of the BT-FSW joints after the BT exits the joints. The BT leaves the joint by a disruption at the end of the welded pass. The arrows on the top (Figure 15a) and bottom (Figure 15b) surfaces show the last vestige of the shoulder on the top and bottom surfaces. The material is transported from the retreating side to the advancing side before starting the weld disruption. However, the stirring mechanism is not able to continue primary stirring from the advancing side to the retreating side, as the stirring material accumulated in the advanced side [28]. Figure 15c,d presents the shape of material flow of the entry regions for the bobbin welds on the upper and lower surface, respectively. The entry defect consists of an ejected tail and is produced on the retreating side. In the early step of the BT entering the workpiece, the material flows from the retreating side to the advancing side are disrupted, and the plasticized mass flows outwards of the joint. The main reason for the material ejection in the entry zone is the free surface at the trailing edge and the absence of the solidified mass to block the material flow outwards [29].

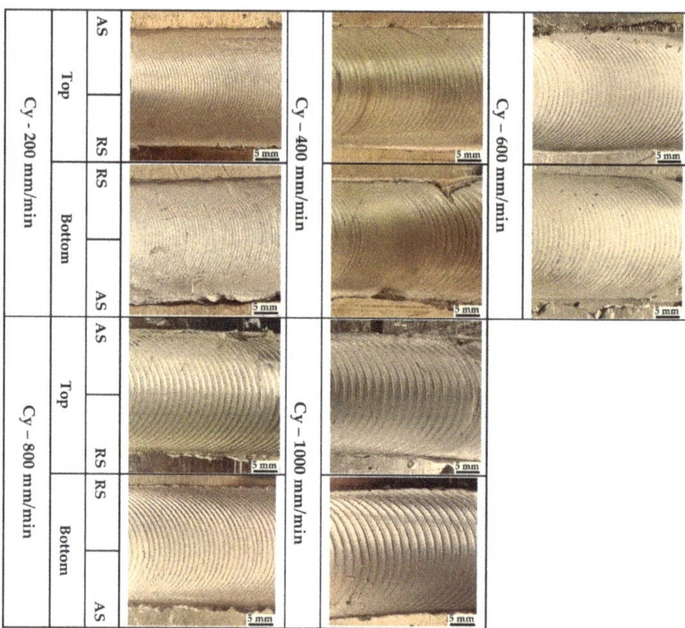

Figure 11. Top and bottom views of BT-FSW joints using Cy pin.

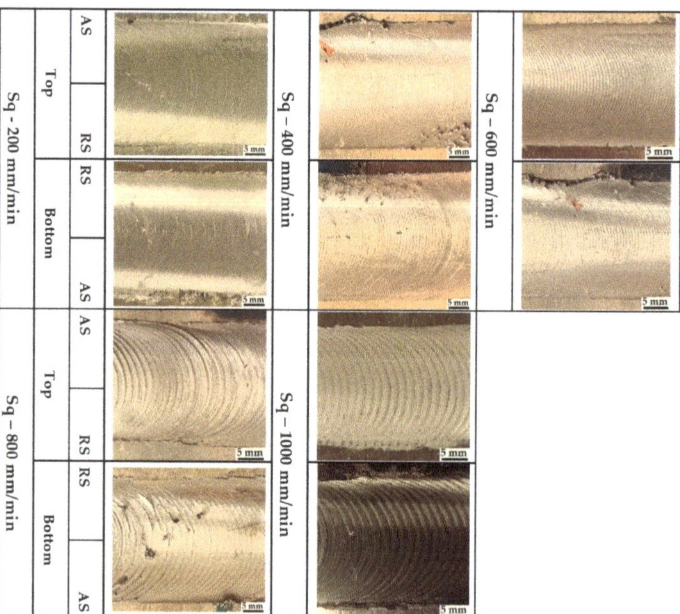

Figure 12. Top and bottom views of BT-FSW joints using Sq pin.

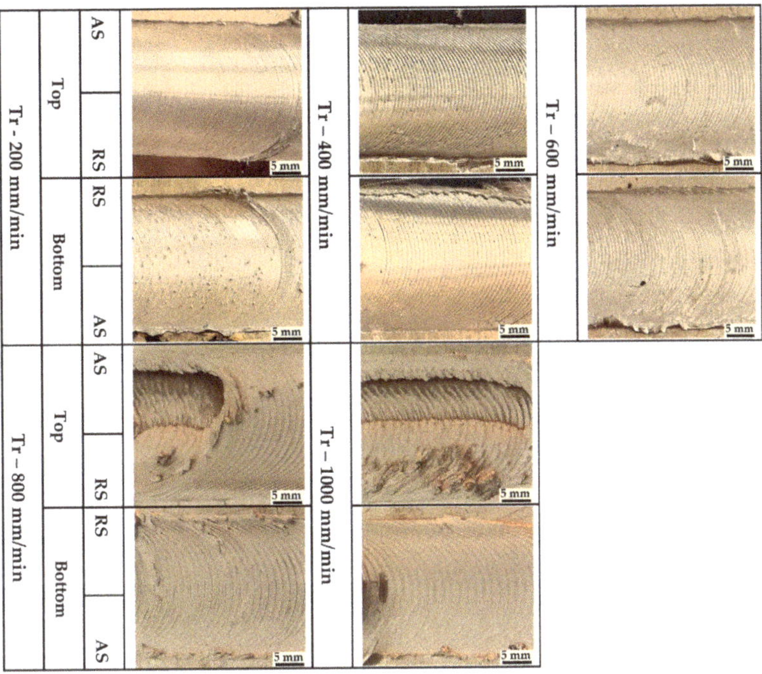

Figure 13. Top and bottom views of BT-FSW joints using Tr pin.

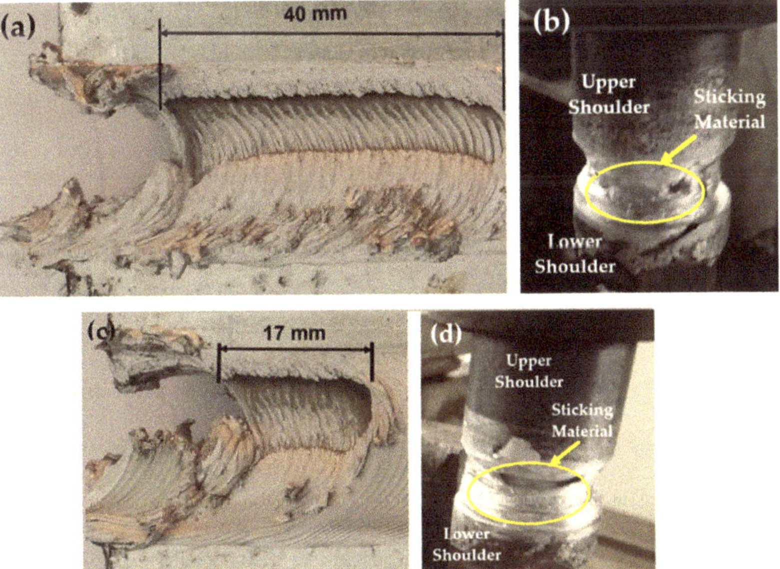

Figure 14. BT-FSW uncompleted joining defect and stick material around the BT using Tr pin at (**a**,**b**) 800 mm/min, (**c**,**d**) 1000 mm/min.

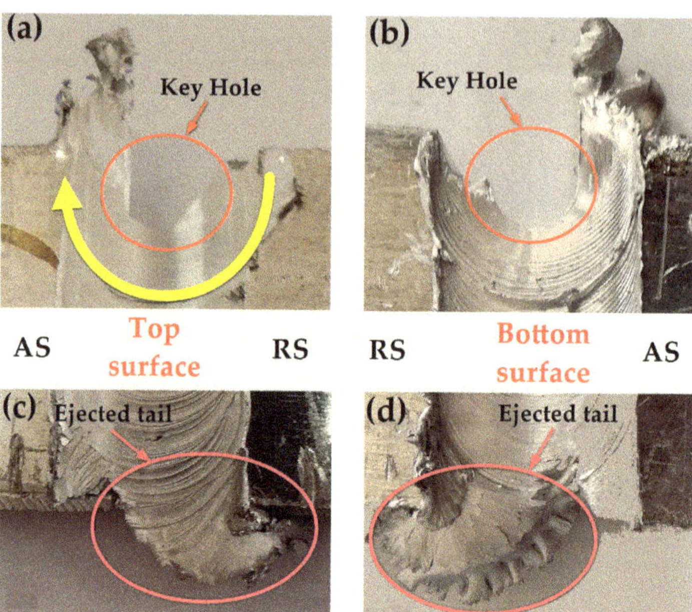

Figure 15. Top view of the BT-FSW of aluminum at start and end region; (**a**,**b**) the exit regions; (**c**,**d**) the entry regions.

3.5. Macrostructure of BT-FSW Lap Joints

Figure 16 shows the cross-sectional macrostructures of the BT-FSW lap joints welded at a rotation speed of 600 rpm and travel speed of 200 mm/min using Cy pin geometry. The different regions in welded joints, BM, HAZ, TMAZ, and SZ, can be easily featured. As seen in Figure 16, a sharp transition between the BM and the SZ occurs on the advancing side, while a more diffuse transient region is present on the retreating side of the welded joints. This has been attributed to the different flow behaviors and the degree of plastic strain/deformation of the material on both sides of the rotating and traveling tool [1]. The macrostructures of the transverse cross-sections of the AA1050-H14 lap joints welded using Cy, Sq, and Tr pins at various travel welding speeds (200, 400, 600, 800, and 1000 mm/min) and at a constant rotation speed (600 rpm) were investigated perpendicular to the welding directions and presented in Figures 17–19, respectively. All cross-sections BT-FSWed at the applied travel speeds using Cy and Sq pin are defect-free, as shown in Figures 17a–e and 18a–e, respectively. The relationship between defects observed in the welds and tool pin geometry is important to analyze. Many researchers [29,30] reported that due to the inappropriate movement and insufficient plasticization of material in the stir zone, the joints show various defects such as kissing bonds, voids, and tunnel defects. It was observed that the BT-FSWed joints obtained using triangular pin geometry showed the presence of tunnel defect, Figure 19. In lap joints produced with Tr pin at 400 and 600 mm/min, separation surface between the upper and lower plates at the joint middle transverse cross-section is shown in Figure 19a,b. Tunnel defects are more pronounced, especially in joints produced at 800 and 1000 mm/min travel speeds. The tunnel defect in the cross-sectional area at 1000 mm/min is larger than that obtained at 800 mm/min, as shown in Figure 19c–e. In general, the defect area increases with increasing travel speed. This is likely due to the decrease of weld temperature, which results in a lack of stirring action, and then promotes the chance of tunnel formation [31,32]. Similar results were also obtained by Hussain et al. [30] by using a triangular tool for FSW of AA6063 butt joints.

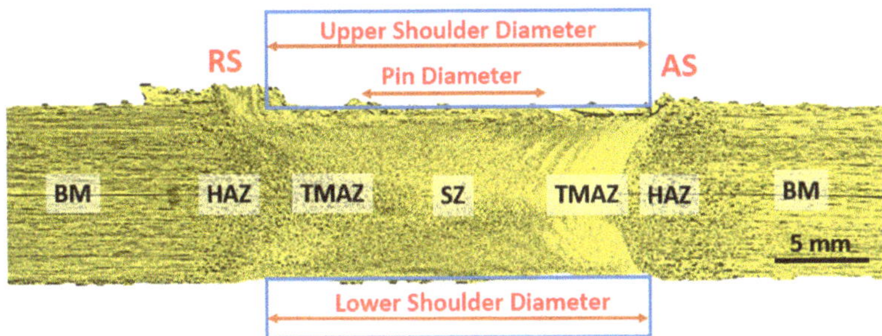

Figure 16. Macrograph analysis of BT-FSW of lap joints welded at 600 rpm and 200 mm/min.

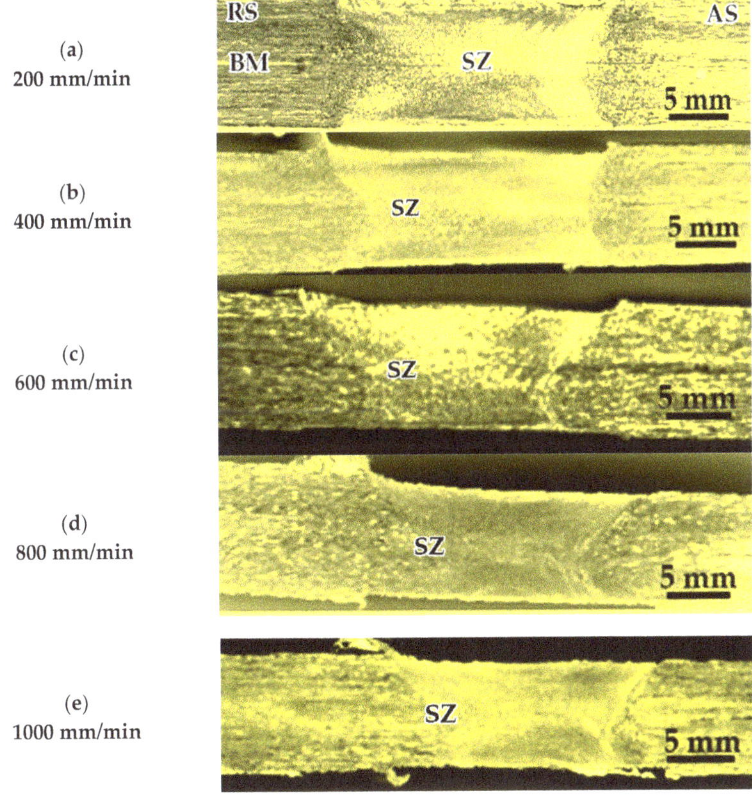

Figure 17. Macrographs of transverse cross-sections of lap joints produced at 600 rpm rotation speed and different travel speeds of (**a**) 200, (**b**) 400, (**c**) 600, (**d**) 800, and (**e**) 1000 using BT with Cy pin.

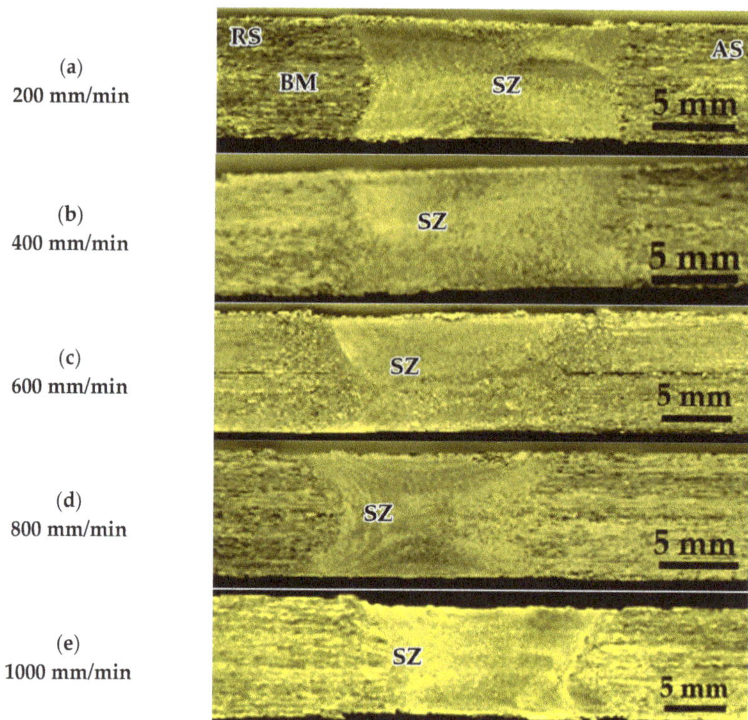

Figure 18. Macrographs of transverse cross-sections of lap joints produced at 600 rpm rotation speed and different travel speeds of (**a**) 200, (**b**) 400, (**c**) 600, (**d**) 800, and (**e**) 1000 using BT with Sq pin.

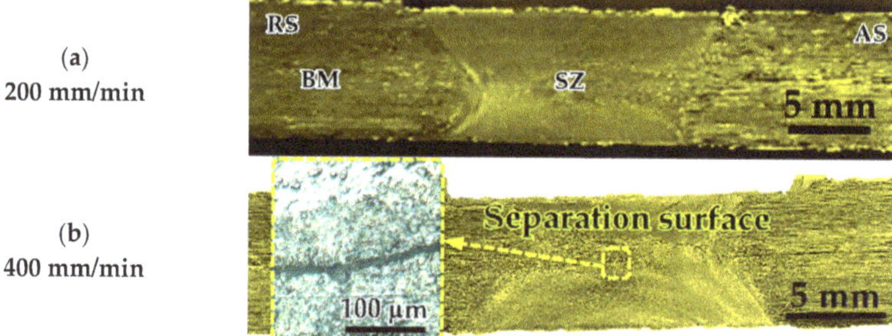

Figure 19. *Cont.*

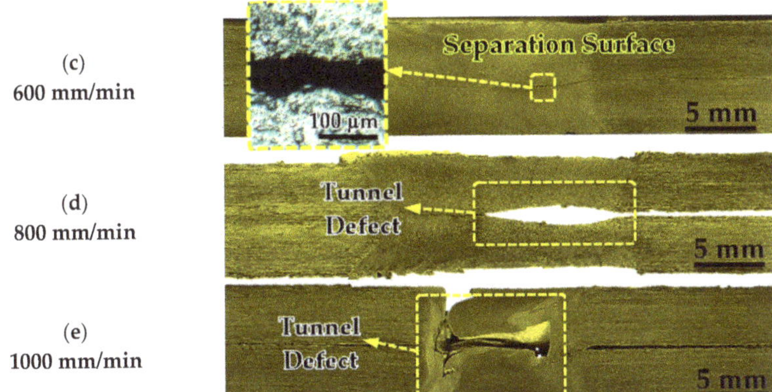

Figure 19. Macrographs of Transverse cross-sections of lap joints produced at 600 rpm rotation speed and different travel speeds of (**a**) 200, (**b**) 400, (**c**) 600, (**d**) 800, and (**e**) 1000 using BT with Tr pin; (**b**,**c**) separation surface defects; (**d**,**e**) tunnel defects.

3.6. Mechanical Properties of BT-FSW Joints

Figure 20 shows the Vickers hardness maps obtained along the transverse cross-sections of the BT-FSW lap joints welded at a rotation speed of 600 rpm and at welding speeds of 200 and 1000 mm/min, using Cy, Sq, and Tr pin geometries. Generally, the hardness profile indicates the typical hardness profile for the strain hardened aluminum alloys that is characterized by a softened region of reduced hardness at the weld nugget; however, a significant effect can be observed in the hardness profile either due to the tool pin geometry or the welding speed. The hardness is mainly affected by the thermal cycle experienced and the maximum temperature. This can be observed in the width of the softened region and the level of hardness reduction in the weld zone. The highest reduction in hardness is occurred at 200 mm/min welding speed and using the Cy BT, as can be observed from the hardness maps that represented in blue color with a hardness of about 24 Hv. Using the same Cy pin at a higher welding speed of 1000 mm/min has reduced the level of hardness reduction, and the hardness reached about 27 Hv at the middle section. The lowest hardness reduction is achieved using the Tr BT at 1000 mm/min, and hardness reached about 29 Hv at the middle section, as can be noted from the orange color of the map. On the other hand, at the same level of hardness, the width of the softened region increases as the thermal cycle reaches higher temperatures for a long time, which occurred here at the low welding speed of 200 mm/min using the Sq tool pin. If, for example, one measures the width at a hardness value of 27 Hv, which is represented by the green color in the maps, it can be observed that the width of the softened region is the maximum at a welding speed of 200 mm/min using the Cy BT and the minimum at the welding speed of 1000 mm/min using the Tr BT. It can also be noted that at all the welding conditions, the hardness is lower at the shoulder surfaces, either top or lower shoulder, due to the maximum heat generated. The results obtained here are consistent with the reported in the literature during FSW of the strain-hardened aluminum alloys such as AA5083 and AA1050.

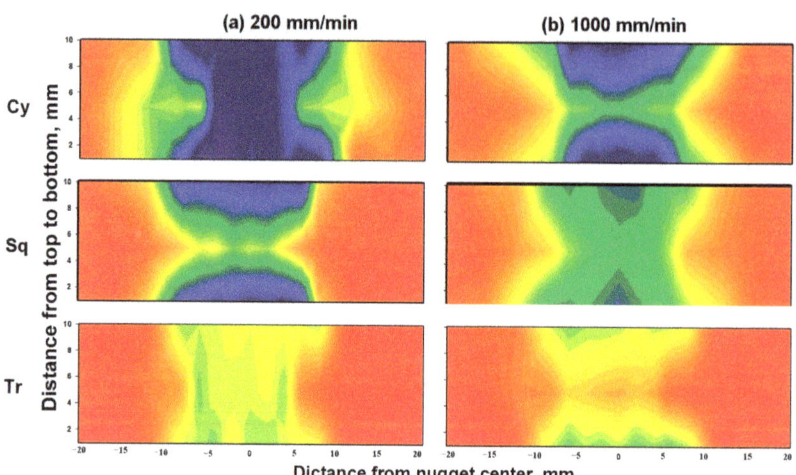

Figure 20. Hardness maps obtained at the transverse cross-section of the BT-FSWed lap joints at (**a**) 200 mm/min and (**b**) 1000 mm/min, using different pin geometries (Cy, Sq, and Tr).

The tensile-shear load of BT-FSW lap joints using the different pins (Cy, Sq, and Tr) at various travel speeds of 200, 400, 600, 800, and 1000 mm/min, at a constant rotation speed of 600 rpm, was plotted in Figure 21. For all the applied pin geometries, the tensile-shear load increases with decreasing travel speed, reaching its peak value at 200 mm/min. The highest tensile-shear load values of the AA1050 similar lap joints were obtained at 200 mm/min welding travel speed. These values are 6491, 5806, and 5419 N for the applied pin geometries: Sq, Cy, and Tr, respectively. In CT-FSW, the pulsating action (pulse/s) (pulse/s = rotational welding speed in one second × number of flat faces) during the stirring process in FSW is a function of rotational speed and the number of flat faces of the pin geometry [29]. Goyal et al. [33] studied the effect of pin geometry on the pulsating action of the material flow around the pin in the stir zone for Al-Mg4.2 at welding conditions of 1200 rpm and 65 mm/min. They concluded that the cylindrical tool does not promote a pulsating action, whilst the triangular and the square pin geometries generate 60 and 80 pulse/s, respectively. Additionally, Elangovan et al. [17] reported that for FSW of AA2219 using different pin geometries, the Sq pin produced more pulse/s compared to the Tr pin. At the rotational welding speed of 1200 rpm, the pulses/sec values were 80 and 60 for the Sq and Tr pin geometries, respectively. The pin geometry of the BT-FSW is mainly responsible for softening and transporting the welded material during the joining process. The movement of plasticized material in the stir zone is controlled by two effects. The first effect is extrusion, where the pin pushes the plasticized material behind it. Additionally, the second effect is the rotary movement of the pin, which promotes the required thrust to ease the material flow [29]. According to the current welding conditions of 600 rpm rotation speed and different pin geometries, the Sq pin geometry produced 40 pulse/s and the Tr pin geometry produced 30 pulse/s, while no such pulsating action was observed in cylindrical pin geometry. The lap joints made with Cy and Tr showed lesser joint strength as compared to other welded lap joints with Sq pin (Figure 21), due to the lack of pulsating action and vertical flow of plasticized material in the stir zone. In the case of Tr pin geometry, improper consolidation of material near the pin base leads to the formation of separation surface 400 and 600 mm/min (Figure 19b,c). Additionally, there is a tunnel defect in the nugget zone at the welding condition of 800 and 1000 mm/min (Figure 19d,e), which consequently deteriorates the joint strength despite hardness increase. In addition, the lap joint welded at 200 mm/min using Tr and Cy pin geometries shows lower tensile-shear load compared with that produced using Sq pin geometry due to the of top and bottom surface roughness of the welded joints and the presence of weld protrusions.

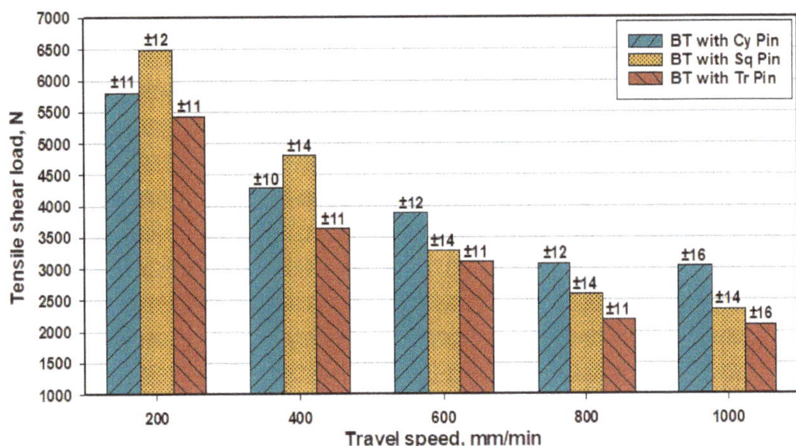

Figure 21. Tensile-shear loads of BT-FSW lap joints at various travel speeds using different pin geometries.

Figures 22 and 23 show typical fracture morphologies of the BT-FSW lap joint using an Sq pin at travel speeds of 200 and 1000 mm/min. It can be observed that the fracture surface of investigated fracture surfaces of the samples consists of mixed brittle and ductile modes. Figure 22 shows the fracture surface of BT-FSW at 200 mm/min, where ductile features as serrations-like areas are shown mixed with small cleavage fracture areas. These features indicate that the fracture mode is more ductile. On the other hand, the fracture mode of the joint produced using 1000 mm/min is mainly brittle, as shown in Figure 23. There is more cleavage faceted area, little serrations, and no dimples on the fracture surface. The above features are confirmed with the FSW heat input and tensile-shear results.

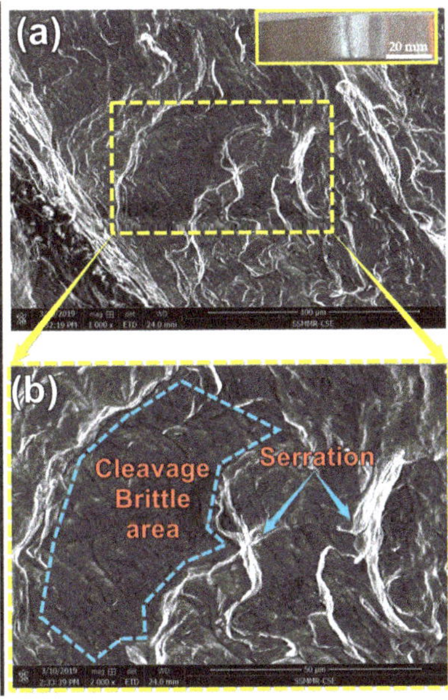

Figure 22. (**a**) The fracture morphology of the BT-FSW lap joint welded at travel speed of 200 mm/min and rotational speed of 600 rpm using Sq pin and (**b**) at a higher magnification.

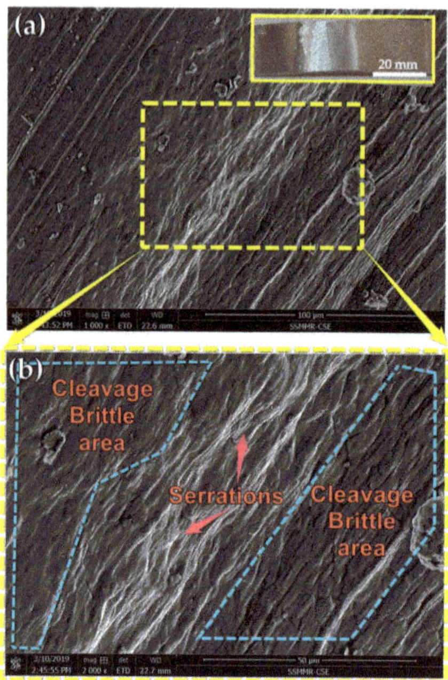

Figure 23. (**a**) The fracture morphology of the BT-FSW lap joint welded at a travel speed of 1000 mm/min and rotational speed of 600 rpm using Sq pin and (**b**) at a higher magnification.

4. Conclusions

BT-FSW of AA1050-H14 similar thick lap joints was carried out using different pin geometries, Cy, Sq, and Tr, at a rotation speed of 600 rpm, and welding speeds from 200 to 1000 mm/min. The following conclusions could be drawn:

1. Pin geometry and travel speed showed significant effect as processing parameters in the BT-FSW process as they affect the temperature at the weld center. Furthermore, the square pin leads to a higher temperature, and the temperature in the advanced side is higher than that in the retreating side at all welding conditions.
2. Increasing the travel speed increases the torque, and the highest torque values are associated with the square pin geometry.
3. BT-FSW using Cy and Sq pins produced defect-free welds at all welding conditions, whereas at travel speeds of 400 and 600 mm/min, using a Tr pin resulted in uncompleted joining. Furthermore, tunnel defects appeared at welding speeds of 800 and 1000 mm/min.
4. At all the welding conditions, the hardness values of the stir zone are lower than those of the BM.
5. The welding conditions of 200 mm/min and 600 rpm using BT with Sq pin produces sound thick lap joints of AA1050-H14 Al alloy with a higher tensile shear load.
6. Mixed brittle-ductile fracture mode is observed on the tensile-fractured FSW samples.

Author Contributions: Conceptualization, M.I.A.H.; Data curation, M.M.Z.A. and M.I.A.H.; Formal analysis, M.M.Z.A., M.M.E.-S.S., S.A., and F.H.L.; Funding acquisition, K.H. and S.A.; Investigation, M.I.A.H.; Methodology, M.I.A.H., M.I.A.H., M.M.E.-S.S., and S.A.; Project administration, M.M.E.-S.S. and S.A.; Resources, M.M.E.-S.S., K.H., and F.H.L.; Software, K.H. and F.H.L.; Supervision, M.I.A.H., M.M.E.-S.S., and A.E.E.-N.; Validation, K.H. and S.A.; Visualization, K.H., S.A., and F.H.L.; Writing—original draft, M.I.A.H.; Writing—review and editing, M.M.Z.A., M.M.E.-S.S., and A.E.E.-N. All authors have read and agreed to the published version of the manuscript.

Funding: This research was funded by the Deanship of Scientific Research at Imam Mohammad Ibn Saud Islamic University (KSA Research Group no. RG-21-12-04).

Institutional Review Board Statement: Not applicable.

Informed Consent Statement: Not applicable.

Data Availability Statement: The data presented in this study are available on request from the corresponding author. The data are not publicly available due to the extremely large size.

Acknowledgments: The authors extend their appreciation to the Deanship of Scientific Research at Imam Mohammad Ibn Saud Islamic University for funding this work through Research Group no. RG-21-12-04.

Conflicts of Interest: The authors declare no conflict of interest.

References

1. Qiao, Y.; Zhou, Y.; Chen, S.; Song, Q. Effect of Bobbin Tool Friction Stir Welding on Microstructure and Corrosion Behavior of 6061-T6 Aluminum Alloy Joint in 3.5% NaCl Solution. *Acta Metall. Sin.* **2016**, *52*, 1395–1402. [CrossRef]
2. Threadgill, P.L.; Ahmed, M.M.Z.; Martin, J.P.; Perrett, J.G.; Wynne, B.P. The Use of Bobbin Tools for Friction Stir Welding of Aluminium Alloys. *Mater. Sci. Forum* **2010**, *642*, 1179–1184. [CrossRef]
3. Goetze, P.; Kopyściański, M.; Hamilton, C.; Dymek, S. Comparison of Dissimilar Aluminum Alloys Joined by Friction Stir Welding with Conventional and Bobbin Tools. *Miner. Met. Mater. Ser.* **2019**, *10*, 3–12. [CrossRef]
4. Ramadan, A.; Essa, S.; Mohamed, M.; Ahmed, Z.; Ahmed, A.Y.; El-nikhaily, A.E. An Analytical Model of Heat Generation For. *J. Mater. Res. Technol.* **2016**, *5*, 234–240. [CrossRef]
5. Ramanjaneyulu, K.; Reddy, G.M.; Rao, A.V. Role of Tool Shoulder Diameter in Friction Stir Welding: An Analysis of the Temperature and Plastic Deformation of AA 2014 Aluminium Alloy. *Trans. Indian Inst. Met.* **2014**, *67*, 769–780. [CrossRef]
6. Peel, M.J.; Steuwer, A.; Withers, P.J.; Dickerson, T.; Shi, Q.; Shercliff, H. Dissimilar Friction Stir Welds in AA5083-AA6082. Part I: Process Parameter Effects on Thermal History and Weld Properties. *Metall. Mater. Trans. A* **2006**, *37*, 2183–2193. [CrossRef]
7. Hamada, A.S.; Järvenpää, A.; Ahmed, M.M.Z.; Jaskari, M.; Wynne, B.P.; Porter, D.A.; Karjalainen, L.P. The Microstructural Evolution of Friction Stir Welded AA6082-T6 Alu- Minum Alloy during Cyclic Deformation. *Mater. Sci. Eng. A* **2015**, *642*, 366–376. [CrossRef]
8. Mirzaei, M.H.; Asadi, P.; Fazli, A. Effect of Tool Pin Profile on Material Flow in Double Shoulder Friction Stir Welding of AZ91 Magnesium Alloy. *Int. J. Mech. Sci.* **2020**, *183*, 105775. [CrossRef]
9. Wen, Q.; Li, W.; Patel, V.; Gao, Y.; Vairis, A. Investigation on the Effects of Welding Speed on Bobbin Tool Friction Stir Welding of 2219 Aluminum Alloy. *Met. Mater. Int.* **2020**, *26*, 1830–1840. [CrossRef]
10. Tonelli, L.; Morri, A.; Toschi, S.; Shaaban, M.; Ammar, H.R.; Ahmed, M.M.Z.; Ramadan, R.M. Effect of FSP Parameters and Tool Geometry on Microstructure, Hardness, and Wear Properties of AA7075 with and without Reinforcing B 4 C Ceramic Particles. *Int. J. Adv. Manuf. Technol.* **2019**, *102*, 9–11. [CrossRef]
11. Ubaid, M.; Bajaj, D.; Mukhopadhyay, A.K.; Siddiquee, A.N. Friction Stir Welding of Thick AA2519 Alloy: Defect Elimination, Mechanical and Micro-Structural Characterization. *Met. Mater. Int.* **2020**, *26*, 1841–1860. [CrossRef]
12. Ahmed, M.M.Z.; Seleman, M.M.E.; Zidan, Z.A.; Ramadan, R.M.; Ataya, S.; Alsaleh, N.A. Microstructure and Mechanical Properties of Dissimilar Friction Stir Welded AA2024-T4/AA7075-T6 T-Butt Joints. *Metals* **2021**, *11*, 128. [CrossRef]
13. Ahmed, M.M.Z.; Wynne, B.P.; Rainforth, W.M.; Addison, A.; Martin, J.P.; Threadgill, P.L. Effect of Tool Geometry and Heat Input on the Hardness, Grain Structure, and Crystallographic Texture of Thick-Section Friction Stir-Welded Aluminium. *Metall. Mater. Trans. A* **2019**, *50*, 271–284. [CrossRef]
14. Fujii, H.; Cui, L.; Maeda, M.; Nogi, K. Effect of Tool Shape on Mechanical Properties and Microstructure of Friction Stir Welded Aluminum Alloys. *Mater. Sci. Eng. A* **2006**, *419*, 25–31. [CrossRef]
15. Chandrashekar, A.; Ajaykumar, B.S.; Reddappa, H.N. Effect of Pin Profile and Process Parameters on the Properties of Friction Stir Welded Al-Mg Alloy. *Mater. Today Proc.* **2018**, *5*, 22283–22292. [CrossRef]
16. Zhao, S.; Bi, Q.; Wang, Y.; Shi, J. Empirical Modeling for the Effects of Welding Factors on Tensile Properties of Bobbin Tool Friction Stir-Welded 2219-T87 Aluminum Alloy. *Int. J. Adv. Manuf. Technol.* **2017**, *90*, 1105–1118. [CrossRef]
17. Elangovan, K.; Balasubramanian, V. Influences of Tool Pin Profile and Tool Shoulder Diameter on the Formation of Friction Stir Processing Zone in AA6061 Aluminium Alloy. *Mater. Des.* **2008**, *29*, 362–373. [CrossRef]
18. Mahmoud, E.R.I.; Takahashi, M.; Shibayanagi, T.; Ikeuchi, K. Effect of Friction Stir Processing Tool Probe on Fabrication of SiC Particle Reinforced Composite on Aluminium Surface. *Sci. Technol. Weld. Join.* **2009**, *14*, 413–425. [CrossRef]
19. Li, G.H.; Zhou, L.; Zhang, H.F.; Guo, G.Z.; Luo, S.F.; Guo, N. Evolution of Grain Structure, Texture and Mechanical Properties of a Mg–Zn–Zr Alloy in Bobbin Friction Stir Welding. *Mater. Sci. Eng. A* **2021**, *799*, 140267. [CrossRef]
20. Wang, F.F.; Li, W.Y.; Shen, J.; Hu, S.Y.; Santos, J.F. Effect of Tool Rotational Speed on the Microstructure and Mechanical Properties of Bobbin Tool Friction Stir Welding of Al—Li Alloy. *Mater. Des.* **2015**, *86*, 933–940. [CrossRef]

21. Ahmed, M.M.Z.M.Z.; Ataya, S.; El-Sayed Seleman, M.M.M.; Ammar, H.R.R.; Ahmed, E. Friction Stir Welding of Similar and Dissimilar AA7075 and AA5083. *J. Mater. Process. Technol.* **2017**, *242*, 77–91. [CrossRef]
22. Ahmed, M.M.Z.; Jouini, N.; Alzahrani, B.; Seleman, M.M.E. Dissimilar Friction Stir Welding of AA2024 and AISI 1018: Microstructure and Mechanical Properties. *Metals* **2021**, *11*, 330. [CrossRef]
23. Ahmed, M.M.Z.; Elnaml, A.; Shazly, M.; Seleman, M.M.E. The Effect of Top Surface Lubrication on the Friction Stir Welding of Polycarbonate Sheets. *Intern. Polym. Process.* **2021**, *36*, 94–102. [CrossRef]
24. Ahmed, M.M.Z.; Seleman, M.M.E.S.; Shazly, M.; Attallah, M.M.; Ahmed, E. Microstructural Development and Mechanical Propertiesof Friction Stir Welded Ferritic Stainless Steel AISI 40. *J. Mater. Eng. Perform.* **2019**, *28*, 6391–6406. [CrossRef]
25. Ahmed, M.M.Z.; Ataya, S.; Seleman, M.M.E.; Mahdy, A.M.A.; Alsaleh, N.A.; Ahmed, E. Heat Input and Mechanical Properties Investigation of Friction Stir Welded AA5083/AA5754 and AA5083/AA7020. *Metals* **2021**, *11*, 68. [CrossRef]
26. Shazly, M.; Ahmed, M.M.Z.; El-Raey, M. Friction stir welding of polycarbonate sheets. In *Characterization of Minerals, Metals, and Materials 2014*; Jon Wiley& Sons: Hoboken, NJ, USA, 2014; ISBN 9781118887868.
27. Tolephih, M.H.; Mahmood, H.M.; Hashem, A.H.; Abdullah, E.T. Effect of Tool Offset and Tilt Angle on Weld Strength of Butt Joint Friction Stir Welded Specimens of AA2024 Aluminum Alloy Welded to Commercial Pure Cupper. *Chem. Mater. Res.* **2013**, *3*, 49–58.
28. Tamadon, A.; Pons, D.J.; Sued, K.; Clucas, D. Formation Mechanisms for Entry and Exit Defects in Bobbin Friction Stir Welding. *Metals* **2018**, *8*, 33. [CrossRef]
29. Fuse, K.; Badheka, V. Bobbin Tool Friction Stir Welding: A Review. *Sci. Technol. Weld. Join.* **2019**, *24*, 277–304. [CrossRef]
30. Azmal Hussain, M.; Zaman Khan, N.; Noor Siddiquee, A.; Akhtar Khan, Z. Effect of Different Tool Pin Profiles on the Joint Quality of Friction Stir Welded AA 6063. *Mater. Today Proc.* **2018**, *5*, 4175–4182. [CrossRef]
31. Zayed, E.M.; El-Tayeb, N.S.M.; Ahmed, M.M.Z.; Rashad, R.M. *Development and Characterization of AA5083 Reinforced with SiC and Al_2O_3 Particles by Friction Stir Processing*; Springer International Publishing: Berlin/Heidelberg, Germany, 2019; Volume 92.
32. Reda, R.; Saad, M.; Ahmed, M.Z.; Abd-Elkader, H. Monitoring Joint Quality: Taguchi Approach and Structure Investigation. *J. Eng. Des. Technol.* **2018**, *16*, 211–232. [CrossRef]
33. Goyal, A.; Garg, R.K. Selection of FSW Tool Parameters for Joining Al-Mg4.2 Alloy: An Experimental Approach. *Metallogr. Microstruct. Anal.* **2018**, *7*, 524–532. [CrossRef]

MDPI
St. Alban-Anlage 66
4052 Basel
Switzerland
Tel. +41 61 683 77 34
Fax +41 61 302 89 18
www.mdpi.com

Materials Editorial Office
E-mail: materials@mdpi.com
www.mdpi.com/journal/materials

www.ingramcontent.com/pod-product-compliance
Lightning Source LLC
LaVergne TN
LVHW070950170726
843515LV00005B/955

* 9 7 8 3 0 3 6 5 4 4 3 8 0 *